W9-CMT-187

(*continued on back*)

Detectors For
Liquid Chromatography

DATE DUE

WD

This item is Due on
or before Date shown.

CHEMICAL ANALYSIS

A SERIES OF MONOGRAPHS ON ANALYTICAL CHEMISTRY AND ITS APPLICATIONS

Editors
P. J. ELVING, J. D. WINEFORDNER
Editor Emeritus: **I. M. KOLTHOFF**

Advisory Board

Fred W. Billmeyer, Jr.	Victor G. Mossotti
Eli Grushka	A. Lee Smith
Barry L. Karger	Bernard Tremillon
Viliam Krivan	T. S. West

VOLUME 89

A WILEY-INTERSCIENCE PUBLICATION

JOHN WILEY & SONS

New York / **Chichester** / **Brisbane** / **Toronto** / **Singapore**

DETECTORS FOR LIQUID CHROMATOGRAPHY

Edited by

EDWARD S. YEUNG

Professor of Chemistry
Iowa State University
Ames, Iowa

A WILEY-INTERSCIENCE PUBLICATION

JOHN WILEY & SONS

New York / Chichester / Brisbane / Toronto / Singapore

Copyright © 1986 by John Wiley & Sons, Inc.

All rights reserved. Published simultaneously in Canada.

Reproduction or translation of any part of this work
beyond that permitted by Section 107 or 108 of the
1976 United States Copyright Act without the permission
of the copyright owner is unlawful. Requests for
permission or further information should be addressed
to the Permissions Department, John Wiley & Sons, Inc.

Library of Congress Cataloging in Publication Data:

Detectors for liquid chromatography.

(Chemical analysis, ISSN 0069-2883; v. 89)
"A Wiley-Interscience publication."
Bibliography: p.
Includes index.
1. Liquid chromatography. I. Yeung, Edward S.
II. Series.

QD79.C454D48   1986   543'.0894   86-11008
ISBN 0-471-82169-1

Printed in the United States of America

10 9 8 7 6 5 4 3 2 1

543.0894
048

87-111

1/21/87— MW 953.60

CONTRIBUTORS

Thomas R. Covey
Equine Drug Testing and Research Program
Cornell University
Ithaca, New York

Jonathan B. Crowther
Equine Drug Testing and Research Program
Cornell University
Ithaca, New York

Robert B. Green
Instrumental Chemical Analysis Branch
Department of the Navy
Naval Weapons Center
China Lake, California

Jack D. Henion
Equine Drug Testing and Research Program
Cornell University
Ithaca, New York

Kiyokatsu Jinno
School of Materials Science
Toyohashi University of Technology
Toyohashi, Japan

Charles N. Kettler
Department of Chemistry
University of Tennessee
Knoxville, Tennessee

Michael D. Morris
Department of Chemistry
University of Michigan
Ann Arbor, Michigan

v

Michael J. Sepaniak
Department of Chemistry
University of Tennessee
Knoxville, Tennessee

Stephen G. Weber
Department of Chemistry
University of Pittsburgh
Pittsburgh, Pennsylvania

Edward S. Yeung
Ames Laboratory-USDOE and Department of Chemistry
Iowa State University
Ames, Iowa

PREFACE

Liquid chromatography (LC) is, for all practical purposes, a hyphenated technique. Physical separation of components is only one part of LC. Except for preparative LC, where physical separation is the entire goal, some technique for quantitation is needed. In principle, any measurement method can be used. In practice, there are several constraints when a measurement method, that is, a detector, is interfaced to LC. First, good sensitivity is needed to deal with the low concentrations of typical analytes, which are often diluted substantially during the separation. This is not simply because of the general rule that the better the detectability, the broader the scope of application, but also because the injected quantity must be kept below the point of saturation of the stationary phase. Second, the volume of the detector must be small to avoid additional band broadening due to extracolumn effects, to preserve the quality of the separation. This volume includes, for example, the connecting tubings to the end of the column, in addition to the actual volume active in producing the signal. Recent developments in microcolumn LC make this problem even more severe. Third, the detector must be able to function in the presence of a large background—that of the eluent molecules. Unlike gas chromatography (GC) where the mobile phase can be an inert material such as helium, the LC mobile phase often affects the actual detector signal. It is necessary to be able to null out this background signal, and to maintain it at a stable level to reduce noise. Fourth, the time response of the detector must be compatible with the chromatographic event. This is even more important for the new high-speed columns. If multidimensional information is to be obtained, for example, recording entire spectra, the detector must cycle rapidly to be useful. Indirectly, time response also affects the detectability, because long time constants cannot be used to average out noise. Last, detector selectivity is much more important in LC compared to GC. Current technology provides separation efficiencies in LC that are inferior to those in GC. The chance of peaks overlapping in LC is then much higher. A selective detector can be used to monitor a subset of all analytes. Effectively, components can thus be resolved without physical separation.

Detectors are not just ancillary techniques for LC. In many situations, analytical measurements can be greatly improved by interfacing the instrument to a liquid chromatograph. What is gained is essentially a simple procedure for sample preparation and cleanup before the measurement. Interferences can then be reduced or eliminated. It is true that one has to worry about additional problems due to dilution and the presence of the eluent. On the other hand, for analytes in solution, LC allows measurement in a well-prescribed solvent—the eluent—rather than the original, often less predictable, solvent. The presence of the eluent can also be turned into an advantage. Because some species are always present at the detector, a signal can be produced in certain detectors even when analytes are absent. When analytes that give no response at the detector are eluted, displacement of the eluent occurs at the detector cell. A negative signal can thus be generated to provide a means for quantitation of the "inactive" analytes. This indirect mode of detection substantially increases the scope of a given measurement technique. Another application resulting from the dependence of the signal on both the eluent and the analyte is that of quantitation without standards. Finally, the time dependence in LC can be used to improve the accuracy and detectabilities of many analytical measurements. This may be called "sample modulation." As long as the noise associated with the analytical measurement is not on the same time scale as the chromatographic event, one can reduce the "noise" compared to static measurements by proper signal averaging or base-line correction.

New books on LC detectors appear on the market at regular intervals. I have tried to assemble here a group of authors (all of whom probably regret having succumbed to my heavy arm-twisting) who are noted experts in the fields that have been "interfaced" to LC, but are not necessarily purebred chromatographers themselves. Perhaps a different viewpoint can be offered. Discussions include the basic principles behind the detector response, the instrumentation, and selected applications for the purpose of comparison and evaluation of the potential.

Many thanks to my friends for their fine contributions, and to my co-workers for hours of argument and for the data that led to many parts of this book. I would like to dedicate this monograph to Professor Philip Elving, without whom the Chemical Analysis Series would certainly have had less impact in the field today.

EDWARD S. YEUNG

Ames, Iowa
October 1986

CONTENTS

Detectors For
Liquid Chromatography

CHAPTER

1

REFRACTIVE INDEX DETECTOR

EDWARD S. YEUNG

Department of Chemistry and Ames Laboratory-USDOE
Iowa State University, Ames, Iowa

1. INTRODUCTION

The refractive index (RI) detector has a unique place in liquid chromatography (LC). It is one of very few universal detectors available. Unless the analyte happens to have exactly the same RI as the solvent, a signal is observed. Thus for the initial survey of samples, when the physical and chemical properties of the analytes are not known, the RI detector can provide useful information. The most severe limitation is the poor limits of detection (LOD) that can be achieved compared to other detectors. Moreover, for very complex samples, when components coelute, there is a possibility of mutual cancellation of the signal because both positive and negative RI changes can occur.

2. QUANTITATIVE RELATIONSHIPS

The RI detector responds to both the eluent and the analyte. The responses from components in a mixture are not additive. Therefore, quantitation is not as straightforward as in the absorbance detector, for example. Many different rules for predicting the RI of a two-component mixture have been suggested. The simplest is a linear interpolation of the RIs of the components (1),

$$n = n_A C_A + n_B C_B \tag{1}$$

where n, n_A, and n_B are the corresponding RIs of the mixture, the component A, and the component B. C_A and C_B are the fractional concentrations of the components, in units of weight, volume, or mole fraction. Because $C_A + C_B = 1$, one also has

$$n = n_A + (n_B - n_A)C_B \tag{2}$$
$$= n_A + kC_B$$

1

The difficulties associated with Eqs. 1 and 2 are obvious from the arbitrary definition of C, because a linear interpolation in one concentration unit implies a nonlinear behavior in the other two units. A more complicated form is derived (2), based on the Gladstone–Dale specific refraction:

$$\frac{100(n - 1)}{d} = \frac{w_A(n_A - 1)}{d_A} + \frac{(100 - w_A)(n_B - 1)}{d_B} \tag{3}$$

where w_A is the weight percent of the solute, and d, d_A, and d_B refer to the densities of the mixture and the two components.

It is also possible to start from first principles to see how RI depends on the components in a mixture. From the basic relations among the electric field, the electric displacement vector, and the polarization, one obtains the Clausius–Mosotti relation (3):

$$\chi = \frac{3}{4\pi} \frac{M}{N_0 \rho} \left(\frac{\varepsilon' - 1}{\varepsilon' + 2} \right) \tag{4}$$

where χ is the susceptibility per molecule, M is the molecular weight, N_0 is Avogadro's number, ρ is the density of the material, and ε' is the dielectric constant, which depends on the frequency of the electric field. It can be shown that, except at high densities or high field intensities, neither of which is true in LC, the susceptibility of a mixture is additive (3):

$$\chi = \sum_i x_i \chi_i \tag{5}$$

where x_i is the mole fraction of component i with susceptibility χ_i. It may be noted that for sufficiently low field strengths or high temperatures

$$\chi_i = \alpha_i + \frac{D_i^2}{3kT} \tag{6}$$

where α_i and D_i are the polarizability and the permanent dipole moment of the species i, k is the Boltzmann constant, and T is the absolute temperature. Now the volume fraction of the component i, C_i, is given by

$$C_i = \frac{x_i(M_i/\rho_i)}{\sum_j x_j(M_j/\rho_j)} \tag{7}$$

The refractive index of component i, n_i, is given by

$$n_i = \sqrt{\mu_i' \, \varepsilon_i'} \tag{8}$$

where μ_i' is the permeability of the medium. But because μ' is very nearly unity (typically deviating by less than 10^{-3}), we can replace ε_i' with n_i^2 in Eq. 4. Combining Eqs. 4, 5, 7, and 8, we have

$$\frac{n^2 - 1}{n^2 + 2} = \sum_i C_i \left[\frac{n_i^2 - 1}{n_i^2 + 2} \right] \tag{9}$$

Now consider the case of a binary mixture, which is composed of an analyte and an eluent. At any particular instant, the measured RI response is determined by the volume fraction of the analyte (of RI n_x) in the flow cell, C_x, and the volume fraction of the eluent in the flow cell $(1 - C_x)$. Using Eq. 9 and the subscript 1 for the eluent, we obtain

$$\frac{n^2 - 1}{n^2 + 2} - \frac{n_1^2 - 1}{n_1^2 + 2} = C_x \left[\frac{n_x^2 - 1}{n_x^2 + 2} - \frac{n_1^2 - 1}{n_1^2 + 2} \right] \tag{10}$$

Combining the two terms on the left of Eq. 10, we get $3(n^2 - n_1^2)/[(n^2 + 2)(n_1^2 + 2)]$, which in turn equals $3\Delta n_1(n + n_1)/[(n^2 + 2)(n_1^2 + 2)]$. Δn_1 is in fact the experimental observable from any of the differential RI detectors. Note that for low concentrations, errors much less than 0.1% are expected if all n's are now replaced by n_1's. Rearranging and redefining terms, we get

$$\Delta n_1 K_1 = C_x(F_x - 1) \tag{11}$$

where $F_i \equiv (n_i^2 - 1)/(n_i^2 + 2)$, and $K_i \equiv 6n_i/(n_i^2 + 2)^2$.

To appreciate the implications of Eqs. 10 and 11, we have plotted the actual dependence of the RI of a mixture of heptane and benzene as a function of the volume fraction of benzene in Fig. 1. The ordinates are shifted on the left compared to the right to facilitate visualization. The solid horizontal line is a linear interpolation of the RIs for the two pure solvents, and is commonly assumed in elementary discussions of applications of refractometry (1). The curved line is the actual dependence from Eq. 10. It can be seen that the linear interpolation does extremely poorly throughout. The dashed lines in Fig. 1 are the tangents to the curve for small concentrations of each of the components. From Eq. 11, the two slopes are given by $(F_x - F_1)/K_1'$ and $(F_1 - F_x)/K_x'$, respectively. Using a standard refractometer, we have independently confirmed linearity and the limiting slope, as specified by Eq. 11, for cases with the minor component present below 5% by volume for a number of binary mixtures involving combinations of hexane, benzene, chloroform, carbon tetrachloride, and carbon di-

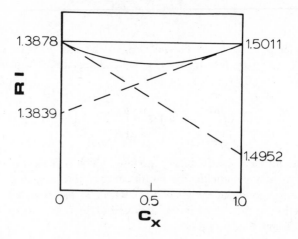

Figure 1. RI dependence on volume fraction for benzene in heptane; solid horizontal line, linear interpolation; solid curve, true dependence; dashed lines, limiting slopes. Reproduced with permission from R. E. Synovec and E. S. Yeung, *Anal. Chem.*, **55**, 1599 (1983). Copyright 1983 American Chemical Society.

sulfide. It can be seen from Fig. 1 that if the true dependence is plotted without a shift in the ordinate, the curvature is not obvious. Presumably this is why a linear interpolation (1) is used routinely.

Equations 10 and 11 do not hold for all situations. It has been pointed out (4) that the Lorentz–Lorenz term (5), $(n^2 - 1)/(n^2 + 2)$, in Eq. 9 can be replaced by the empirical Eykman term, $(n^2 - 1)/(n + 0.4)$. Good fit to experiments is also reported using the relationship

$$\frac{n - 1}{d} = \frac{1}{d_A - d_B}\left[n_A - n_B + \frac{(n_B - 1)d_A - (n_A - 1)d_B}{d_B + C_A(d_A - d_B)}\right][1 + C_A(1 - C_A)K] \tag{12}$$

where K is an empirical constant and C is the concentration in mole fraction. There is also a problem in the case of nonideal solutions. Nonideal mixing changes the implication of the volume fraction C_x, and in a sense makes the horizontal scale in Fig. 1 nonlinear. Some examples are mixtures of water–methanol and water–acetonitrile (6). In general, polar species interact with each other and nonideal mixing occurs. So Eqs. 10 and 11 are applicable only to mixtures of nonpolar species that do not differ greatly in molecular size.

3. INSTRUMENTATION

There are essentially four main types of RI detectors. They are grouped based on the physical principles used in the instruments, although variations in the optical arrangements are common, depending on the manufacturer. They all share a common feature, such that the output reflects the *difference* in refractive index between a sample flow cell and a reference flow cell. This is because (in contrast to an absorption detector) the chromatographic solvent also contributes to the signal. A differential measurement can then compensate for the solvent contribution to enhance detection of the analyte. Thus even though the Abbe refractometer is commonly used for RI measurements, it is of little value as an LC detector.

3.1. Deflection Type

Snell's law governs the angles of incidence and refraction at an interface, so that

$$n_1 \sin \theta_1 = n_2 \sin \theta_2 \tag{13}$$

where θ_1 is the angle of the beam with respect to the normal of the interface in the medium with RI of n_i. Figure 2 shows the dual-cell arrangement used in the commercial instruments. There are actually three glass cell walls to contain the liquids. If these are thin, their contributions to beam deflection can be neglected. Even if they are thick, they will provide a constant deviation if the cell walls

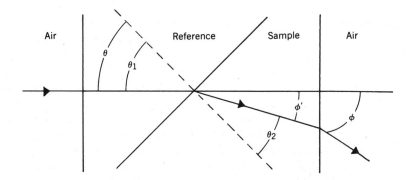

Figure 2. Propagation of light beams through the reference and sample flow cells in a deflection-type RI detector.

are pairs of parallel faces. For simplicity, we have assumed that the light beam enters normal to the first cell wall. n_2 is simply $n_1 + \Delta n$, where Δn is the contribution due to the presence of the analyte, according to Eq. 11. Equation 13 then becomes

$$n \sin \theta = (n + \Delta n) \sin(\theta - \phi') \tag{14}$$

At low concentrations, that is, those typical of LC after the column, Δn and ϕ' are both small, so

$$\phi' = \frac{\Delta n}{n} \tan \theta \tag{15}$$

If we further assume that the RI of air is unity, Snell's law gives

$$\phi = (n + \Delta n)\phi' \tag{16}$$
$$\simeq \Delta n \tan \theta$$

Actually, the commercial instrument uses a rear mirror to fold the light beam through the same cell a second time, doubling the net deflection. If the deflection is measured as a displacement of Δx at a distance X away from the cell, then

$$\Delta x = 2 \tan \theta \, \Delta n \, X \tag{17}$$

It can be seen that the sensitivity is best when θ is maximized. In practice, a compromise of $\theta = 45°$ has to be used to maintain a small cell volume. If X is 13 cm, then $\Delta x = 26 \, \Delta n$ cm. Because the dual-diode position sensor can detect a change of about 10^{-4} cm on its most sensitive setting (full scale), the maximum RI sensitivity is 3.8×10^{-6} units full scale.

The limitations on the commercial unit include beam collimation, sensitivity of the position sensor, and mechanical rigidity of the system. A slit formed by a mask defines the light beam from a tungsten lamp. If a laser is used instead, the beam collimation is improved and a longer optical lever can be used. Position sensors are now available that can detect deflections to 10^{-9} rad, or three orders of magnitude better than the ones used in commercial RI detectors. However, the pointing stability of the laser and the mechanical rigidity of the system will have to be improved substantially to take advantage of these new position sensors.

Finally, an interesting feature of the commercial RI detector is a mechanical null adjustment. Even though Fig. 2 shows no deflection whenever Δn is zero for any solvent RI, there is actually a deflection that depends on n because the incidence angle on the cell is not zero degrees. So, a glass plate is put in the optical path to compensate for the deflection by rotation in the incidence plane. If a true mechanical null is achieved when both reference and sample flow cells are filled with the chromatographic solvent, the *sensitivity* as well as the *detectability* of the instrument should remain constant regardless of n. The compensation plate must be rigidly mounted, however, so that additional drift and fluctuations are not introduced into the system.

3.2. Reflection Type

The behavior of light incident on a dielectric interface (Fig. 3) is described by both Snell's law of refraction, Eq. 13, and Fresnel's laws of reflection. Fresnel's laws describe the reflectivity and transmittance of light at an interface for the two types of linearly polarized light. The polarizations are p when the electric vector of light is parallel to the incidence plane made by the normal and the incident light ray, and s when the electric vector is perpendicular to the incidence plane. The reflectivity of both polarizations are described by the following equations:

$$R_s = \left| \frac{\cos\theta - \sqrt{n^2 - \sin^2\theta}}{n^2\cos\theta + \sqrt{n^2 - \sin^2\theta}} \right|^2 \tag{18}$$

$$R_p = \left| \frac{-n^2\cos\theta + \sqrt{n^2 - \sin^2\theta}}{n^2\cos\theta + \sqrt{n^2 - \sin^2\theta}} \right|^2 \tag{19}$$

R_s and R_p are the reflectivities for the s and p polarized light, respectively, n is the ratio n_2/n_1, and θ is the incident angle (7). Thus the measurement of Δn

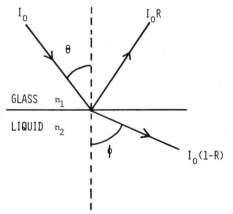

Figure 3. Reflectivity and transmittance of light at an interface: I_0, incident light intensity; θ, incident angle; ϕ, angle of refraction; n_1, refractive index before interface; n_2, refractive index after interface; R, reflectivity.

becomes a measurement of ΔI due to the change in reflectivity. The intensity change can be monitored in the reflected beam or the transmitted beam. However, the latter is preferred because the transmitted intensity is lower and background noise is reduced. To maximize the signal, one should work as close to the critical angle, $n = \sin \theta_c$, as possible. In practice, the light beam is not perfectly collimated so that one cannot approach the critical angle closer than the beam divergence. Also, absorption of light by the medium will cause the discontinuity in Eqs. 18 and 19 to become gradual transitions (8). In these situations, the point of highest sensitivity is for an angle slightly smaller than the critical angle, and one has to contend with a larger background of transmitted light.

To approach the critical angle at the liquid–glass interface, the light beam is coupled through a prism, as in Fig. 4. The transmitted light is backscattered by a metal plate that forms the opposite side of the cell. The metal plate is grained to scatter light mainly in the incidence plane. The scattered light is collected by a lens in a direction different from that of the reflected light at the first interface, which must be rejected by an aperture to maintain a low light background. The grooves in the scattering surface are randomly spaced, so that no interference fringes appear that are due to the grating-like structure. The cell region is defined by a gasket placed between the prism and the metal plate, and is properly coupled to the photodetector and the light source by masks and collimating lenses. A two-cell design is used, to allow compensation of the solvent contribution in the reference and sample flow cells. Light from the common light source is detected by separate photocells, the difference in which reflects Δn. The fraction of light backscattered is found empirically to be

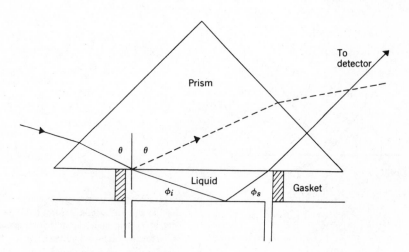

Figure 4. Propagation of light beams in a reflection-type RI detector.

$$S = [4.33 \times 10^{-4}(45 - \phi_s)] + 0.0187\phi_i^{0.0097(45 - \phi_s) + 0.705} \qquad (20)$$

where the angles are defined in Fig. 4. Thus no absolute RI scale can be established, and calibration is necessary.

The problems with the commercial system are that the cell volume is still quite large (9 μL), the light beam is not well collimated, the collected light is only a fraction of the total transmitted intensity, and there is difficulty in matching the two photocells, a lack of monochromaticity in the light beam, and nonlinearity as predicted by Eq. 20. Many of these problems can be avoided if a true transmission arrangement is used, by a second prism to couple the exiting light beam, and if a laser is used. One such arrangement (9) is shown in Figs. 5 and 6. In Fig. 5, a helium–neon (HeNe) laser is modulated between the two flow cells by a Bragg cell that is driven by a signal generator at 100 kHz. A 50-cm focal-length lens focuses the beam into the optical cell. An optical flat is used to balance the intensities of the two beams reaching the detector, based on the variation in natural reflection with angle and with polarization. After passing through the cell light is detected by a photodiode. The output from the diode is sent into either an oscilloscope or a lock-in amplifier. The oscilloscope was used

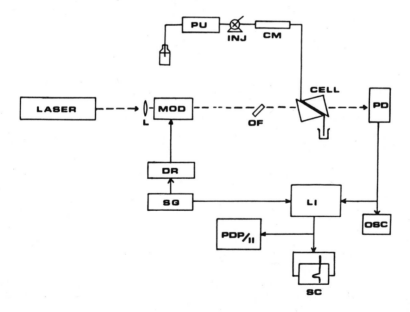

Figure 5. Refractive index detector and chromatographic system: PU, pump; INJ injector; CM, microbore column; LASER, He–Ne laser; L, lens; MOD, Bragg cell modulator; OF, optical flat; CELL, optical cell; PD, photodiode; DR, driver; SIG, signal generator; LI, lock-in amplifier; PDP/11, minicomputer; SC, strip-chart recorder; OSC, oscilloscope.

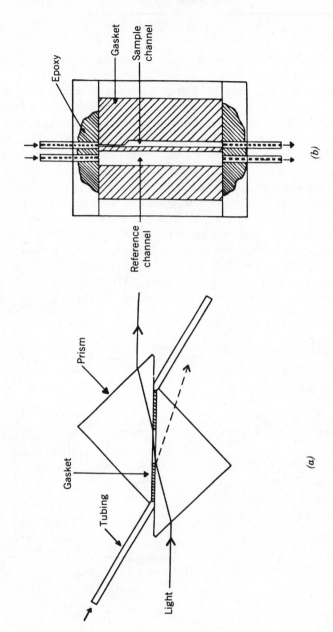

Figure 6. (*a*) Side view and (*b*) top cutaway view of optical cell.

10

to optimize the modulation of the Bragg cell and the lock-in amplifier converts the modulated signal to a DC signal with a 3-s time constant, which is in turn sent to both a strip-chart recorder and a minicomputer for data collection.

The 100 kHz high-frequency modulation allowed us to improve the intensity stability of the laser by lowering the "flicker noise" from a noise-to-signal (N/S) ratio of 5×10^{-3} to 2×10^{-5}. With the cell placed in the beam and adjusted to 20° less than the critical angle, we still have a N/S ratio of 2×10^{-5}, but with the cell adjusted near the critical angle (allowing 10% of the light to be transmitted) we see an increase in noise, making the ratio 2×10^{-4}. This increase in noise is caused by the beam being partially clipped by the optical beam aperture of the glass–liquid interface.

The cell consists of a Teflon tape gasket squeezed between two high-quality right-angle prisms, as shown in Fig. 6a. The Teflon tape defines the flow channels for both the reference and sample chambers as shown in Fig. 6b. The inlet and outlet tubes are made of 1/16-in. OD stainless steel tubing and are filed down to a wedge at the cell end to minimize dead volume. The connecting tubing between the column and the optical region is 4 cm long with an inside diameter of 0.004 in. The other three tubes (outlet to the sample chamber and inlet and outlet to the reference chamber) have inside diameters of 0.010 in. Epoxy seals the tubes to the prisms, and braces hold the tubes rigidly and minimize the danger of breaking the seal owing to accidental bumping. The prisms are pressed together by additional braces placed at the apex of each. This homemade assembly was mounted in a rotation stage with a resolution of 10^{-3} deg. For the sample channel in Fig. 6b (1-cm optical pathlength, a 80-μm-thick gasket, and a 1-mm-wide flow channel) the cell has a volume of 0.8 μL.

The test mixture was chosen to be benzene diluted in acetonitrile. Solutions were prepared by successive dilutions no greater than 100-fold at each step to minimize dilution errors. Figure 7 shows a chromatogram of benzene being eluted by the acetonitrile eluent at a flow rate of 30 μL/min. The dip before the peak is due to an unidentified impurity and does not add to the peak height or area. To obtain this peak, 0.5 μL of 5×10^{-4} (v/v) benzene in acetonitrile was injected. The RI of the test solution was calculated by using Eq. 11 with the RI of benzene and acetonitrile being $n_D = 1.501$ and $n_D = 1.344$, respectively. Taking into account the dilution to a peak volume of 3 μL, this peak shows a detectability of 2.0×10^{-7} RI units (S/N = 3, with noise = 1 standard deviation). This is comparable to the detectability of commercial RI detectors. The detectable mass of injected benzene is only 6 ng, however, which shows the advantage of using microbore columns for increased mass detectability. The chromatogram also shows the high number of plates possible with a microbore column, 78,000 plates/m in this case, with $K' = 1.2$. It should be noted that because the peak has a half-width of only 3 μL and commercial RI detectors have cells with optical volumes of approximately 10 μL (with total volumes

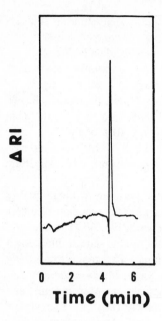

Figure 7. Refractive index chromatogram of benezene: peak is 370 ng of benzene.

much larger), the peak could not have been detected properly with any presently available RI detector without severe band broadening.

When optimizing the detector, questions arise such as what is the best angle at which to set the liquid–glass interface with respect to the critical angle, and how critical is the placement of the cell with respect to the focal point of the lens. Although Fresnel's equations predict the change in transmitted light per change in RI to increase as the critical angle is approached, the fractional flicker noise in the transmitted beam also increases. We found the optimum setting of ΔRI detection to be that with 10% of the incident beam being transmitted and occurring 0.004 deg from the critical angle. At this angle the transmitted intensities for s versus p polarized light differ by less than 1 percent; therefore one loses very little sensitivity by having randomly polarized light. Although the radius of curvature of the wave front of a focused Gaussian beam goes to infinity at the focal point, we did not see a noticeable difference in S/N ratio when working at the focal point versus slightly away from it. Our choice of a 50-cm focal-length lens is a compromise between the beam waist, which increases with increased focal lengths (thus requiring larger cell volumes), and the curvature of the wave front, which decreases with increasing focal lengths, to give a greater S/N ratio.

An interesting feature of this particular optical arrangement is that the light beam travels a full centimeter before exiting at the second interface. The geometry is favorable for the detection of absorption and fluorescence. This is in contrast to other small-volume optical cells which are based on transverse excitation or a Z-shaped flow path. For a solvent of low RI, the eluting analytes cause an increase in the RI and thus an increase in the transmitted light intensity (Eqs. 18 and 19). However, an absorbing analyte causes a decrease in the light intensity. The two types of events can then be readily distinguished from each other. By mounting a photomultiplier tube at one of the prism faces above the flow region, fluorescence can also be monitored. Therefore the same cell can be used to monitor RI, absorbance, and fluorescence in the LC effluent simultaneously. Because the same 1-μL region is used, no band broadening is expected compared to using the three types of detectors in a series arrangement.

3.3 Interference Type

When two light beams from the same source travel different optical paths and then recombine, constructive and destructive interference occur, depending on whether the optical path difference is an integer multiple or a half-integer multiple of the wavelength, respectively. Optical path length, however, is the physical path length multiplied by the RI. Thus the shift in the interference pattern can be used to measure changes in RI, if the physical path length is kept constant. This is the basis for a commercial interference type RI detector, as shown in Fig. 8. The conventional light source used makes this a "white light" interferometer. If the optical paths are identical, then constructive interference holds regardless of the wavelength of light. As the optical path changes, destructive interference occurs to different extents for different wavelengths. Thus the intensity at the detector does not follow a true cosine behavior as the RI changes. Perfect constructive and perfect destructive interference occur only when the optical pathlengths are identical. Fortunately, the light source is still narrow in spectral output compared to its wavelength; therefore useful interference fringes are still observed. To maximize the sensitivity, one must be close to the optical

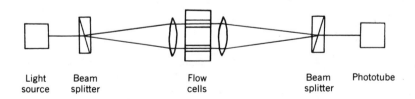

| Light | Beam | Flow | Beam | Phototube |
| source | splitter | cells | splitter | |

Figure 8. Optical arrangement in a commercial interference type RI detector.

null between the reference and the sample paths. This is accomplished by rotating an optical flat in one of the beams, artificially introducing an additional optical delay in that path.

The quality of the interference depends on the degree of collimation of the light. Also, because the intensity of light is translated directly to the output signal, shot noise is a limiting factor. Normally, only for phase differences within a small range of values can the output be considered linear with RI changes. In principle, nonlinearity here can be corrected by a mathematical function, for example, a damped cosine dependence, if the "zero" is well defined relative to the point of total constructive interference, but most likely shot noise will degrade such an attempt. Therefore, interchanging the existing lamp with a laser can bring some improvements in detectability.

To achieve significantly better detection, major changes in the optical scheme above seem inevitable. The most promising approach is probably Fabry–Perot interferometry. It is known that for plane-parallel mirrors, maximum constructive interference occurs whenever

$$m\lambda = 2dn \qquad (21)$$

where m is any integer (the order of interference), λ is the wavelength of light in vacuum, d is the separation of the mirrors, and n is the RI inside the interferometer. The interferometer is usually operated in the differential mode; that is, Δd rather than d is determined. The interferometer is scanned by a piezoelectric crystal, which can be quite linear in the resulting displacement with respect to applied voltage, or can be calibrated independently. Figure 9 shows the trans-

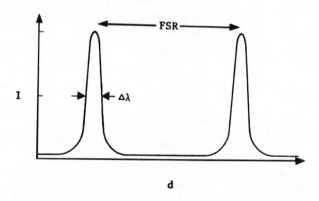

Figure 9. Transmission characteristics of a Fabry–Perot interferometer for a monochromatic source as a function of the mirror spacing d: FSR, free spectral range; $\Delta\lambda$, width of the interference peak.

mission characteristics of a monochromatic light source as d changes. One notes from Eq. 21 that successive interference peaks are separated by displacements of $\lambda/2n$. The equivalent change in λ to shift the interference peak to the same displacement is called the free spectral range (FSR) of the interferometer, and is equal to $\lambda/2nd$. Of importance is the width of the observed peaks, $\Delta\lambda$, which determines the resolution of the interferometer. This is described by a quantity known as the finesse, F, such that

$$F = \frac{\text{FSR}}{\Delta\lambda} \tag{22}$$

Interferometers of this type have been used to determine laser wavelengths ($\Delta\lambda/\lambda$) to an accuracy of 3 parts in 10^{11} (10) when n is held constant (when used in air). It is clear from Eq. 21 that $\Delta\lambda/\lambda = \Delta n/n$ for the same mirror separation, so that potentially the same level of detectability in RI changes can be achieved, if λ can be kept stable to the same level.

The main difference between the Fabry–Perot interferometer and others, which include the commercial detector (11), the Michelson interferometer (12), the Jamin interferometer (13), and the Mach-Zender interferometer (14), is the finesse F and thus the ultimate resolution. The Fabry–Perot device is a multiple-beam interferometer as opposed to the others, which are all two-beam interferometers. The latter have transmission properties following a cosine dependence on d, so that the finesse is limited to a value of 2. The former, however, can have F in the order of 200 depending on the alignment, the reflectivity of the mirrors, and the flatness of the mirrors. The interference peak is thus much sharper, and the sensitivity for measuring RI changes increases accordingly, even if the photoelectric detection scheme used is the same as in the commercial detector. With the increased resolution, one can instead use the *location* of the interference peak as the experimental observable. In practice, this translates to the position on the voltage ramp used to scan the interferometer. The peak can be located with a resolution much better than $\Delta\lambda$, which is the width of the peak. Using this observable, the RI measurement is then independent to first order of shot noise from the light source. Also, any degradation of the finesse due to flow inhomogeneities, mechanical instabilities, and alignment seriously affects the shape of the interference peak, but not so much the peak position. Furthermore, if by chance the analyte or any coeluting material absorbs light, any intensity-based detector is affected but the peak position detector is not. The Fabry–Perot interferometer is thus ideal for RI detection.

To function as an RI detector for LC, an appropriate flow cell must be placed inside the interferometer cavity. This must be accomplished without introducing additional losses, which degrade the reflectivity finesse. The first model of this (15) is based on having Brewster angle windows, and is shown in Fig. 10. For

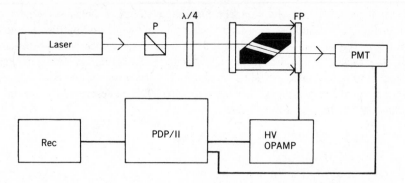

Figure 10. A RI detector based on Fabry–Perot interferometry: LASER, He–Ne laser; linear polarizer; λ/4, quarter-wave plate; FP, interferometer; PMT, photomultiplier tube; REC, chart recorder; PDP/II, minicomputer. Reproduced with permission from S. D. Woodruff and E. S. Yeung, *Anal. Chem.*, **54**, 1175 (1982). Copyright 1982 American Chemistry Society.

a frequency stability of a few megahertz, a single-frequency HeNe laser is used. To avoid feedback from the interferometer, the laser is isolated by a quarter-wave plate and a linear polarizer. Under computer control, the interferometer is scanned with the help of a high-voltage operation amplifier. The photomultiplier tube then provides a signal similar to that in Fig. 9. The interference peak is located by a simple point-by-point comparison algorithm and is then plotted on a chart recorder. In principle, even better determination of the peak position should be possible by fitting the experimental points with the known Airy function, but this has not been found to be necessary. Using a low-quality interferometer and a cell path length of 1 cm, one obtains a detectability of $\Delta n = 10^{-7}$ (S/N = 3). By using a longer cell (10 cm) and a more stable interferometer, the detectability is improved to $\Delta n = 1.5 \times 10^{-8}$ with some smoothing of the short-term noise, which is primarily due to acoustic noise in the room and fluctuations in the solvent delivery pump.

To improve further on this concept, a dual-beam arrangement must be used. Also, the liquid must occupy as large a fraction of the physical length of the interferometer as possible, because the effective RI that is measured is prorated between the liquid, the windows, and the air space. The air space contributes to instabilities through acoustic noise in the laboratory. The second-generation arrangement (16) is shown in Fig. 11. The laser is split into the two parallel beams by an optical flat for the reference and the sample beams. The feedback from the interferometer to the laser is of the order of 1%, and no further optical isolation is necessary. A normal incidence is chosen for the cell windows, so that the air space is minimized. To avoid optical losses, the windows are anti-

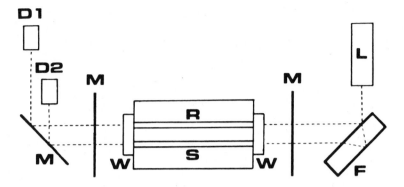

Figure 11. A double-beam RI detector: L, laser; F, optical flat; M, interferometer mirrors; W, cell windows; R, reference flow cell; S, sample flow cell; D1, phototube for reference beam; D2, phototube for sample beam; dashed lines, optical paths. Reproduced with permission from S. D. Woodruff and E. S. Yeung, *Anal. Chem.*, **54**, 2124 (1982). Copyright 1982 American Chemical Society.

reflection coated at the air interface. This geometry also avoids the need for rotation of the cell window when an eluent of very different RI is used. Two phototubes are used to obtain separate interference peaks from the two beams. The electronic controls are similar to the earlier version (15), except that the difference in the peak positions in the two beams is plotted. Temperature drifts, mechanical vibrations, laser frequency instabilities, and misalignment can then be partially compensated for. The computer keeps track of how many FSR the interference peak has shifted, so that a large dynamic range is possible. This is shown in Fig. 12a, where each chromatographic peak is derived from shifts of several FSR in the interferometer. Good linearity is seen in Fig. 12b, where both the injected concentration and the vertical scale have been decreased by a factor of 20. The detectability of the system is $\Delta n = 7 \times 10^{-9}$ for water as the eluent and $\Delta n = 4 \times 10^{-9}$ for acetonitrile.

A few variations on the above designs are possible. Data collection and scanning of the interferometer can be accomplished by analog circuitry (17). This allows scanning at 100 Hz to improve the base-line stability. Also, the photoelectric detector can be a simple photodiode rather than a photomultiplier tube, because the transmission characteristics of available Fabry–Perot mirrors are very good. Finally, confocal Fabry–Perot interferometers have twice the free spectral range as plane ones for the same physical length. The optical beam is also naturally focused at the center. These two properties can lead to smaller detector volumes at the same level of detectability.

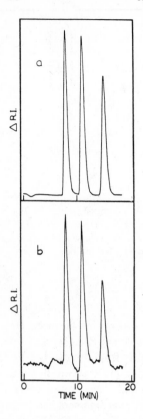

Figure 12. Refractive index chromatograms for a mixture of α-glucose, sucrose, and raffinose, in the order of elution: (*a*) 14.4 μg of each injected; (*b*) 0.72 μg of each injected, with the scale expanded by a factor of 20. Column, C_{18} 10-μm, 25 cm × 4.6-mm ID; eluent, water at 0.5 mL/min. Reproduced with permission from S. D. Woodruff and E. S. Yeung, *Anal. Chem.*, **54**, 2124 (1982). Copyright 1982 American Chemical Society.

3.4. Christiansen Effect Type

The Christiansen effect is based on natural reflection and refraction of light by Snell's law and Fresnel's laws, summarized in Section 3.2. When light travels through a sample cell filled with solid particles of RI identical to that of the eluent, it propagates without disturbance through the cell. However, any slight change in the RI of the eluent, for example, when analyte is eluted, causes reflection and refraction at the face of each particle. The intensity of light exiting the cell thus decreases, and can be used as a measure of Δn. For example, isooctane and lithium fluoride have almost identical RI around 530 nm. A narrow-band light source is therefore used to assure a match in RI. The commercial instrument has a second flow cell used as a reference to compensate for intensity fluctuations in the light source and temperature fluctuations in the eluent. The need for solid particles limits the size of this detector, and interface to micro-columns is unlikely. The match in RI limits the choice of solvents, and thus the versatility of LC. Also, absorbing species cause a signal that cannot be distinguished from the RI effect. Finally, the particles themselves may interact with the analyte and change the separation process. Compared to the other types of RI detectors, the Christiansen effect type is the least attractive alternative.

4. APPLICATIONS

It is beyond the scope of this chapter to survey the many applications of RI detectors in LC. Instead, several unique features of RI detection are discussed here. These features result in distinct advantages for LC detection for special analytical problems. The implementation of these ideas does not involve extensive modifications of standard instrumentation. Thus they have the potential for broad applications.

4.1. Sequential Differential Detection

The RI detector is a number density detector, according to Eq. 5. Therefore, even when no analyte is eluted from the column, any fluctuations in the number density of the eluent in the optical region appear as a background signal. Eventually, these may be the limiting factors in achieving good detectability. The effect of temperature can be determined from the coefficient of thermal expansion of the liquid. The temperature dependence, dn/dT, ranges from -1×10^{-4} K^{-1} for water to -6.4×10^{-4} K^{-1} for benzene at room temperature (4). We note that water at 4°C has a zero coefficient of thermal expansion, and provides an extremely favorable condition for RI measurements if technical problems for operating at 4°C can be solved. The pressure dependence of RI can be determined from the isothermal compressibility of the liquid. Thus dn/dp ranges from 0.16×10^{-9} m^2/N for water to 0.90×10^{-9} m^2/N for pentane (6). Because most often RI detectors are connected to the outlet end of the chromatographic column, where the pressure is essentially at atmospheric pressure, one does not expect much contribution from pressure to the baseline in the chromatogram, even considering changes in atmospheric pressure. This is not true if the outlet of the RI detector is restricted, for example, by devices to control back pressure or by narrow-bore connections to other detectors. The pressure drop proportionately reflects the pressure fluctuations in the solvent delivery system to add to baseline noise. Even when no intentional restrictions are used at the detector outlet, solvent viscosity may produce pressure changes at the detector. The viscous liquid responds slowly to the strokes of the pump in the solvent delivery system, synchronously causing pressure pulses at the optical region. This effect decreases at low flow rates and if, for example, syringe pumps are used.

Of these density fluctuations, the most serious problem is temperature effects.

For a stability of $\Delta n = 10^{-7}$, the temperature has to be held to $10^{-3}°C$ for water or $1.6 \times 10^{-4}°C$ for benzene. Therefore most RI detectors depend on a reference/sample arrangement to offset room temperature changes partially. Then only the temperature *difference* in the two cells has to be controlled, not the absolute temperature. Some instruments even use an active thermostated heater to provide temperature stabilization.

In practice, however, it is very difficult to achieve a true cancellation of the contributions of the LC eluent even using a sample and a reference chamber. The reason is that the two chambers are effectively isolated from each other so that the environment in one chamber is always slightly different from the other. The inability to reproduce the conditions in the two chambers is then the primary limitation in differential detectors. Because the reference cell in the RI detector is typically used in the static mode, complete equilibration of temperature and pressure relative to the sample flow cell is unlikely. It is in principle possible to split the eluent before the point of injection to be used in the reference cell, with or without an intervening matching column. Even so, the elution process in the analytical column can itself lead to differences in temperature and pressure in the sample flow cell. The extreme case is that of gradient elution, when one must balance the sample and the reference cells to the order of the RI change of the analyte.

One solution to this problem is to use differential detectors in series along the chromatographic effluent stream (19). It is clear that because the two chambers are connected (by a delay loop of volume V), pressure and temperature differences can be minimized. A response in the sample cell at any given elution volume, V', during the separation will give an equal but opposite response in the reference cell (opposite because of the differential nature of the detector output) at an elution volume $V' + V$. If the eluent flow rate is F (mL/min), this corresponds to a time delay of V/F (min) between the equal but opposite signals of the same event to register at the detector. Therefore, to produce chromatograms that resemble conventional detection methods, one can simply add to the detector signal at any given instant the particular detector signal recorded exactly V/F min earlier in the separation. The conversion is simple enough to be handled by a microprocessor in real time. The concept of sequential differential detection thus allows proper equilibration between the sample and the reference cells of a differential detector, while retaining all of the chromatographic information.

A closer examination shows that there are in fact some guidelines toward the choice of the delay volume, V. If V is too large, band broadening within the delay loop destroys the simple correlation between the responses in the sample cell and the reference cell. If V is too small, the net signal obtained for any chromatographic event is reduced, because the response then becomes a true derivative of the normal chromatographic signal. Sensitivity in the measurement is then sacrificed. A reasonable choice is then a volume on the order of half the elution volume of a typical chromatographic peak ($\frac{1}{2}$ the width of the base) under the particular conditions for separation. In fact, because the differential detector is now properly balanced, detectability may even be improved.

The main reason for the series arrangement is to balance the sample and the reference flow cells. It is therefore important to compare the noise level there to that of a conventional arrangement. Using the series arrangement and a commercial RI detector, we obtained noise level comparisons owing to a reciprocating pump with chloroform as the eluent, as shown in Fig. 13. The peak-to-peak pressure fluctuations due to the stroking of the pump are four times worse using the conventional, static reference cell (Fig. 13a) compared to the series arrangement (Fig. 13b). The residual fluctuations in the latter are probably due to a slight mismatch in the inlet tubes to the two cells, and a slight pressure drop across the delay loop. A more careful optimization of the length versus diameter of the delay loop may bring further improvements.

To show that this scheme works even for complex chromatograms, we studied the separation of benzene, cumene, and 1,2,3,5-tetramethylbenzene using acetonitrile as the eluent. The untreated response is shown in Fig. 14a and the calculated response (by adding the response V/F min earlier) is shown in Fig.

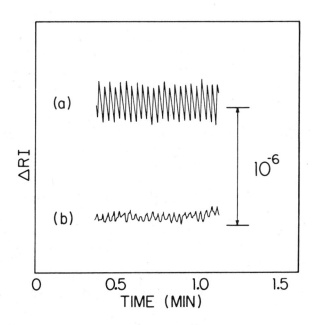

Figure 13. Noise level due to pressure fluctuations in a conventional arrangement (a) and a series arrangement (b). Reproduced with permission from Ref. 19.

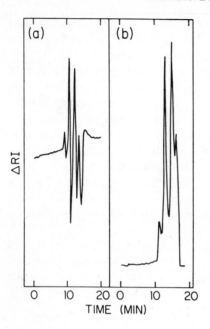

Figure 14. RI response (*a*) and reconstructed chromatogram (*b*) of the separation of a mixture of benzene, cumene, and tetramethylbenzene in acetonitrile. Flow rate, 0.33mL/min; delay, 88 μL; full scale, 10^{-7} RI units; column, C_{18} 10-μm, 25-cm × 4.6-mm ID; mobile phase, acetonitrile. Reproduced with permission from Ref. 19.

14*b*. The results are again as expected, and reproduce even an impurity peak eluted before benzene. For each species, the peak width, shape, and height are as expected in a normal chromatogram. It should be noted that for a changing baseline, only the signal superimposed on the baseline should be used in correcting the response at a time V/F min later. Otherwise, the calculated chromatogram will show an incorrect baseline. One can see that the calculated response at any time is derived from all times $m(V/F)$ min earlier in the separation, where m includes every integer from 1 to V'/F. The resulting chromatogram thus provides an averaging effect on any base-line noise, as long as the noise is random.

Perhaps the worst case for balancing the sample and the reference flow cells is gradient elution in LC. A typical gradient may involve an RI change of 0.01–0.10 units. The difficulty is to try to reproduce the eluent composition at any arbitrary time at the reference cell. It can be seen that if a gradient is used with a linearly varying volume fraction, a series arrangement will produce a constant difference signal. The magnitude of this difference signal is simply the total RI change multiplied by the ratio of the delay volume and the total volume of the gradient. If desired, this constant difference can even be nulled out optically by the detector. A test of this concept using a commercial system programmed to generate a linear gradient from pure chloroform to pure carbon tetrachloride at 5%/min and at 1.0 mL/min flow is shown in Fig. 15. We note that the null point for this detector (as seen from the extreme left and the extreme right of the chromatogram) varies with the RI because of the inherent design of the optics. Taking this into account, the shape of the baseline is as predicted. In fact, the difference signal is about 1/100 of the difference in RI between chloroform and carbon tetrachloride, as predicted by a 200-μL delay loop and a 20-mL gradient. Normally, the chromatogram will drift off-scale. In Fig. 15, useful information can still be obtained.

4.2. Quantitation Without Standards

The details of quantitation without standards is discussed in Chapter 9. For the RI detector, it can be seen that Eq. 11 relates the two unknown quantities—the concentration and the RI of the analyte—to the observed chromatographic peak size through all known parameters (the RI of the eluent). If the same sample is eluted with a second eluent, identical in elution strength to the first but with a different RI, a different peak size will be observed. The two chromatographic signals then give rise to two independent equations of the form of Eq. 11, which can be solved to determine the concentration of the analyte, C_x, and the RI of the analyte (via F_x).

An example of such an application is in the characterization of crude oils (20). The same sample of crude oil is eluted in a gel-permeation (GPC) column using first toluene and then chloroform as the eluent. Because the elution order

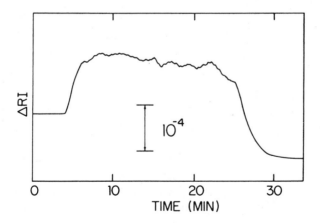

Figure 15. Base-line response of a linear gradient from chloroform to carbon tetrachloride at 5% per min. Flow rate, 1.0 mL/min; delay, 200 μL. Reproduced with permission from Ref. 19.

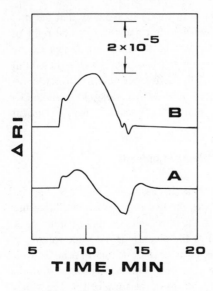

Figure 16. RI chromatograms of North Slope crude oil using GPC. A, toluene eluent; B, chloroform eluent; flow rate, 0.8 mL/min.

of components is expected to be constant, one can divide up the individual chromatograms in Fig. 16 into 1-s time segments. A pair of simultaneous equations according to Eq. 11 can then be obtained for each time segment in the two chromatograms, and a concentration can be calculated. A concentration chromatogram can then be constructed from these calculations, and is displayed in Fig. 17a. This obviously contains meaningful information for characterization of the crude oil, unlike the original chromatograms. The absorbance chromatogram, Fig. 17b, also gives an incorrect picture of size distribution because most of the crude oil components do not absorb. As mentioned above, the RI for materials eluted in each time segment can also be calculated, as shown in Fig. 17c. The extension of this example to the study of biological materials with high molecular weights using GPC is straightforward.

It is well known that the retention times of components in GPC are related to their molecular sizes, that is, the molar volume V_m (mL/mol). For a given chromatographic column, this relationship can be calibrated using model compounds. By calculating the moles of solute at each slice in time using the V_x chromatogram, the total moles of the crude oil can be calculated. V_x is the volume fraction (at time T) with respect to the injected volume. By multiplying V_x by the injection volume V_I and dividing by the effective molar volume of the solute (at time T), the moles x, of the solute, per slice can be calculated by

$$C_x = \text{moles } x = \frac{V_x}{V_m} V_I \qquad (23)$$

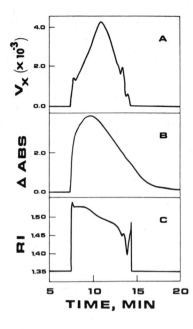

Figure 17. Characterization of North Slope crude oil by size-exclusion chromatography. A, volume fraction of materials eluted each second; B, absorbance of materials at 365 nm; C, RI of the materials; volume injected, 10 μL.

Both the V_x and the C_x chromatograms for the North Slope crude oil and an Arun crude oil are shown in Fig. 18. It is interesting to note that the two crude oils are quite different with respect to molecular size distributions. It is known that the Arun crude oil was a condensate (i.e., somewhat refined), and was sampled directly from an ocean tanker. The North Slope crude oil, on the other hand, was similar to a "true" crude oil in appearance, and was obtained by direct sampling from the Alaskan pipeline. Thus it was expected that the Arun crude oil would contain fewer components, at higher number densities per component, and at molecular sizes somewhat smaller than the North Slope crude oil because of the condensation processing.

A unique application of this quantitation scheme is for testing the purity of a chromatographic peak in LC. Following the same procedure as above, one can calculate the RI at various points in a chromatographic peak. This is similar to that in Fig. 17c, but the time segments are much shorter, down to the limit of the time response of the detector. If a chromatographic peak is pure, a constant RI is obtained across the peak. RI variations during a peak are indicative of incomplete resolution of components. A scheme based on the absorption ratio at two wavelengths has been suggested along the same lines (21). However, the RI method is more universal because it is also applicable to analytes that are nonabsorbing.

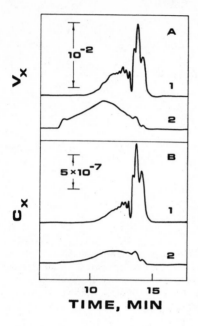

Figure 18. Concentrations of materials in crude oils separated by size-exclusion chromatography. (*a*) Volume fraction eluted each second; (*b*) Moles eluted each second; 1, Arun crude oil; 2, North Slope crude oil.

Another application of this quantitation scheme is the determination of elution orders of analytes in LC without identification. This is often needed to optimize the separation. The correlation scheme (22) for determining the elution orders for a sample consisting of N analytes in two different eluents (regardless of the columns used) is based on obtaining the chromatogram of the same sample in a third eluent (regardless of the column used). To use the RI detector, this third eluent should have an RI different from those of the first two eluents. It is assumed that the N chromatographic peaks obtained in each of the three eluents are resolved well enough to obtain areas for each peak, by deconvolution if necessary. We have already shown above that if the correct elution orders are known, then any two of these chromatograms will provide enough information to predict the peak areas in the third, or any other chromatogram. This is because the two chromatograms can be used to calculate, without analyte identification, the concentration and the RI of each component in the sample. This information can then be used to predict the response of each of the components in an arbitrary third eluent. Thus the unknown elution order in the third eluent can be determined simply by matching the peak areas found experimentally and the predicted peak areas. However, if the elution orders are not known in either of the first two eluents, there can then be a total of $N!$ possible combinations of predictions from the first two chromatograms. The idea then is to find the one combination out of the $N!$ combinations that best predicts the third chromatogram with respect to the individual peak areas, with the help of some least-squares criterion. Thus one relies on achieving consistency in the quantitative information among the assigned elution orders in the three chromatograms to arrive at the best choice, without requiring analyte identification.

5. SUMMARY

The RI detector, despite many experimental difficulties, including the lack of sensitivity and background contributions, is currently the only routine universal detector for LC. For substances that do not absorb in a convenient spectral region, the RI detector can at least provide useful information. Any future developments in improving the LOD or stabilizing the background signal will greatly enhance the utility of RI detection.

ACKNOWLEDGMENTS

I thank the many co-workers in my laboratory who have contributed to the work described here, particularly S. D. Woodruff, R. E. Synovec, S. A. Wilson, and G. Y. Chen, and the U.S. Department of Energy, Office of Basic Energy Sciences, Division of Chemical Sciences, for partial research support through the Ames Laboratory, Iowa State University, under contract No. W-7405-eng-82.

REFERENCES

1. S. Z. Lewin and N. Bauer, in I. M. Kolthoff and P. J. Elving, Eds., *Treatise on Analytical Chemistry*, Part I, Vol. 6, Wiley-Interscience, New York, 1965, Chapter 70.

2. N. Bauer and S. Z. Lewin, in A. Weissberger, Ed., *Physical Methods of Organic Chemistry*, Vol. 1, Part 2, Wiley-Interscience, New York, 1960, p. 1156.

3. J. O. Hirschfelder, C. F. Curtiss, and R. B. Bird, *Molecular Theory of Gases and Liquids*, Wiley-Interscience, New York, 1964, p. 852.

4. J. A. Riddick and W. B. Bunger, in A. Weissberger, Ed., *Techniques of Chemistry*, Wiley-Interscience, New York, 1970, p. 27.

5. M. Born and E. Wolf, *Principles of Optics*, 5th ed., Pergamon, New York, p. 100.

6. M. N. Munk, in T. M. Vickrey, Ed., *Liquid Chromatography Detectors*, Dekker, New York, 1983, p. 165.

7. G. R. Fowles, *Introduction of Modern Optics*, 2nd ed., Holt, Rinehart and Winston, New York, 1968, p. 44.

8. G. H. Meeten, A. N. North, and F. M. Willmouth, *J. Phys. E: Sci. Instrum.*, **17**, 642 (1984).

9. S. A. Wilson, Ph.D. thesis, Iowa State University, Ames, IA, 1985.

10. P. T. Woods, K. S. Shotton, and W. R. C. Rowley, *Appl. Opt.*, **17**, 1048 (1978).

11. For example, Optilab 902 refractometer, Vallingby, Sweden.

12. R. P. W. Scott, *Liquid Chromatography Detectors*, Elsevier, Amsterdam, 1977, p. 95.

13. D. A. Cremers and R. A. Keller, *Appl. Opt.*, **21**, 1654 (1982).

14. C. C. Davis, *Appl. Phys. Lett.*, **36**, 515 (1980).

15. S. D. Woodruff and E. S. Yeung, *Anal. Chem.*, **54**, 1175 (1982).

16. S. D. Woodruff and E. S. Yeung, *Anal. Chem.*, **54**, 2124 (1982).

17. G. Y. Chen and E. S. Yeung, unpublished results.

18. Gow-Mac Instrument Co., *Operating Manual, Christiansen Effect Detector Model 89-100*, Bound Brook, NJ.

19. S. D. Woodruff and E. S. Yeung, *J. Chromatogr.*, **260**, 363 (1983).

20. R. E. Synovec and E. S. Yeung, *J. Chromatogr. Sci.*, **23**, 214 (1985).

21. A. C. J. H. Drouen, H. A. H. Billiet, and L. D. Galan, *Anal. Chem.*, **56**, 971 (1984).

22. R. E. Synovec and E. S. Yeung, *Anal. Chem.*, **56**, 1452 (1984).

23. R. E. Synovec and E. S. Yeung, *Anal. Chem.*, **55**, 1599 (1983).

CHAPTER

2

ABSORPTION DETECTORS FOR HIGH-PERFORMANCE LIQUID CHROMATOGRAPHY

ROBERT B. GREEN

*Instrumental Chemical Analysis Branch, Department of the
Navy, Naval Weapons Center, China Lake, California*

Almost every review of high-performance liquid chromatography (HPLC) will concede that the lack of a detector comparable in sensitivity to those available for gas chromatography (GC) has been a limiting factor. This deficiency is a result of the similarity of the mobile and stationary phases in HPLC. Both the solute and the solvent are liquids, very often with nearly identical physical and chemical properties. This book examines many approaches for addressing this dilemma.

This chapter concentrates on absorption detectors for HPLC. These include the detection of ultraviolet–visible (UV–VIS) molecular absorption, a relatively mature approach driven by new technology, along with more recent approaches utilizing UV–VIS atomic absorption and infrared (IR) absorption. Finally, intracavity absorption spectrometry is addressed as a possibility for HPLC detection.

The coverage of this chapter is not comprehensive, particularly in the area of applications. The information presented here was drawn primarily from citations compiled from a computer search of the Chemical Abstracts data base from 1967 through 1984. Also included is discussion of commercially available instrumentation where appropriate (1). Although products of specific manufacturers are mentioned as examples, instrumentation with similar capabilities may be available from other sources. The choices for discussion were obtained from information generated by responses to a general mailing in 1984. The mailing list was drawn from vendors listed in *Analytical Chemistry*'s 1983–1984 Lab-Guide (2). This biennial publication is an excellent source of information on commercial manufacturers of analytical instruments.

1. ULTRAVIOLET–VISIBLE ABSORPTION DETECTORS

Ultraviolet–visible detectors may be divided into two categories for discussion: atomic absorption detectors and molecular absorption detectors. The latter are discussed first because they maintain an overwhelming advantage in terms of current usage when considering optical detectors for HPLC. Historically, UV–VIS molecular absorption detectors have played a prominent role in the development of HPLC and have been commercially available since the earliest days of HPLC research.

The absorption of ultraviolet or visible light by an atom or a molecule is the result of an electronic transition; that is, a valence electron is promoted to a higher, bound excited state (Fig. 1a). The wavelength of light absorbed depends on the discrete energy levels involved in the transition and the quantity of light absorbed is proportional to the number of absorbing centers or the concentration of the absorber. Because the energy level distribution is specific for a particular atom or molecule, the former condition is a sufficient criterion for selectivity and the latter relationship governs sensitivity.

This energy absorption, measured as transmitted light, is codified as Beer's law (3), which states that $\log P_0/P = \log 1/T = A = \varepsilon bC$, where P_0 = incident radiant power, P = transmitted radiant power, T = transmittance, A = absorbance, ε = molar absorptivity (L/mole · cm), b = path length (cm), and C = concentration (moles/L). The value of ε is characteristic of the absorbing substance at a specific wavelength in a particular solvent and is independent of the concentration and path length. Absorbance is linear with concentration whereas the relationship of concentration to transmittance is logarithmic. Using modern data systems, it is often possible to compensate for nonlinearity by point-to-point calibration curves or curve-fitting routines. These techniques are approx-

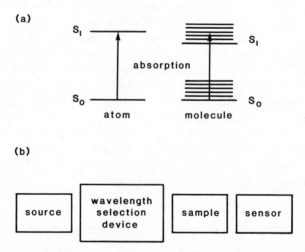

Figure 1. (*a*) Simplified energy level diagrams. The molecular diagram shows vibrational levels superimposed on the electronic states, S_0 and S_1. (*b*) Block diagram of a UV–VIS absorption detector.

imations and lead to uncertainty in the results. A linear system is always preferable.

Beer's law assumes that the incident radiation is monochromatic, absorption occurs in a volume of uniform cross-section, and absorbing species behave independently of each other in the absorption process. This last assumption implies that the absorbance of a multicomponent mixture at any wavelength is the sum of the absorbances of the individual components at the same wavelength. This relationship is the basis of quantitative methods for determining the individual concentrations of absorbing substances in mixtures. Equations written for the total absorbance of all components at the same wavelength may be solved simultaneously to yield the unknown concentrations (3). As the number of the components and hence the number of equations increase, solutions become increasingly difficult. Significant spectral overlap of the components also reduces the accuracy of the results.

In simplest terms, a UV–VIS absorption detector consists of a source, a wavelength selection device, a sample, and a sensor (see Fig. 1b). The source emits radiation in the UV–VIS region of the spectrum. The wavelength selection device isolates and passes the appropriate wavelength(s) from the source and rejects the rest of the light. The wavelength selection device may precede the sample as indicated or it may follow it. The sensor responds to the transmitted light and generates a measurable electrical signal.

Noise or unwanted fluctuations of the signal may originate in the detector system or the separation system (4). The source can generally be diagnosed by determining the effect of changing the solvent flow rate on the noise. If the noise depends on flow, the separation system is likely responsible. Early UV–VIS molecular absorption detectors were very sensitive to flow rate. This led to the development and use of constant flow-rate pumps to minimize fluctuations in the detector signal. The primary cause of flow rate sensitivity was temperature-induced changes in the index of refraction (5,6). This effect can be reduced by thermal equilibration of the solvent with the detector cell. Bubbles in the solvent stream due to dissolved gases upset UV–VIS detectors and can be eliminated by degassing of solvents or pressurizing detector cells with a suitable flow restrictor downstream. Drift may be a problem where the solvent composition changes with time, for example, gradient elution. Reproducible drift may be compensated for by electronic means (7) or the use of matched-flow sample and reference columns with a dual-beam detector (8).

Noise originating within the detector system may be traced to any of the electronic components. Johnson noise results from thermally induced motion of electrons in resistive circuit elements. The average number of electrons moving in a given direction in a resistor element changes continuously owing to this motion. Johnson noise occurs in all transducers and electronic components of photometers and spectrophotometers, even in the absence of current. Johnson noise is insignificant relative to other sources of noise in phototubes, photo-multipliers, and amplifiers but it is important in thermal detectors for infrared radiation (3). Flicker or $1/f$ noise $(f =$ frequency) in photometers and spectro-photometers is principally associated with the radiation source. The magnitude of flicker noise cannot be predicted from theory but it can be eliminated by appropriate instrument design. In atomic absorption detection schemes, the flame atom reservoir into which eluents are aspirated may contribute to optical flicker noise. Environmental noise such as 60-Hz pickup can also be avoided by good instrument design, incorporating proper shielding and grounding. Sources of UV–VIS radiation degrade with time and consequently the intensity of the light decreases. As the source degrades, the relative importance of shot noise increases. Shot noise arises from the transfer of electrons or other charged particles across junctions, such as in phototubes. The current is the result of a series of random events and therefore can be treated statistically (3). Shot noise cannot be entirely eliminated from an instrument. The uncertainty introduced into absorbance mea-surements by imprecision in cell positioning may be ignored because HPLC detectors generally have fixed cells. With all these qualifications, shot noise is usually the limiting factor in modern, well-designed UV–VIS absorption detec-tors.

1.1. Molecular Absorption Detectors

In the face of competition from more sensitive and selective detection techniques, why does the popularity of UV–VIS molecular absorption detectors persist? The pressure of the marketplace is one reason. Many companies now manufacture UV–VIS detectors; about 20,000 units were produced in the first decade of HPLC (4). More compelling reasons also exist for the continued use of UV–VIS detectors. Many solute molecules of interest contain chromophores, making them amenable to detection. The complex molecules typically separated by HPLC have large molar absorptivities, permitting sensitive detection as well. Quanti-fication of solutes is straightforward using well-characterized mathematical re-lationships. Just as importantly, solvents that are commonly used for HPLC are generally transparent in the UV–VIS spectral region. Another factor is that UV–VIS detectors are relatively inexpensive to purchase and maintain and are simple to use. UV–VIS spectrometry itself benefits from prior separation of a sample into its component molecules. UV–VIS spectra of molecules in solution tend to be broad and featureless, making qualitative identification of the com-ponents in a mixture difficult without separation. In some cases, quantitative determinations by UV–VIS spectrometry alone are equally complicated by spec-tral overlap.

It is possible to convert nonabsorbing species into compounds that absorb UV–VIS radiation. The addition of a chromophore to a nonabsorbing molecule is called "derivatization" (9,10). Initially, derivatization was performed on a sample prior to injection into the HPLC or after elution from the HPLC. Off-line techniques contribute to sample contamination and the introduction of artifacts into the overall analysis as well as requiring additional time, materials, and equipment. On-line methods are preferred for both pre- and postcolumn derivatization to minimize these problems (11). Of the two, postcolumn derivatization has received more emphasis. Automated instrumentation has been introduced for performing postcolumn reactions (12). It should be recognized that better selectivity may result from derivatization as well as improved sensitivity (11). Further discussion of derivatization is beyond the scope of this chapter.

Recently, the introduction of the self-scanning photodiode and other array devices has caused a resurgence in interest in UV–VIS spectroscopy, which lends itself to multichannel detection. An entire UV–VIS spectrum may be simultaneously acquired as the solute elutes from the HPLC column by replacing a conventional single-channel sensor with an array detector. Even with multichannel detection capability, UV–VIS multicomponent analysis presents a challenge. Chemometrics suggests that it is possible to extract additional valuable information from what may seem to be an overdetermined system (13). Using microprocessor technology with multivariate analysis, in some cases it is possible and practical to overcome the inadequacies of the sensor through data processing. The data of interest may be extracted from the information matrix that is generated for each eluting molecule with photodiode array detection. The rapid acceptance and commercialization of UV–VIS molecular absorption with multichannel detection along with the development of microcomputer hardware and software for data analysis has had a significant impact on modern HPLC.

UV–VIS molecular absorption detectors may be classified as follows: (1) photometric detectors, sometimes referred to as fixed-wavelength detectors; (2) spectrophotometric/single-channel detectors, sometimes referred to as variable-wavelength detectors; and (3) spectrophotometric/multichannel detectors, sometimes referred to as multiwavelength detectors. Each of these is discussed in terms of instrumentation and applications.

1.1.1. Photometric Detectors

These detectors utilize atomic vapor lamps as sources of radiation. In these lamps, a discharge produces electrons whose collisions with the atoms result in line spectra superimposed on a continuum, even at high pressures. The wavelength(s) of interest is isolated with an appropriate filter. The resolving power of a monochromator is unnecessary because the narrow atomic lines emitted by the source are widely separated and unwarranted because of the additional expense. A photovoltaic cell, photodiode, phototube, or photomultiplier may be

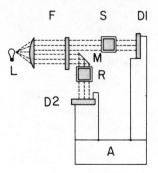

Figure 2. Schematic diagram of a double-beam photometer. L = light source, F = filter, S = sample, D1 and D2 = photodiodes, M = mirror, R = reference, A = amplifier.

used as the sensor. Figure 2 is a schematic diagram of a typical photometric detector using a double-beam optical design. Virtually all photometric detectors employ the double-beam configuration to compensate for source intensity fluctuations due to voltage variations. Photometric detectors are commonly referred to as "fixed-wavelength" detectors because the source emission is not continuously tunable. In spite of this limitation, it is usually possible to isolate several different wavelengths, depending on the atomic vapor contained within the source, by inserting an appropriate filter. A low-pressure mercury lamp is the most common photometric source although lamps containing zinc, cadmium, and magnesium are now available from HPLC absorption detector manufacturers (see Table 1). Besides increasing the total number of available lines, these newer detectors all have emission lines at shorter wavelengths than provided by the mercury lamp. This permits the detection of a wider range of chromophores. Many molecules have larger molar absorptivities below 254 nm. Shorter wavelengths permit the detection of molecules such as peptides or proteins, which have little or no absorptivity at 254 or 280 nm. The zinc lamp, in particular, produces an almost universal response from approximately 90% of all organic molecules at 214 nm (14). However, as the wavelength decreases, background absorbance from all eluents often increases, placing constraints on the choice of buffers used. It is also possible to use a phosphor-coated element that will absorb UV light from the source and reemit it at longer wavelengths. Emission intensity from phosphors is generally lower than from the source. This may result in lower S/N ratios and consequently higher limits of detection, but the flexibility of the line source is greatly increased (see Table 1).

The technology for photometric HPLC detectors has existed for many years but these detectors still retain a sizable share of the market. This is because in many cases photometric detectors provide adequate selectivity without the additional expense and complexity of spectrophotometric detectors. Almost every manufacturer includes photometric detectors in their product line. The features of a few of the currently available detectors are discussed.

**Table 1.   Table of Typical Wavelengths for HPLC
Photometric Detectors (4)**

Source	Emission Lines (nm)	Phosphor (nm)
Mercury	254	280
	313	300
	365	320
	405	340
	436	470
	546	510
	578	610
		660
Cadmium	229	
	326	
Zinc	214	
	308	
Magnesium	206	

Few currently marketed molecular absorption detectors are limited to operation with a mercury lamp alone. All but the least expensive detectors offer a variety of source lamps. Cadmium and zinc lamps are most commonly available in addition to mercury but LKB (Bromma, Sweden) also offers a 206-nm lamp along with the more or less standard complement of metal vapor lamps. LKB sources utilize a low-power, high-intensity electrodeless design that is proprietary (15). The 206-nm line is probably magnesium emission. A 226-nm lamp is also offered. This wavelength was not listed by other manufacturers who responded to the same survey. Isco (Lincoln, NB) currently markets a dual-beam absorbance/fluorescence photometer (UA-5) for HPLC which offers a selection of 20 absorbance wavelengths. The additional source wavelengths that are now commercially available have greatly improved the utility of photometric HPLC detectors.

Flow-cell volumes in commercial instruments range from a few tenths to tens of microliters, depending on the path length desired. An 8-μL cell is common because a cell with a convenient 10-mm path length and 1-mm diameter has this volume. Because of the growing interest in microbore HPLC, many small-volume flow cells are being offered; for example, LKB has a 0.8-μL cell with a 3-mm path length. In HPLC detectors, if the source light strikes the cell walls increased sensitivity to flow and refractive index effects may result. Therefore, the source light is usually focused, particularly with small-volume cells. Most manufacturers have designed flow cells that are specific for their detectors and address the common flow problems associated with HPLC. For example, Beckman Instruments, Inc. (Fullerton, CA) boasts a flow-cell design that eliminates base-line fluctuations due to refractive index effects and flow changes. LDC/Milton Roy (Riviera Beach, FL) cites an integral, low-volume heat exchanger for temperature equilibration as a remedy for base-line drift. High-purity HPLC-grade solvents must be used; otherwise detector performance may be obscured by impurities.

Double-beam detectors can monitor the differential absorbance beween the two liquid streams but there are also detector designs that permit detection at two wavelengths in the same cell. One design uses a single mercury lamp coated with a phosphor at one end with 254 and 280 nm filters isolating the appropriate emission lines from the lamp (16). The sample cell is irradiated at right angles and separate photodetectors are used. Gilson International (Middleton, WI) uses a similar scheme to produce a dual-wavelength UV detector for HPLC. There are some applications where monitoring at a single wavelength is inadequate. Proteins are regularly detected at 254 and 280 nm and the ratio of the response at the two wavelengths is used to characterize the eluting bands. (Response ratios are discussed in more detail in the next section.) These detectors have demonstrated higher flow sensitivity and lower S/N ratios, apparently because of the off-axis illumination (4).

1.1.2. Spectrophotometric/Single-Channel Detectors

These detectors utilize continuum sources. Wavelength selection is accomplished with a monochromator except in a few cases where high light throughput is essential. A diffraction grating rather than a prism is commonly the dispersive element. The exit slit of the monochromator isolates a narrow band of wavelengths that illuminate the sample. As with the photometric detector, the sensor of transmitted light is photoelectric. Because a photoelectric sensor is sensitive to a wide range of wavelengths, without the monochromator it would introduce little wavelength selectivity. Figure 3 is a schematic diagram of a typical double-beam spectrophotometer with single-channel detection. The arrangement of the monochromator before the sample minimizes sample photodecomposition due to irradiation of the sample with high-energy UV emission from the source.

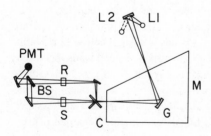

Figure 3. Schematic diagram of a double-beam spectrophotometer with single-channel detection. PMT = photomultiplier, BS = beam splitter, R = reference, S = sample, C = chopper, G = grating, M = monochromator, L1 and L2 = light sources.

The spectrophotometric/single-channel detector is commonly referred to as a "variable-wavelength detector," although it is clear that photometric detectors are capable of operation at a variety of wavelengths. Because the absorption wavelength is continuously tunable with spectrophotometric detectors, the detector response for virtually any molecule may be optimized. The user does not depend on finding a coincidence with an available source line. This feature may also allow greater selectivity by increasing the opportunity for operation at interference-free wavelengths. If several components are to be detected in a single chromatogram and the wavelengths of their absorption maxima vary, it is often better to use a relatively large monochromator bandwidth (17). This approach trades maximum sensitivity for the components that absorb strongly at the monochromator setting for improved sensitivity for other components. On the other hand, it may be necessary to reduce the bandwidth to improve linearity for quantitative analysis of a wide range of concentrations.

Deuterium lamps are generally used as sources in spectrophotometric detectors. The intensity of a deuterium lamp is three to five times greater than the intensity of a hydrogen lamp of comparable design and wattage. High-pressure deuterium gas excited by electron impact in an electrical discharge produces a continuum emission in the ultraviolet down to about 165 nm. Detection at wavelengths lower than the 254-nm mercury line is routinely possible with deuterium sources. (Quartz windows that absorb below 200 nm may be the limiting factor for deuterium lamp emission at short UV wavelengths.) The short-wavelength emission available with a deuterium lamp was more of an advantage for spectrophotometric versus photometric detectors prior to the commercial availability of a variety of alternate line sources. The deuterium lamp's intensity drops off toward the visible. The upper limit of usefulness is generally considered to be about 400 nm for a standard deuterium lamp.

A novel approach for producing UV photons for HPLC utilizes the Cerenkov effect (18). The Cerenkov effect occurs when electrons travel through a medium with a velocity greater than the velocity of light (in that medium) and lose energy by generating photons. In this instance, the energetic electrons result from the beta decay of strontium-90. Interference filters limit the continuum Cerenkov photon flux to wavelengths of 210, 254, or 280 nm.

Deuterium lamps are the most widely used source for commercial spectrometric detectors. The wavelength range most commonly specified by manufacturers is from 190 to approximately 350 nm. In one case, a commercial detector (LDC/Milton Roy) has been used successfully at 185 nm, below the rated lower wavelength limit for the deuterium lamp (19). As is true for short-wavelength UV line sources, the rationale for using continuum-source absorption wavelengths below the traditional 254 nm (i.e., mercury lamp) is to achieve more universal response from functional groups. HPLC detection in this spectral regime leads to a new set of problems. The absorption of the solvents as well as the solutes in the 200-nm region becomes more important in determining whether this strategy is a practical one. The choice of a mobile phase is clearly more limited at these low UV wavelengths. It may be necessary to use sequential isocratic steps under microprocessor control rather than conventional gradient elution at wavelengths as low as 190 nm (20). Deuterium lamp emissions at 195 (21) and 206 nm (22) have also been reported in reverse-phase HPLC with near "universal" detection at 210 nm for rapid survey analyses of unknown mixtures (23). A variety of techniques for minimizing base-line problems is discussed.

Visible operation is often but not uniformly available as an option from photometric detector manufacturers. Generally, complete coverage of the UV–VIS spectral regions requires a tungsten lamp in addition to the deuterium source. The spectral distribution from a tungsten filament is typical of an incandescent source or blackbody. The emission peaks at around 600 nm at the normal operating lamp temperature. Although most of the emission is in the infrared spectral region, the radiation at shorter wavelengths is sufficient for tungsten lamps to be widely used as soures of visible radiation.

Several companies offer UV–VIS detectors that use deuterium lamps well into the visible, in spite of a factor of 10 lower emission from 350 to 700 nm. A variety of approaches is used to compensate for the relative increase in noise above 350 nm (Spectra-Physics, San Jose, CA; Varian, Palo Alto, CA; Beckman Instruments, Inc.; Kontron Analytical, Zurich, Switzerland; LKB Instruments, Inc.). The quantum efficiency of a silicon photodiode, a commonly used detector, increases with wavelength. The visible noise is generally within a factor of 2 of that achieved in the ultraviolet (14), permitting acceptable S/N ratios with a deuterium source. Spectra-Physics uses a high-intensity deuterium lamp and single-beam optics. The high light throughput permits compensation for reduced signal (i.e., amplification) without being overwhelmed by noise (24). Kontron Analytical uses a high-intensity deuterium lamp with filters for wavelength selection, rather than a monochromator, to maximize light throughput. In any case, these instruments offer the advantages of a single source for the entire usable UV–VIS range.

A few manufacturers offer a quartz–halogen lamp for visible operation which extends the visible range to near-IR wavelengths (Kontron Analytical; TCS, Southampton, PA). The TCS instrument is a dual-wavelength model intended for biomedical applications.

Most HPLC detectors with continuum sources use monochromators for wavelength selection, and holographic diffraction gratings are offered by many manufacturers as standard equipment. With sources operating below about 200 nm, it is prudent to purge the monochromator with an inert gas to eliminate the atmosphere. Molecular oxygen, which begins to absorb at around 195 nm, is the worst problem. Gilson and Jasco (Tokyo, Japan) both indicate that the monochromators for their spectrometric detectors are purgeable. Kontron Analytical offers several detectors with deuterium lamps and multiple filters instead of the more common monochromator. One model employs a double-beam optical system with an eight-position filter wheel. These filters are centered at wavelength intervals from 200 to 570 nm.

With some spectrometric detectors it is possible to monitor the absorbance at more than one wavelength during a chromatogram. This practice has been referred to as "wavelength" or "detector programming." Of course, multiple detectors can be used to achieve this goal but it is also possible to use a single detector under microprocessor control in a time-shared mode. The latter approach avoids loss of chromatographic resolution by peak broadening in multiple detector cells.

A relatively low-cost detector using a rotating disk with four narrow bandpass interference filters has been developed that allows up to four wavelengths to be monitored during an analysis (25). Another approach to achieving sequential multiwavelength capability employs stepwise cyclic programs to slew the monochromator rapidly between the wavelengths of interest over the 190–360-nm range (26). The Varian UV-100 permits programming of detector parameters such as wavelength, attenuation, or time constant (27). Figure 4 illustrates the separation of several hydrocarbons. The early section of the chromatogram contains compounds best detected near 254 nm; the intermediate section is monitored at 240 nm. The final peaks to elute are detected at 270 nm. The signal was autozeroed at each wavelength and the recorder range was programmed to give the best presentation of the data.

The low rate of diffusion in the HPLC mobile phase allows stop-flow wavelength scanning for eluting components without unacceptable band broadening (28). A UV–VIS spectrum is sometimes helpful in characterizing compounds. Spectra-Physics offers a spectrophotometric/single-channel detector for HPLC that features automatic stop-flow peak scanning. The HPLC pump start/stop functions are integrated into the program software. When the automatic stop-flow scanning mode is in operation, the mobile phase is interrupted and the monochromator wavelength scanned each time a peak exceeds a preset absorbance threshold. The peak scans are corrected for the mobile phase background by first storing the mobile phase spectrum in memory and subtracting it from subsequent peak scans. The slew rate is 190–600 nm in 12 s.

There are several instruments available that scan rapidly enough to acquire an eluent spectrum on-the-fly. These rapid-scanning instruments eliminate any chance of band broadening by interruption of the mobile phase flow and speed up the overall analysis time. There are some advantages to sequential scanning as opposed to simultaneous spectral acquisition, which is discussed in the next section. The latter technique using available photodiode array technology is less sensitive than a photomultiplier, which is used in single-channel detection. Also, a compromise has to be made between spectral resolution and wavelength cov-

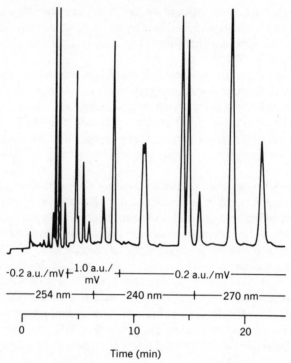

Figure 4. Chromatogram obtained with a Varian UV-100 detector. The detector was time-programmed to employ three different wavelengths and to achieve the best sensitivity for all compounds. Two different ranges were used during the chromatogram. Courtesy of Varian, Palo Alto, CA.

erage when using arrays. One approach to rapid scanning uses a monochromator that is repeatedly slewed over the instrument's entire wavelength range during the progress of the chromatogram (29). Deuterium and tungsten lamps are automatically alternated in the appropriate wavelength regions, giving this instrument wide wavelength coverage. Full-wavelength scanning with this instrument imposes constraints on the maximum flow rate. Other versions of rapid-scanning spectrophotometric detectors use an oscillating optical element in the monochromator to increase the scanning rate. The Harrick Rapid Scan Spectrometer (Ossining, NY) employs an oscillating mirror with scan rates up to 4.25 kHz and a repetition rate of 218 Hz. The spectra can be continuously monitored on a storage oscilloscope. In addition to the increased qualitative information acquired with the rapid-scanning spectrometer (RSS), it is possible to maximize the sensitivity for all components by using the optimum wavelengths from the RSS three-dimensional chromatogram for subsequent separations (30). Beckman

Instruments, Inc. also offers a rapid-scanning UV–VIS detector with sophisticated microprocessor control and data processing. An oscillating grating provides wavelength scanning. To scan the intrument's entire wavelength range, the span of the monochromator's oscillation is rapidly incremented to provide an on-the-fly scan.

Peak ratio techniques whose goal is component identification become a practical possibility with variable-wavelength detectors. Rapid scanning permits the monitoring of eluent absorbance at several wavelengths during the progress of the chromatogram. The ratio of the absorbance measured at any two preselected wavelengths is defined as the "response ratio." Beer's law prescribes that the absorbance ratio at two different wavelengths is concentration independent and defined as long as the absorbance at either wavelength is not zero. Therefore, the absorbance ratio measured at two different wavelengths should be characteristic of a solute (31,32). When a pure component elutes, the absorbance ratio at two wavelengths must be constant. A ratio change indicates that the peak of interest contains a chromatographically unresolved second component. If the absorption spectrum of both components is known, a quantitative interpretation is possible. When coupled with retention times, the measurement of response ratios can reduce the uncertainty in identifying separated components in a chromatogram to 5% (4). Although response ratios are not reliable for the identification of a complete unknown, in some cases they do appear to be more useful than wavelength scanning for qualitative analysis (33).

In double-beam arrangements, such as pictured in Fig. 3, part of the beam is diverted to a reference photocell by a beam splitter. Because the measurement of P and P_0 is made simultaneously or nearly simultaneously, most short-term electrical fluctuations are compensated for, as well as other time-dependent variations in the source, detector, and amplifier. For these reasons, a double-beam instrument is essentially a necessity for wavelength scanning. Continuous double-beam measurement with two optical sensors is favored over a chopped double-beam approach with a time-shared sensor to avoid modulation signal loss. This provides double-beam stability with monitoring of 100% of the transmitted light, which maximizes the S/N ratios. As with photometric detectors, a variety of sensors from photovoltaic cells to photomultipliers is used in commercial spectrometric/single-channel detectors. The additional components and complexity associated with a double-beam instrument translates to higher cost than single-beam photometric detectors.

As with photometric detectors, the basic technology of spectrometric/single-channel detectors is mature and many models are commercially available with a variety of refinements. Most recent publications using spectrophotometric/single-channel detection for HPLC cite commercial rather than laboratory-built detectors.

1.1.3. Spectrophotometric/Multichannel Detectors

Photodiode array technology has been embraced by both users and manufacturers of HPLC molecular absorption detectors (34,35). With the exception of the photodiode array itself, the rest of the detector components are similar to those described in the preceding section; for example, deuterium lamps are commonly used as sources. The optical system is reversed to permit simultaneous multi-wavelength detection after the light has been dispersed (see Fig. 5).

Simultaneous multiwavelength detection of HPLC eluents has long been a goal of chromatographers. The value of multiwavelength detection (albeit sequential) was demonstrated with a rapid-scanning spectrometer which used an oscillating mirror (30), but noise levels below 250 nm were high and the flow cell was large (87 μL). Some of the advantages of multichannel detection for HPLC were mentioned in the introduction to Section 1. Many of the benefits that accrue to simultaneous multiwavelength detection are a consequence of the "multiplex advantage." Because all the resolution elements may be viewed simultaneously, S/N ratios may be improved over single-channel data acquisition by averaging the entire spectrum for the conventional scanning time. Conversely, multiwavelength spectra with the same S/N ratios as spectra acquired channel by channel may be obtained in much less time. Acquisition of spectra of eluents as they exit the HPLC column without compromising flow rates and possibly resolution is important. Three-dimensional chromatograms with axes of wavelength, time, and absorbance may be generated. With multichannel data acqui-

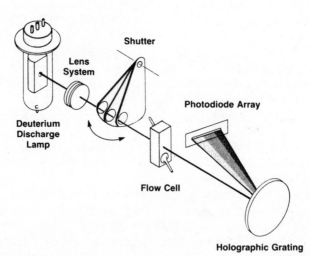

Figure 5. Schematic diagram of a spectrophotometric detector with photodiode array detection. Courtesy of Hewlett-Packard Company, Palo Alto, CA.

sition, an option exists to return to the stored data and to extract information at different wavelengths than originally interrogated without rechromatographing the sample. This is not possible with other detectors with the possible exception of rapid-scanning spectrometers. Applications that require monitoring several wavelengths, such as response ratioing, may be easily implemented. All of this is accomplished by array detection with no moving parts, and thus no chance for irreproducible monochromator positioning.

A variety of optoelectronic devices exists for multichannel detection (36,37) but the photodiode array has been the most widely adopted (38). Silicon vidicons, which were first explored for HPLC detection (39–42), generally have poor UV response and are more expensive than photodiode arrays. Incomplete readout of the silicon target and charge blooming between the diode elements limit the capacity of the vidicon to follow fast events (38). Charge-coupled devices respond rapidly enough for HPLC detection and have low noise and dark charge levels, but their UV response is low. In addition, charge-coupled devices have poor signal collection aspect ratios (38). Other optoelectronic devices also suffer from similar deficiencies that make them less suitable for HPLC detectors than photodiode arrays.

The development of the photodiode array as a multichannel detector for HPLC began around 1976 (43–48). The photodiode array consists of several hundred photosensitive diodes generally configured in a linear pattern to mimic the focal plane of a spectrometer (44). Each photodiode is connected in parallel to a storage capacitor. Transmitted light from the sample is spatially dispersed by the spectrometer so that when it illuminates the photodiode array, wavelength as well as intensity information is encoded. The photocurrent generated by light striking individual photodiodes discharges their respective capacitors. Each capacitor is discharged according to the amount of light that falls on the corresponding photodiode during the integration time. Each photodiode is successively interrogated by a digital shift register. The entire array readout process is repeated every few microseconds to prevent saturation of the photodiodes at high incident light levels. An analog signal is generated that is representative of the state of all of the photodiodes in the array; that is, an absorbance versus wavelength spectrum is produced. A microprocessor network is required to manage the data acquisition tasks and process the data in real time.

Instrument manufacturers have contributed significantly to the development and acceptance of spectrometric/photodiode array detectors for HPLC. There are several commercial versions currently being marketed. The Hewlett-Packard 1040A detector (Palo Alto, CA) was introduced in 1982 and is probably one of the most thoroughly documented instruments on the market. In addition to the manufacturer's literature, several other publications describe the design and operation of the optical system, the photodiode array, and the microprocessor network in detail (49–53). The device covers a spectral range of 190–600 nm with an optical resolution of 4 nm. The photodiode array consists of 211 diodes, each 50 μm wide. The raw data generated by the photodiode array are reduced by a 16-bit microprocessor in real time and the spectrum is displayed on a cathode ray tube. The detector system can also perform data treatment, in addition to acquisition, making data processing a useful adjunct or sometimes a substitute for chromatographic resolution.

The Shimadzu UV–VIS spectrometric/photodiode array detector (SPD-M1A, Columbia, MD) utilizes a double-beam optical configuration with rotating sector mirror to divert the beam to sample and reference cells alternately. Deuterium and tungsten lamps permit absorbance measurements over the entire UV–VIS region (200–699 nm) with automatic changeover. A three-dimensional chromatogram can be displayed on a cathode ray tube or an XY-T recorder.

In addition to the capability for full spectra, the PU 4021 Multichannel UV/VIS Detector (Phillips/Pye Unicam Ltd., Cambridge, UK) provides the flexibility of selecting the output from one to nine different photodiodes. The selected photodiode (i.e., wavelength) or photodiodes can be preprogrammed to be monitored at specific times during the chromatogram. A spectral storage facility may be used to store spectra manually or automatically during the run. Any or all of the spectra may be plotted after the chromatogram has been completed.

LKB's Rapid Spectral Detector (2140) has a built-in 56K dedicated microprocessor that provides stand-alone capability with single or multichannel potentiometric recorders. It is also compatible with the IBM Personal Computer. The additional computing power permits extensive data manipulation. In addition to displaying data as a three-dimensional chromatogram, a two-dimensional contour plot may be generated. The effectiveness of this data presentation is increased by the use of color graphics keyed to specific wavelengths to indicate the contour region of identical absorbances for the separated components (54). The colored "isograms" serve as chromatographic "fingerprints" for comparison of samples. An ink-jet printer–plotter allows the screen to be dumped to a full-color hard copy.

The LDC/Milton Roy CMX-50 photodiode array HPLC detector also is designed to use another manufacturer's computer, a Digital Equipment Corporation PDP/11. Compatibility with a laboratory computer that is not dedicated only to a HPLC detector provides more access to software and consequently allows more sophisticated data manipulation.

In addition to the above detectors, Jasco Incorporated introduced a multiwavelength detector photodiode array detector at the 1984 Pittsburgh Conference (55). Several companies, such as Tracor Northern, Inc. (Middleton, WI) and EG&G PARC (Princeton, NJ), also offer multipurpose photodiode arrays with associated electronics that can be adapted to HPLC detection (56).

Recently, a new type of UV–VIS photodiode array spectrometer adapted to HPLC detection has been demonstrated (57). Groton Technology, Inc. (Waltham,

MA) has introduced a relatively inexpensive Fourier transform spectrometer for the UV–VIS spectral regions. An interferometer encodes the spectral information at all wavelengths simultaneously and uses a mathematical transformation to resolve the data into a conventional frequency or wavelength spectrum. (Fourier transform spectroscopy is discussed in more detail in Section 2.) The "common-path" interferometer used in this instrument encodes the signal in a spatial domain rather than the time domain used by a Michelson interferometer. A common-path interferometer was chosen over the Michelson interferometer, normally used in infrared instruments, because the precision required by the latter device is much greater at shorter wavelengths, making the instrument prohibitively expensive. In the Groton LC/S spectrometer, the spatially encoded information is sensed by a 512 element photodiode array and transmitted to a computer for Fourier transformation into a frequency domain spectrum. The current price of the LC/S is $25,000, about the same as one of the top-of-the-line grating photodiode detectors with an associated computer. Although there is no price advantage, this new approach could stimulate the continued revitalization of molecular UV–VIS detector technology that photodiode array technology sparked a few years ago.

Spectrometric/photodiode array detectors and associated computer hardware and software have improved the implementation of standard techniques such as response ratioing and peak monitoring at several wavelengths to optimize response for all separated components. They have also permitted some unique applications (49). A few of these applications will illustrate the potential of photodiode array detectors for HPLC.

A photodiode array detector coupled with an HPLC can be used for purity analysis (58). Samples can be rapidly screened for UV–VIS absorbing impurities. Because the entire spectrum may be monitored, there is no chance of overlooking the wavelength that yields the best sensitivity, even if the components are unknown. In addition, a spectrum taken at any wavelength within the absorption envelope of a component should exhibit only amplitude changes if the fraction is pure. If a chromatographically unresolved component is present, shifts in maxima or minima can be expected (59). Multicomponent analysis routines also may permit quantification of multicomponent mixtures without altering chromatographic conditions.

One of the most intriguing aspects of multichannel detection is the ability to collect large amounts of data that can then be processed to extract the meaningful information. The chromatogram need be run only once, but the resulting data may be manipulated in a variety of ways to produce the most desirable result. For example, digital algorithms have been examined for peak recognition and spectral deconvolution of data obtained with photodiode array detectors (60). Although the deconvolution of unseparated peaks for quantification still requires that the spectral properties of the overlapping components be defined and sufficiently different, numerical methods for multicomponent analysis hold considerable promise for the future.

It has been suggested that HPLC with photodiode array detection is a low-cost alternative to mass spectrometry for qualitative analysis (38,49). Although photodiode array detection presents some interesting and useful alternatives, there seems to be no danger that either photometric or spectrophotometric/single-channel detectors will be eliminated from the marketplace. As with any analytical problem, the situation will dictate the choice of detector.

Table 2 summarizes Section 1.1. The cost of a molecular absorption detector for HPLC can be linked, in at least a superficial way, to the detector's versatility; that is, how many things can it do? Specific comparisons of detectors for applications of limited scope may indicate that a photometric detector will perform better than a spectrophotometric/photodiode array detector, particularly in terms of cost efficiency. A photodiode array detector will be preferred for screening analyses but once the optimum wavelengths have been determined, a single-channel photomultiplier may be used to obtain the lowest detection limits. In any case, it is clear that molecular absorption detectors will remain competitive with more exotic techniques for HPLC detection in the near future.

1.2. Atomic Absorption Detectors

The increasing use of atomic absorption detectors for HPLC has been stimulated by several factors. Determination of metals and their chemical forms (i.e., speciation) is becoming increasingly important for environmental, clinical, biological, and agricultural studies. For example, organic mercury is much more toxic to biological organisms than elemental mercury (61). One of the deficiencies of atomic absorption spectrometry is that the common methods for sample introduction reduce most compounds to the atomic state or at least alter their composition. The memory of the original form of the element is lost; that is, only the total concentration of an element is measured. Chromatographic separation prior to atomic absorption permits the identification and quantification of the molecules that contain the detected element. In addition, metal labeling of molecules to introduce more selective detection into an HPLC separation is also a possibility.

Atomic absorption is an ideal technique for element-specific detection because it provides good detection limits (ng/mL), atomic absorption spectrometers are readily available, and it is often simple to interface a LC column to an atomic absorption spectrometer. HPLC as well as conventional LC has benefited from atomic absorption detectors. Atomic absorption spectrometry has been used as a specific metal detector for reverse-phase liquid [e.g., (62)], size exclusion [e.g., (63)], ion [e.g., (64)], ion-exchange [e.g., (65)], and affinity chromatog-

Table 2.   UV–VIS Molecular Absorption Detectors for HPLC

Type	Wavelength Coverage	Wavelength Selector	Sensor	Comments
Photometric (fixed wavelength)	Discrete lines (Table 1)	Filter	Single-channel	High throughput. Least expensive. Sample must absorb at source wavelengths
Spectro-photometric (variable wavelength)	Continuous	Monochromator	Single-channel	Stop-flow spectral acquisition. Source wavelength can be optimized. Wavelength programming and response ratioing possible
Spectro-photometric (multi-wavelength)	Continuous	Monochromator with rapid scanning	Single-channel	On-the-fly spectra with flow rate limits
	Continuous	Polychromator	Multichannel	Simultaneous spectral acquisition. Choice of maximum resolution or wavelength coverage. Advanced data analysis possible. Less sensitive than a photomultiplier

raphy [e.g., (66)], with both isocratic and gradient elution. Several reviews have discussed the use of atomic absorption spectrometers for HPLC detection (67–72).

Atomic absorption is the result of an electronic transition induced by a radiant source (3). Figure 6 is a schematic diagram of a typical atomic absorption spectrometer. A resonance line source, commonly a hollow cathode or electrodeless discharge lamp containing the element of interest, is used in an atomic absorption spectrometer rather than a continuum lamp because atomic absorption

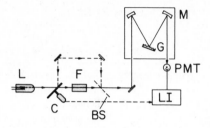

Figure 6. Schematic diagram of a double-beam atomic absorption spectrometer. L = light source, C = chopper, F = flame, BS = beam splitter, G = grating, M = monochromator, PMT = photomultiplier, LI = lock-in amplifier.

line widths are very narrow. Continuum-source atomic absorption spectrometers have been developed but they have been used to a lesser extent and commercial versions are not available.

In an atomic absorption spectrometer, a dilute gaseous solution of the atomized sample replaces the cuvette used for molecular absorption. The simplest approach to sample atomization is a premixed burner (Fig. 7). The liquid sample is aspirated into a chamber where it is mixed with the fuel and oxidant. A few percent of the aerosol is introduced into the flame while the large droplets condense on baffles within the mixing chamber and are removed via the drain. A slot burner head several centimeters long provides a laminar flow flame and a long absorption path length. Acetylene–air is the most widely used fuel–oxidant combination but atomic absorption detection of elements that form refractory oxides requires a higher temperature acetylene–nitrous oxide mixture (73).

There are also nonflame methods for sample atomization. A common alternative to flame atomic absorption involves thermal atomization with a furnace (3). Several types of graphite furnaces are commercially available. Figure 8 shows a simplified schematic diagram of a Perkin-Elmer furnace (Norwalk, CT). A few microliters of sample are deposited with a micropipette through a small hole in the center of the graphite cylinder. The graphite furnace is then resistively heated to evaporate the solvent, char the sample to remove organics, and finally atomize the sample. A transient measurement is required because the atomic vapor has a finite residence time in the graphite tube. This is in contrast to flame spectometry in which the sample may be continuously introduced into the flame.

Graphite furnace atomizers generally produce several orders-of-magnitude lower detection limits than flames for several reasons. The sample is confined to the measurement volume for a longer period of time in the furnace, graphite is a strongly reducing medium, and the furnace is usually purged with helium so the atmosphere is less reactive than the flame (68). On the negative side, graphite furnace atomic absorption is limited to very small, discrete samples, and elements that form refractory carbides give reduced sensitivity. Flame analysis is more precise than atomic absorption with electrothermal atomization and is much simpler to implement, particularly as a detector for HPLC.

Atomic absorption spectrometers may be single or double beam. Flame back-

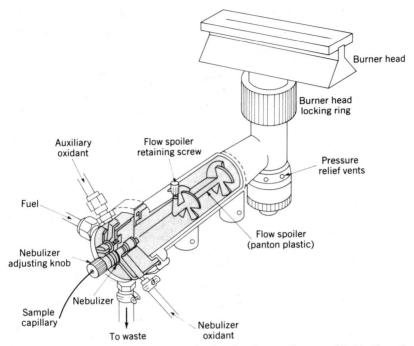

Figure 7. Schematic diagram of a premixed, laminar flow burner. Courtesy of Perkin Elmer Corporation, Norwalk, CT.

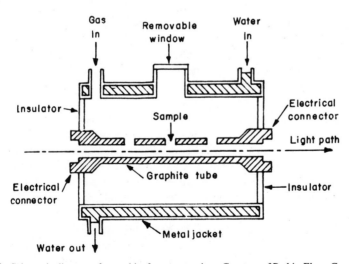

Figure 8. Schematic diagram of a graphite furnace atomizer. Courtesy of Perkin-Elmer Corporation, Norwalk, CT.

ground is discriminated against by modulating the source with a mechanical chopper. Background correction for nonspecific absorption or scattering of source radiation is available for both flame and graphite furnace atomic absorption spectrometers. It is generally considered essential for the latter.

Although atomic absorption spectroscopy is basically a single-element technique when used with line sources, automated sequential "multielement" analysis (usually for up to six elements) is available with some commercial instruments. Simultaneous multielement analysis is an important feature of continuum-source instruments but compromise flame conditions must be used, reducing sensitivity for some elements.

The rest of the components used in an atomic absorption spectrometer are similar to those described for molecular absorption spectrometers. Although the source is imaged in atomic absorption spectrometry, a low- to medium-resolution monochromator is generally sufficient with line sources because analyte atomic lines are widely separated and the source line widths fall within the flame-broadened analyte atom line widths. With a continuum-source instrument, a high-resolution polychromator is necessary to extract the signal in the presence of radiation falling outside the analyte absorption line width. An echelle grating with an order-sorting prism fulfills the resolution requirements while providing a simultaneous presentation of the wavelengths at the focal plane of the polychromator (74). Photomultipliers are used almost exclusively as sensors of transmitted light in atomic absorption spectrometers. A photodiode array may be substituted for a matrix of photomultipliers in a continuum-source/multiwavelength instrument.

Fraction collection may be used to accumulate predetermined volumes of HPLC effluent for atomic absorption spectrometry sequentially, but direct coupling of the column to the atom cell is desirable (72). When the effluent is directly monitored, there is no opportunity for sample contamination. In addition, determination of elution times or retention volumes may be less precise with off-line detection. Only on-line detection schemes are discussed further.

The interfacing of an ion-exchange column to an atomic absorption burner was reported in 1973 (75). Atomic absorption was used as a metal specific detector for the determination of chelating agents as copper complexes. In 1975, an atomic absorption detector was coupled directly to a gel permeation chromatograph for the determination of condensed phosphate anions measured as magnesium complexes (76). The magnesium complexes were produced by eluting the phosphate anions through a column preequilibrated with a magnesium chloride solution.

Atomic absorption spectrometry has been applied to HPLC detection for the separation of organochromium compounds (77). Using a mobile phase of 0.5% pyridine in toluene and an air–acetylene flame, a detection limit of 40 ng was reported for the chromium compounds investigated. In this work, the column exit was connected directly to the atomic absorption burner with a short length of small-bore tubing. This is a common approach for interfacing the HPLC with the burner, for the effluent flow or the burner solution uptake rate can often be made compatible.

There are several possibilities for manipulating flow rates and aspiration rates. The two flow rates may be balanced (78) or the nebulizer can be "starved" by decreasing the column flow rate (79). In one case, the burner aspiration rate was adjusted to be slightly less than the HPLC flow rate so that compatibility could be achieved without a postcolumn low-pressure region (62). Nebulizers can also be adjusted to eliminate aspiration entirely by creating a slight back pressure to the HPLC column (80). A HPLC solvent pump can easily overcome this back pressure because it is designed to deliver a constant volume of liquid regardless of the impedance. Depending on the type of nebulizer, "back-pressure" operation may provide the highest S/N ratios owing to improved droplet characteristics and transport efficiencies (80). The internal baffles may be removed from the premixing chamber of a standard burner to increase the fraction of the effluent to enter the flame (81).

The use of organic solvents may also require some modification of standard procedures. The heat of combustion of most flammable organic solvents is less than that of acetylene. Although this lowers the flame temperature and consequently the atom fraction, the flame temperature is probably higher than with aqueous solvents. Noncombustible solvents affect flame temperature similarly to the addition of an aqueous solvent. A three-slot burner head may produce better results than a standard single-slot head when organic solvents are used (79).

When it is not possible to interface the HPLC column and the nebulizer directly, there are several alternatives. In one case, a Teflon funnel was attached to the nebulizer of the premixed burner (82). The effluent droplets from the column exit were caught in the funnel and nebulized one at a time. The chromatogram consisted of a series of spikes but there was no loss of sensitivity because a 100-μL drop was sufficient to give a steady-state atomic absorption signal. If dilution of the sample is acceptable, another possibility is to augment the HPLC flow whose rate is generally less than the nebulizer requirements with additional solvent (69). A flow-injection sample manipulator has been used to sample a HPLC effluent, dilute the aliquot appropriately, and present it to the atomic absorption spectrometer at a flow rate compatible with the nebulizer (83).

A method of calculating detection limits based on known chromatographic and spectrophotometric parameters has been devised (84). Calculated and experimental detection limits for copper chelates of (ethylenedinitrilo)tetraacetate and nitrilotriacetate were compared for continuous sample aspiration and detection after separation. Not surprisingly, the latter detection limits were poor by comparison. In addition to the differences that may be attributed to discrete

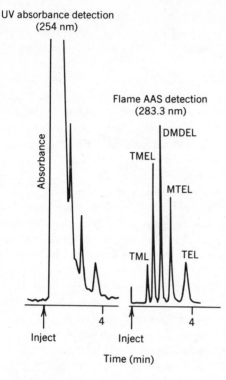

Figure 9. Separation of tetraalkyllead compounds in leaded gasoline by reverse-phase HPLC with atomic absorption detection (62). TML-tetramethyllead, TMEL-trimethyllead, DMDEL-dimethyldiethyllead, MTEL-methyltriethyllead, TEL-tetraethyllead.

versus continuous sampling, the loss of sensitivity was due to solute peak broadening on the column.

In spite of an anticipated reduction of sensitivity by coupling HPLC with an atomic absorption spectrometer, the combination provides good detection limits with impressive selectivity (62). Figure 9 illustrates the HPLC separation of tetraalkyllead compounds in gasoline using both UV–VIS absorption at 254 nm (Hg vapor lamp) and flame atomic absorption at 283.3 nm (Pb hollow cathode lamp). The HPLC/UV–VIS molecular absorption chromatogram is not very useful because of the strong absorption of coeluting unsaturated and aromatic compounds in the gasoline sample at 254 nm. The HPLC/atomic absorption chromatogram at the 283.3-nm lead resonance line shows all of the tetraalkyllead compounds permitting accurate quantitation by peak height or peak area methods. Detection limits were approximately 10-ng Pb for each tetraalkyllead compound.

In some cases, flame atomic absorption detection limits are not adequate for trace metal determinations and the greater sensitivity available with graphite furnace is required. Because of the need for discrete sample atomization and the multiple steps involved, the graphite furnace or similar electrothermal device is more difficult to interface to a HPLC than the standard atomic absorption burner. Basically, all the approaches to interfacing involve automation of some type of fraction collection (85,86) and transfer of the samples to the furnace. Autosamplers are available from manufacturers of atomic absorption spectrometers and work well for transferring a sample of the HPLC effluent from the column to the graphite furnace (66,87–94). Effluent is collected in sampling cups that are arranged on a carrousel and sequentially injected into the furnace. It is also possible to sample periodically a single cup that is modified to allow excess effluent to overflow to a drain (68). Another approach involves valving systems, sample loops, and usually automation by microprocessor control (95–101).

Other nonflame methods of sample atomization for HPLC–atomic absorption spectrometry have also been reported. An automated detection system for ion chromatography used continuous generation of volatile arsine derivatives from arsenic-containing environmental samples followed by a heated quartz cell for atomization (102). None of the techniques reported previously could individually determine all of the arsenic species that were present. A cold vapor atomic absorption technique has been used for determining mercuric compounds separated by HPLC (103). The mercury vapor that was produced by addition of a reducing solution to the column effluent was swept into the absorption cell where absorption measurements were made at 253.7 nm.

Some of the graphite furnace atomic absorption spectrometers utilize Zeeman effect background correction (94–101). If the atomizer is placed between the poles of a strong magnet, the absorption line can be split into at least three components: a central component at the resonance wavelength and two side bands (104). The light absorbed by the sample can be varied by changing the polarization of the source light with a rotating polarizer. The central component absorbs light parallel to the magnetic field and the side bands are perpendicularly polarized. Because the background absorbs both polarizations equally, subtraction of the signal at the perpendicular polarization from the signal at the parallel polarization results in background correction. Other implementations of Zeeman background correction produce similar results. This technique has the advantage that the correction is performed at the resonance line. Standard background correction techniques for line source atomic absorption use a continuum lamp measurement at wavelengths slightly off resonance. An important restriction is that the element of interest must have a resonance line that demonstrates a suitable Zeeman effect.

Components separated by HPLC with flame atomic absorption detection yield continuous Gaussian curves, indicating the concentration of the detected species as a function of time (87). With discrete sampling, which is characteristic of most nonflame atomic absorption techniques, a series of histograms indicates the analyte concentration at specific time intervals. The chromatogram is a series of spikes, the sum of which is related to the total amount of analyte eluted from the column. Two methods of quantification have been used successfully: the peaks due to the individual measurements may be connected and the resultant area computed, or the areas of the individual peaks may be summed. Both methods yield comparable results (72).

Although there are no commercially available interfaces for HPLC–atomic absorption spectrometry and certainly no dedicated instruments, the technique will remain an important one for samples containing metals, because of the relative ease of coupling the two techniques and the excellent selectivity afforded, as well as good sensitivity.

2. INFRARED ABSORPTION DETECTORS

The desire to improve the selectivity of absorption measurements has prompted investigations into the possibilities for adapting IR spectrometry to HPLC detection. Infrared absorption in solutions corresponds to transitions between the quantized vibrational levels that reside within electronic states of molecules (see Fig. 1a). Most molecules have characteristic absorptions in the IR region and the absorption bands are narrow. Interference from molecules other than the analyte is possible but the IR is so information-rich that measurements in another region of the spectrum may be sufficient for an identification. This is in contrast to UV–VIS absorption spectometry where spectra are broad and featureless.

Although IR absorption follows Beer's law, deviations are common because the absorption peaks are usually narrow and slit widths may be opened up to compensate for low source intensities. Beer's law deviations can often be overcome by using analytical calibration curves.

The generic elements of an IR spectrometer are similar to those of a UV–VIS instrument but the specific components and their arrangement is very different. A common IR source is an inert solid, heated electrically to temperatures between 1500 and 2000 K. Continuum radiation approximating a blackbody emitter results. Detection of radiation in the IR is more difficult than in the UV–VIS region of the spectrum because source intensities are low and IR photons are not energetic enough to use sensitive photoemissive devices. Thermal detectors, such as thermocouples, have been widely used. Pyroelectric detectors, for example, lithium tantalate or triglycine sulfate, are receiving increased usage because of their more rapid response characteristics. These detectors are based on the temperature sensitivity of the crystal's dipole moment.

IR spectrometers are of two types, dispersive and nondispersive. Dispersive spectrometers employ gratings to separate the IR radiation into its component wavelengths. Dispersive spectrometers have been used primarily for qualitative analysis and are generally double-beam instruments because a spectral scan is necessary (Fig. 10). Double-beam design relaxes the demands on source and

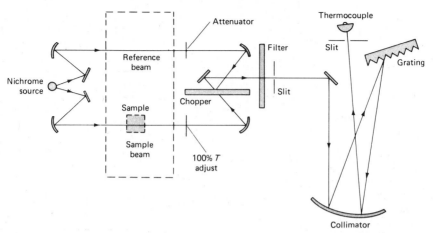

Figure 10. Schematic diagram of a double-beam IR grating spectrometer. Courtesy of Beckman Instruments, Inc., Fullerton, CA.

detector stability in addition to compensating for their wavelength dependence. Chopping rates are limited to low frequencies because of the slow response of most IR sensors. Double-beam IR spectrometers usually employ an optical null design with a comb or absorbing wedge for beam attenuation. The monochromator is located immediately before the sensor to shield it from the heat generated by the mechanical chopper and other stray radiation.

Jasco markets a double-beam infrared spectrometer designed specifically for HPLC detection. Any solvent that has at least 5% transmittance may be used because the transmittance scale has been expanded in the 0–10% range (105). A sensitive vacuum thermocouple is used as the detector. A variety of standard double-beam IR spectrometers has also been adapted for HPLC detection.

Double-beam IR spectrometers are less satisfactory for quantitative work than single-beam instruments because the additional electronic and mechanical components contribute to increased noise. In addition, the double-beam instrument is subject to errors arising from zero overshoot inherent with an optical null design (3). Simple, single-beam instruments have been designed for quantitative IR spectrometry as a result. Foxboro/Wilks (Foxboro, MA) markets an automated spectrometer for HPLC detection (106). The dispersing element of the MIRAN IR detector consists of three filter wedges mounted on a wheel with a slit mask to define the optical beam (Fig. 11). The motor drive and the potentiometric control permit rapid computer-controlled wavelength selection from 2.5 to 14.5 μm. Analyses can be conducted from a few percent down to parts per million. The IR source is a heated nichrome wire and the detector is lithium tantalate.

Most dispersive IR detectors are used to monitor absorption at a fixed wave-

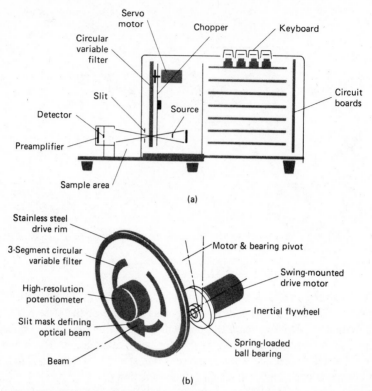

Figure 11. An IR instrument for quantitative analysis: (*a*) schematic of instrument; (*b*) circular variable filter wheel. Courtesy of The Foxboro Company, Foxboro, MA.

length [e.g., (107–112)] because the slow scanning rates available are not compatible with on-the-fly spectral acquisition. A computer-controlled double-beam, ratio-recording spectrometer has been used in a stop-flow mode to record spectra of eluted compounds (113). The typical time from stopped flow to pump restart was less than 4 min. The data were collected on a floppy disk for subsequent processing.

The sensitivity of a number of spectroscopic techniques may be enhanced by Fourier transformation of spectral data. In a Fourier transform infrared (FTIR) spectrometer, the most widely used nondispersive instrument, a Michelson interferometer, replaces the monochromator used in a dispersive instrument. An in-depth discussion of FTIR techniques for HPLC detection is presented in Chapter 3. A bench-top FTIR spectrometer is available for approximately $35,000. This includes a computer, which is necessary for utilization of the instrument. A general-purpose grating IR spectrometer costs less by a factor of 3. In addition to the dedicated computer, the price difference may be attributed to the interferometer and the peripheral instrumentation necessary for its precise and accurate operation. The sources and detectors for FTIR spectrometers are similar to those previously discussed except there is a preference for pyroelectric detectors because of their rapid response characteristics. Larger FTIR spectrometers with more options and advanced computational capability cost well over $100,000.

Of the absorption techniques described to this point, IR spectrometry is the most difficult to adapt to HPLC and the least studied. Only a few instrument manufacturers have entered the market, apparently because it is not clear that the best approach for interfacing HPLC with IR spectrometry has been developed or the interest is sufficient within the HPLC user community. Mobile phase elimination, complicated as it can be, may represent the best hope for more widespread application of IR spectrometry to HPLC detection. FTIR spectrometers are ideally suited to HPLC detection but even the lower-priced models are relatively expensive. As FTIR instruments become more common, their use for HPLC will surely increase.

3. INTRACAVITY ABSORPTION DETECTORS

Placement of the sample within the resonant cavity of a laser is a technique for enhancing the sensitivity of absorption measurements. Depending on the mirror separation and the gain medium, one or more longitudinal modes oscillate within the laser cavity (114). Gas lasers have several longitudinal modes whereas hundreds of modes may be oscillating within the gain curve of a dye laser. Two types of interaction between an absorber and the laser modes have been described (115). In type I, the laser mode falls within the absorption line width of the sample. This case applies to a single-mode laser or an inhomogeneously broadened laser with one of its modes overlapped by the absorption line. Insertion of an absorber into the cavity results in reduction of the laser power because of increased cavity loss. Enhancement over an extracavity absorption (i.e., a transmission) measurement is primarily due to the extended effective path length. When several of the modes of a homogeneously broadened laser fall outside the absorption line width, type II intracavity absorption prevails. In type II, the absorption interacts with relatively few modes and the laser power is shifted into the remaining modes. Enhancement factors of 10–100,000 have been observed and attributed to this mode competition.

Intracavity absorption spectrometry may be implemented for both atomic and molecular absorption measurements in the UV–VIS and the IR but it has not been applied to HPLC detection. A HeNe laser operating simultaneously at 3.39 (IR) and 0.63 μm (VIS) has been used as a selective detector for hydrocarbons in the effluent of a gas chromatograph (116). The IR and VIS laser transitions

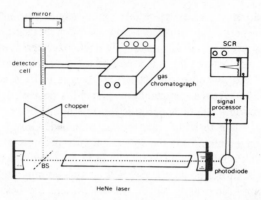

Figure 12. Schematic diagram of a double-beam HeNe intracavity absorption detector (116).

originate at the same excited energy level and are competitive. When a hydro-carbon enters the laser's resonant cavity, the 3.39-μm energy is absorbed because of the C–H stretching vibration, and the VIS emission is enhanced. The VIS laser emission is monitored as a quantitative measurement of the concentration of the absorbing molecule (see Fig. 12). The minimum detectable concentration for propane using a double-beam configuration was 20 pg/mL (117), which is 25 times lower than the best value reported for a thermal conductivity detector.

In practice, the detector's selectivity for hydrocarbons is modified by various substituents. The detector responds to aliphatic and aromatic hydrocarbons with aliphatic side chains, except for those substituted with halogens. It does not respond to water or carbon dioxide.

The HeNe laser intracavity absorption detector might be used for HPLC. The main limitation would be the choice of solvent. Further investigation of the constraints on selectivity imposed by operating with a condensed rather than gas phase absorber are necessary. It is clear that it would be possible to use halocarbon solvents for HPLC with this detector. There are other possibilities for application of intracavity absorption to HPLC detection. Perhaps the increased sensitivity and selectivity available with intracavity absorption spectrometry will justify the additional complexity and expense of using a laser source.

4. CONCLUSIONS

Absorption detectors offer a variety of approaches for introducing sensitivity and selectivity into the detection of eluents from a HPLC column. No detector is a panacea. Ultraviolet–visible molecular absorption detectors are firmly established and will continue to be used extensively by themselves and in conjunction with other detectors. Photodiode array detection of molecular absorption has justified early optimism as an HPLC detector and has promise for the future, particularly where advanced computational methods can be applied to data analysis. Advances in microcomputer technology will continue to stimulate growth in this area. The other absorption detectors discussed will never have the same impact as UV–VIS molecular absorption detectors but they will fill a need for increased specificity in HPLC detection. Atomic absorption and infrared spectrometers are widely used as "stand-alone" instruments and their role as HPLC detectors will increase if suitable interfaces are available. The engine that drives HPLC detector research is the need for more sensitive and selective detectors. Absorption detectors will continue to play a prominent role in fulfilling this need.

REFERENCES

1. Identification of a commercial product by a manufacturer's name or label in no instance implies endorsement by the Naval Weapons Center, nor does it imply that particular products or equipment are necessarily the best for that purpose.

2. 1983–84 LabGuide, *Anal. Chem.*, **55**, 82 (1983).

3. D. A. Skoog and D. W. West, *Principles of Instrumental Chemical Analysis*, 2nd ed., Saunders College, Philadelphia, 1980, pp. 149–166.

4. R. L. Stevenson, *Chromatogr. Sci. (Liq. Chromatogr. Detect.)*, **23**, 23 (1983).

5. F. Grum and R. J. Becherer, *Optical Radiation Measurements*, Vol. 1, Academic, New York, 1979.

6. G. Booker, *Anal. Chem.*, **43**, 1095 (1971).

7. M. W. Hunkapiller and L. E. Hood, *Science*, **207**, 24 (1980).

8. R. L. Stevenson and C. A. Burtis, *Clin. Chem.*, **17**, 774 (1971).

9. R. W. Frei and J. F. Lawrence, Eds., *Chemical Derivatization in Analytical Chemistry*, Vol. 1, *Chromatography*, Plenum, New York, 1981, Chapter 3.

10. D. R. Knapp, *Handbook of Analytical Derivatization Reactions*, Wiley, New York, 1979.

11. I. S. Krull and E. P. Lankmayr, *Am. Lab.*, **14**(5), 18 (1982).

12. Post-Column Reaction System Technical Literature, Kratos Analytical Instruments, Inc., Westwood, NJ, 1981–82.

13. T. Hirschfeld, J. B. Callis, and B. R. Kowalski, *Science*, **226**, 312 (1984).

14. S. R. Abbott and J. Tusa, *J. Liq. Chromatogr.*, **6**, 77 (1983).

15. Private communication, K. Lohse, LKB Instruments, Inc., Gaithersburg, MD, Jan. 11, 1985.

16. G. Munktell, *Protides Biol. Fluids, 1979*, **27**, 735 (1980).

17. J. E. Stewart, *J. Chromatogr.*, **174**, 283 (1974).

18. D. J. Malcome-Lawes and P. Warwick, *J. Chromatogr.*, **240**, 297 (1982).

19. Sj. Van Der Waal and L. R. Snyder, *J. Chromatogr.*, **255**, 463 (1983).

20. V. V. Berry, *J. Chromatogr.*, **199**, 219 (1980).

21. B. J. Compton and W. C. Purdy, Anal. Chim. Acta, **142**, 13 (1982).

22. W. S. M. Geurts Van Kessel, W. M. A. Hax, R. A. Demel, and J. De Gier, *Biochim. Biophys. Acta*, **486** 524 (1977).

23. V. V. Berry, *J. Chromatogr.*, **236**, 279 (1982).

24. Private communication, J. Dukes, Spectra Physics, San Jose, CA, Jan. 28, 1985.

25. T. Catterick, *J. Chromatogr.*, **259**, 59 (1983).

26. G. I. Baram, M. A. Grachev, N. I. Komarova, M. P. Perelroyzen, Yu. A. Bolvanov, S. V. Kuzmin, V. V. Kargaltsev, and E. A. Kuper, *J. Chromatogr.*, **264**, 69 (1983).

27. G. L. Burce and K. Klotter, *Am. Lab.*, **14** (3), 74 (1982).

28. M. A. Tompkins, *Am. Lab.*, **16** (2), 110 (1984).

29. K. Saitoh and N. Suzuki, *Anal. Chem.*, **51**, 1683 (1979).

30. M. S. Denton, T. P. DeAngelis, A. M. Yacynych, W. R. Heineman, and T. W. Gilbert, *Anal. Chem.*, **48**, 20 (1976).

31. P. A. Webb, D. Ball, and T. Thornton, *J. Chromatogr. Sci.*, **21**, 477 (1983).

32. A. C. J. H. Drouen, H. A. H. Billiet, and L. DeGalan, *Anal. Chem.*, **56**, 971 (1984).

33. Anonymous, Perkin-Elmer Publication L-579, Norwalk, CT, Aug. 1979.

34. A. F. Fell, *Anal. Proc.*, **17**, 512 (1980).

35. A. F. Fell and H. P. Scott, *J. Chromatogr.*, **273**, 3 (1983).

36. Y. Talmi, *Anal. Chem.*, **47**, 658A (1975).

37. Y. Talmi, *Anal. Chem.*, **47**, 697A (1975).

38. S. A. Borman, *Anal. Chem.*, **55**, 842A (1983).

39. L. B. Rogers, *Chem. Eng. News*, 18 (Apr. 15, 1974).

40. E. McDowell and H. L. Pardue, *Anal. Chem.*, **48**, 1815 (1976).

41. E. McDowell and H. L. Pardue, *Anal. Chem.*, **49**, 1171 (1977).

42. L. N. Klatt, *J. Chromatogr. Sci.*, **17**, 225 (1979).

43. R. E. Dessy, W. G. Nunn, and C. A. Titus, *J. Chromatogr. Sci.*, **14**, 195 (1976).

44. R. E. Dessy, W. D. Reynolds, W. G. Nunn, C. A. Titus, and G. F. Moler, *J. Chromatogr.*, **126**, 347 (1976).

45. M. J. Milano, S. Lam, and E. Gruska, *J. Chromatogr.*, **125**, 315 (1976).

46. M. J. Milano and E. Gruska, *J. Chromatogr.*, **133**, 352 (1977).

47. M. J. Milano, S. Lam, M. Savonis, D. B. Pautler, J. W. Pav, and E. Gruska, *J. Chromatogr.*, **149**, 599 (1978).

48. T. Amita, M. Ichise, and T. Kojima, *J. Chromatogr.*, **234**, 89 (1982).

49. J. C. Miller, S. A. George, and B. G. Willis, *Science*, **218**, 241 (1982).

50. G. E. James, *Can. Res.*, 39 (Dec./Jan. 1980/81).

51. H. Elgass, A. Maute, R. Martin, and S. George, *Am. Lab.*, 71 (Sept. 1983).

52. J. Leyrer, G. E. Nill, D. Harbawnik, G. Hoschele, and J. Diekmann, *Hewlett-Packard J.*, **35**, 31 (1984).

53. S. A. George and A. Maute, *Chromatographia*, **15**, 419 (1982).

54. K. L. Lohse, R. Meyer, W. Lin, I. Clark, and R. Hartwick, *LC* (LKB Instruments, Inc., Gaithersburg, MD), **2**, 226.

55. Private communication, J. D. Bambling, Jasco Incorporated, Easton MD, Mar. 26, 1984.

56. J. A. Haas, L. J. Perko, and D. E. Osten, Tracor Northern, Inc. Publication Form No. DA-3018, Middleton, WI.

57. S. A. Borman, *Anal. Chem.*, **57**, 276A (1985).

58. B. F. H. Drenth, R. T. Ghijsen, and R. A. De Zeeuw, *J. Chromatogr.*, **238**, 113 (1982).

59. C. J. Warwick and D. A. Bagon, *Chromatographia*, **15**, 433 (1982).

60. A. F. Fell, H. P. Scott, R. Gill, and A. C. Moffat, *J. Chromatogr.*, **282**, 123 (1983).

61. M. Fujita and E. Takabatake, *Anal. Chem.*, **55**, 454 (1983).

62. J. D. Messman and T. C. Rains, *Anal. Chem.*, **53**, 1632 (1981).

63. K. T. Suzuki, H. Sunaga, and T. Yajima, *J. Chromatogr.*, **303**, 131 (1984).

64. J. M. Pettersen, *Anal. Chim. Acta*, **160**, 263 (1984).

65. J. Treit, J. S. Nielsen, B. Kratochvil, and F. F. Cantwell, *Anal. Chem.*, **55**, 1650 (1983).

66. J. W. Foote and H. T. Delves, *Analyst*, **108**, 492 (1983).

67. F. F. Fernandez, *At. Abs. Newsl.*, **16**, 33 (1977).

68. J. C. Van Loon, *Anal. Chem.*, **51**, 1139A (1979).

69. J. C. Van Loon, *Am. Lab.*, 47 (May 1981).

70. J. C. Van Loon, *Can. J. Spectrosc.*, **26**, 22A (1981).

71. M. Ya. Bykhovskii and A. Yu. Braude, *Zh. Anal. Khim.*, **38**, 2236 (1983).

72. K. L. Jewett and F. E. Brinckman, *Chromatogr. Sci. (Liq. Chromatogr. Detect.)* **23**, 205 (1983).

73. R. M. Cassidy, M. T. Hurteau, J. P. Mislan, and R. W. Ashley, *J. Chromatogr. Sci.*, **14**, 444 (1976).

74. J. M. Harnly, T. C. O'Haver, B. Golden, and W. R. Wolf, *Anal. Chem.*, **51**, 2007 (1979).

75. S. E. Manahan and D. R. Jones, *Anal. Lett.*, **6**, 745 (1973).

76. N. Yoza and K. Kouchiyama, *Anal. Lett.*, **8**, 641 (1975).

77. D. R. Jones and S. E. Manahan, *Anal. Lett.*, **8**, 569 (1975).

78. N. Yoza and S. Ohashi, *Anal. Lett.*, **6**, 595 (1973).

79. D. R. Jones, H. C. Tong, and S. E. Manahan, *Anal. Chem.*, **48**, 7 (1976).

80. J. A. Koropchak and G. N. Coleman, *Anal. Chem.*, **2**, 1252 (1980).

81. C. Botre, F. Cacace, and R. Cozzani, *Anal. Lett.*, **9**, 825 (1976).

82. W. Slavin and G. J. Schmidt, *J. Chromatogr. Sci.*, **17**, 610 (1979).
83. B. W. Renoe, C. E. Shideler, and J. Savory, *Clin. Chem.*, **27**, 1546 (1981).
84. D. R. Jones and S. E. Manahan, *Anal. Chem.*, **48**, 1897 (1976).
85. P. T. Tittarelli and A. Mascherpa, *Anal. Chem.*, **53**, 1466 (1981).
86. G. Becher, G. Oestvold, P. Paus, and H. M. Seip, *Chemosphere*, **12**, 1209 (1983).
87. F. E. Brinckman, W. R. Blair, K. L. Jewett, and W. P. Iverson, *J. Chromatogr. Sci.*, **15**, 493 (1977).
88. E. J. Parks, F. E. Brinckman, and W. R. Blair, *J. Chromatogr.*, **185**, 563 (1979).
89. R. H. Fish, F. E. Brinckman, and K. L. Jewett, *Environ. Sci. Technol.*, **16**, 174 (1982).
90. E. A. Woolson, N. Aharonson, and R. Iadevaia, *J. Agric. Food Chem.*, **30**, 580 (1982).
91. R. H. Fish and J. J. Komlenic, *Anal. Chem.*, **56**, 510 (1984).
92. E. J. Parks, R. B. Johannesen, and F. E. Brinckman, *J. Chromatogr.*, **255**, 439 (1983).
93. R. H. Fish, J. J. Komlenic, and B. K. Wines, *Anal. Chem.*, **56**, 2452 (1984).
94. H. Koizumi, T. Hadeishi, and R. McLaughlin, *Anal. Chem.*, **50**, 1700 (1978).
95. T. M. Vickrey, M. S. Buren, and H. E. Howell, *Anal. Lett.*, **A11**, 1075 (1978).
96. T. M. Vickrey and W. Eue, *J. Auto. Chem.*, **1**, 198 (1979).
97. R. A. Stockton and K. J. Irgolic, *Int. J. Environ. Anal. Chem.*, **6**, 313 (1979).
98. D. Chakraborti, D. C. J. Hillman, K. J. Irgolic, and R. A. Zingaro, *J. Chromatogr.*, **249**, 81 (1982).
99. F. E. Brinckman, K. L. Jewett, W. P. Iverson, K. J. Irgolic, K. C. Ehrhardt, and R. A. Stockton, *J. Chromatogr.*, **191**, 31 (1980).
100. H. Koizumi, R. D. McLaughlin, and T. Hadeishi, *Anal. Chem.*, **51**, 387 (1979).
101. T. M. Vickrey, H. E. Howell, G. V. Harrison, and G. J. Ramelow, *Anal. Chem.*, **52**, 1743 (1980).
102. G. R. Ricci, L. S. Shepard, G. Colovos, and N. E. Hester, *Anal. Chem.*, **53**, 610 (1981).
103. M. Fujita and E. Takabatake, *Anal. Chem.*, **55**, 454 (1983).
104. S. D. Brown, *Anal. Chem.*, **49**, 1269A (1977).
105. S. Mori, A. Wada, F. Kaneuchi, A. Ikeda, M. Watanabe, and K. Mochizuki, *J. Chromatogr.*, **246**, 215 (1982).
106. S. Day, *Eur. Spectrosc. News*, **26**, 1 (1979).
107. E. Papadopoulou-Mourkidou, Y. Iwata, and F. A. Guther, *Agri. Food Chem.*, **29**, 1105 (1981).
108. K. Payne-Wahl, G. F. Spencer, R. D. Plattner, and R. O. Butterfield, *J. Chromatogr.*, **209**, 61 (1981).
109. M. Cooke and N. R. Godfrey, *J. Chromatogr.*, **237**, 151 (1982).
110. C. Fujimoto and K. Jinno, *J. High Res. Chromatogr. & Chromatogr. Commun.*, **6**, 374 (1983).

111. K. Jinno and C. Fujimoto, *Chromatographia,* **17,** 259 (1983).

112. S. Shi-Hua and A. Y. Kou, *J. Chromatogr.,* **307,** 261 (1984).

113. S. L. Smith and C. E. Wilson, *Anal. Chem.,* **54,** 1439 (1982).

114. E. H. Piepmeier, "Atomic Absorption with Low-Intensity Lasers," in N. Omenetto, Ed., *Analytical Laser Spectroscopy,* Wiley-Interscience, New York, 1979, pp. 132–134.

115. R. A. Keller and J. C. Travis, "Recent Advances in Analytical Laser Spectroscopy," in N. Omenetto, Ed., *Analytical Laser Spectroscopy,* Wiley-Interscience, New York, 1979, pp. 496–509.

116. J. D. Parli, D. W. Paul, and R. B. Green, *Anal. Chem.,* **54,** 1969 (1982).

117. R. B. Green, *Anal. Chem.,* **55,** 20A (1983).

CHAPTER

3

FTIR DETECTION

KIYOKATSU JINNO

School of Materials Science, Toyohashi University of Technology, Toyohashi, Japan

1. INTRODUCTION

High-performance liquid chromatography (LC) has become an extremely powerful tool, enough to separate complex mixtures into their components. With LC, compounds of poor thermal stability and low volatility, which cannot be separated by gas chromatography (GC), can often be readily separated for either quantitative or qualitative analysis. As a general rule, the amount of information that can be obtained from any chromatographic separation, however effective, depends on the detector. In GC, the thermal conductivity and flame ionization detectors (TCD and FID) have found wide use as universal detectors. No comparable LC detectors have yet appeared because of the large variety of mobile phases and their greater mass. In addition, commercially available LC detection systems such as ultraviolet–visible (UV–VIS) spectrometry and refractive index (RI) measurements suffer to some degree from limitations resulting from lack of sensitivity, selectivity, and/or versatility. None of these gives structural information on compounds eluted from a LC column. As the field of application for LC has increased, the limitations of these conventional detectors have become increasingly restrictive to the growth of the technique. This has led many scientists to look in new directions for LC detectors that can provide more information.

One detection method that has great growth potential is Fourier transform infared spectrometery (FTIR). The large number of absorption bands present in an IR spectrum offers the possibility of having both universal and chemically selective or specific detection capabilities. FTIR has the features of very short scan times (less than 1s) and signal averaging along with spectral subtraction capabilities. Because of the multiplex character of FTIR, IR spectrum is available for any point of interest on the chromatogram. Thus chemical identity of chromatographic peaks can be confirmed by comparison with library spectra for interpretation. The well-developed and analytically powerful application of various GC–FTIR interfaces (1) has led many analysts to believe that a similarly

64

effective and simple hybrid system may be developed for interfacing LC and FTIR.

Fundamental problems of compatibility have been encountered, however, and the development of the LC–FTIR interface has proved to be quite difficult. Unfortunately, most practical LC solvents have strong absorption bands in the mid-IR region, so that some spectral regions are opaque. In order to maintain a sufficiently high transmittance to allow solute absorption bands to be observable at most wavelengths across the spectrum (around 40% at the frequency of interest), the path length of the flow cell must be kept short, typically 100 μm or less. If the path length of the flow cell is increased, the detection limit may be increased in the spectral windows, but the proportion of the spectrum lost through absorption by the organic solvent is increased. Therefore, a compromise to allow maximum cell thickness consistent with solvent transmission over all the bands of interest is sought. This is a very difficult problem that has to be solved.

In spite of these disadvantages, several approaches have been proposed to add a new dimension to the analysis by the use of LC and FTIR in combination while preserving the capabilities of the separate techniques. In this chapter, recent approaches in LC–FTIR interface are reviewed and discussed based on their performance in practical situations, after the basics of FTIR spectrometers are introduced.

2. FTIR

Instead of using a monochromator attached with conventional IR spectrometers, the IR radiation from a source, before passage through a sample (in a very rare case, after passage through a sample), can be analyzed by means of a scanning interferometer. For FTIR spectrometry, the most commonly used device is the Michelson interferometer, although other types of two-beam interferometers such as the lameller grating interferometer have also been used. The Michelson interferometer, which is shown schematically in Fig. 1, consists of a movable mirror, a fixed mirror, and a beam splitter. Radiation from the infrared source is collimated by a mirror and the resultant beam is divided at the beam splitter, half of the beam passing to the fixed mirror and half reflected to the movable mirror. After reflection the two beams recombine at the beam splitter and, for any particular wavelength, constructively or destructively interfere depending on the difference in optical paths between the two arms of the interferometer. The intensity of the emerging radiation at any one particular wavelength modulates in a regular sinusoidal manner with a constant mirror velocity. The emerging beam is a complex mixture of modulation frequencies, which, after passing through the sample compartment, is focused onto the detector. The detector

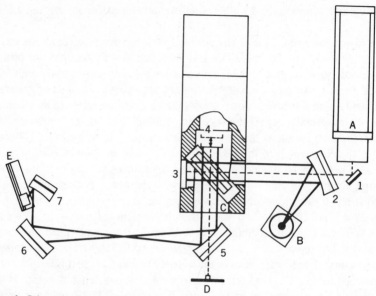

Figure 1. Schematic diagram of Michelson interferometer used in FTIR spectrometers. 1, 2, 5, 6, 7, mirrors; 3, fixed mirror; 4. movable mirror; A, HeNe laser; B, IR source; C; beam splitter; D, detector for reference signal; E, detector for IR beam.

signal is sampled at very precise intervals coinciding with the mirror scan. Both the sampling rate and the mirror velocity are controlled by a reference signal from another detector produced by modulation of the beam from a HeNe laser at the different positions. The signal obtained is known as an interferogram and contains all the information required to reconstruct the spectrum by a mathematical treatment known as Fourier transformation. Because dispersion or filtering is not required in FTIR, energy is not lost at the slits. The use of a HeNe laser as a reference results in very high frequency accuracy, better than 0.01 wave number.

There are two significant advantages of FTIR over dispersive spectrometers. The first is "Fellgett's advantage," that is, multiplex advantage. Measurements of equal signal-to-noise (S/N) ratio and resolution will take M times less time on a FTIR than on a dispersive spectrometer where M is the number of resolution elements, given as the spectral range divided by the desired resolution. This assumes that data are taken on instruments with equal optical throughput and efficiency with identical source. For measurements taken with equal data acquisition time, the S/N ratio of measurements taken by FTIR will be $M^{1/2}$ times better than that of the same measurement taken on a dispersive spectrometer.

The second advantage is "Jacquinot's advantage"; that is, the optical throughput of an interferometer should be greater than that of a monochromator operating at the same resolution. This advantage is typically between one and two orders of magnitude for mid-IR regions. The decreased throughpout of the slits of a monochromator in a dispersive spectrometer is the cause of Jacquinot's advantage, whereas the lack of an "exit slit" in a FTIR spectrometer is the cause of Fellgett's advantage.

Because the detectors used in commercial FTIR and grating spectrometers are different, it is very difficult to compare their performances. However, it might be thought that the performance of FTIR for the mid-IR regions would be greater than is actually found in practice, if the total gain in S/N ratio given by the combination of two advantages is calculated.

It is apparent that the combination of Fellgett's advantage and Jacquinot's advantage in FTIR spectrometry can lead to superior performance compared to conventional IR spectrometry. There is, of course, nothing intrinsically different between a detector on a FTIR instrument and one on a conventional dispersive IR spectrometer. However, the differences in the modulation frequency and in the S/N ratio for two types of instruments require that some care be taken in the choice of detectors for FTIR spectrometry. The most commonly used detector for mid-IR regions is the triglycine sulfide (TGS) pyroelectric bolometer with a KBr window. At the modulation frequencies necessary to prevent the interferogram from becoming digitization-noise limited (about 1 kHz) when incandescent continuous sources are measured, it is found that the specific detectivity (where this value is independent of the dimensions of the detector; the greater the value, the more sensitive the detector) of conventional thermal detectors, such as the thermocouple, falls off drastically compared to their specific detectivities at about 15 Hz. On the other hand, the TGS detector is more sensitive than the thermocouple for radiation modulated at 1 kHz, and thus has become the standard detector for FTIR.

Photoconductive detectors, such as lead sulfide, lead selenide, mercury cadmium telluride, and lead tin telluride, have better sensitivity than that of TGS, but these photoconductive detectors have a limited frequency range and cannot be used to cover all the mid-IR spectrum from wave number 4000 to 400. The sensitivity of photoconductive detectors increases as the modulation frequency increases, and this characteristic makes these detectors much suitable for use with rapid-scanning interferometers.

These significant advantages of FTIR as stated above offer several merits as an LC detector. First, several IR absorption bands can be simultaneously monitored in real time. Second, complete spectra corresponding to every point on the chromatogram are available. Third, the computer system for Fourier transformation is readily available for further treatment of the data generated.

In IR measurements, there are two types of spectroscopic techniques, absorption and reflection. Both can be used for LC–FTIR interfacing. Briefly the differences of both techniques are described below for a better understanding of LC–FTIR interfaces.

For absorption spectroscopy the sample is held at a focus that is commonly formed using a long-focal-length off-axis paraboloidal mirror. By an appropriate choice of the focal length of the paraboloidal mirror, sampling accessories that have been designed for conventional IR spectrometers can be used with FTIR spectrometers. The sample focus is commonly formed in the path between the interferometer and the detector (rather than between the source and the interferometer) for two reasons. First, the beam splitter can sometimes act as a filter for undesirable wavelengths. Second, by placing the sample after the interferometer, any unexpected radiation emitted from the sample is not modulated by the interferometer and is therefore not measured as part of the interferogram.

Reflection measurements can be performed on most instruments designed for absorption spectroscopy merely by placing an accessory for specular reflectance or attenuated total reflectance in the beam at the sample focus, that is, normal sample compartment. The measurements of diffuse reflectance spectra, that is, reflection spectra from samples that scatter the incident radiation over a wide angle, can perhaps best be accomplished on a special-purpose instrument using an integrating sphere. The sample is held at the surface of the sphere that is coated with a layer of a material of very high reflectance. Radiation reflected from the sample is then "integrated" over the surface of the sphere. A common detector is placed somewhere in that sphere. Even though only a small portion of the radiation reflected from the sample is measured, it has been shown that very accurate measurements of reflectance spectra can be obtained. Those spectra are obtained from calculations, such as by using the Kubelka–Munk function. Accessories for diffuse reflectance spectroscopy using an integrating sphere are commercially available from many FTIR suppliers. More details on FTIR spectroscopic measurements can be found in many books and reviews on the subject (2–4).

3. BASICS OF FTIR DETECTION IN LC

Vidrine stated the desired characteristics of an IR detector for LC as follows (2):

1. High sensitivity (as high as possible).
2. Compatibility with all samples chromatographed.
3. Compatibility with all solvents used as the mobile phase.
4. Nondestructive to samples.
5. Real-time output of chromatogram.
6. Capacity for monitoring several IR absorptions simultaneously.

7. Storage of complete spectra for later use.

8. Usability with gradient elution.

To fulfill the above requirements, IR analysis of LC fractions has been accomplished via two alternative approaches: (1) solvent removal and (2) direct flow-cell detection. Solvent removal for LC–FTIR eliminates the solvent automatically prior to measurements of the IR spectrum of the solute, but is inherently unsuitable for volatile or sensitive compounds. Direct IR detection via flow cell offers advantages in simplicity, quantitation, and universal applications. However, it is often rather difficult to use flow cells for FTIR measurements of samples separated by LC, where the choice of solvent is critical to the separation and gradient elution techniques often must be applied, because of IR absorption bands of the solvents used. In gradient elution, the composition of the mobile phase changes continuously throughout the run, and the spectral subtraction routines for solvent absorptions are difficult, in fact, usually impossible. Although both approaches have some intrinsic problems in satisfying the requirements described above, several systems for LC–FTIR measurements have appeared in the last few years. In the following sections typical attempts to combine LC with FTIR are introduced and evaluated on the basis of their performance. Before we begin the detailed discussions, some recent trends in LC should be noticed.

Over the past several years, much effort has been directed toward increasing speed and efficiency in LC. An especially recent development in LC in that direction is the use of low-volume columns. There are two ways for the reduction of the column volume: (1) reduction of the column length, using smaller particle size as stationary phases to compensate for the loss of efficiency, for example, the high-speed LC approach; (2) reduction of the column diameter, because of the essential independence of efficiency on column diameter with well-packed columns, for example, micro LC. The former is along the same lines as conventional LC, but the latter has completely different system requirements. Small column diameters (less than 1 mm) and flow rates (on the order of microliters per minute) are characteristics of micro LC. These features offer the possibility of solving certain problems associated with conventional LC–FTIR techniques. Therefore, separate conventional LC–FTIR and micro LC–FTIR interfaces are discussed separately.

3.1. Conventional LC–FTIR

3.1.1. Solvent Removal Approach

The significant difference in sensitivity between GC–FTIR and conventional LC–FTIR systems is the severe interference caused by the strong IR absorption of the mobile phase in LC. In a flow configuration for conventional LC–FTIR, the effluent from the column is passed through a flow cell, and interferograms are continuously measured and stored during the entire chromatographic run. At the end of this run, the solution spectra are computed and the absorption bands due to the mobile phase solvent are subtracted out. In order to keep solvent bands from "blacking out" excessively wide regions of the spectrum, the path length of the flow cell has to be kept quite small—typically about 100 μm for most solvents used as mobile phases in normal-phase LC, and less than 30 μm for solvents in reverse-phase LC. This means only a small fraction of each eluting sample is present in the cell at any moment during the measurement of the spectrum of the solute of interest, thereby drastically degrading the detectability.

This intrinsic disadvantage of the flow cell configuration in LC–FTIR can be overcome by eliminating the solvent prior to the measurements of the IR spectrum of the solute. The most successful of this approach was reported by Kuehl and Griffiths (5) in 1979. Their method involved a device in which the effluent from the LC column is concentrated and then dropped onto KCl powder. The remaining solvent is rapidly evaporated, leaving only the desired solute on the KCl, and the diffuse reflectance spectrum (DRIFT) of the solute is then measured. In 1980, Kuehl and Griffiths proposed a microcomputer-controlled interface between LC and DRIFT along this line (6).

The automated system for collecting LC peaks is based upon a sample carrousel that supports more than 32 equally spaced sample cups. The sample wheel is machined from 0.5-in. aluminum stock to assure high accuracy of sample positioning. This wheel is rotated by an AC chart drive motor via belt drive and a reduction pully. The sample cups are 5 mm in diameter and 19 mm high, and the sample well is 4.5 mm by 2.5 mm deep. A fine-mesh wire screen is used to support the KCl powder in the sample cups, which are hollowed to assist in solvent evaporation (Fig. 2).

The effluent from the liquid chromatograph is passed through the UV detector and sprayed into a concentration tube. The sprayer is constructed from glass capillary tubes that are drawn out to a fine tip, and the capillaries are held in place by a Teflon holder. The concentrator is constructed from 2-mm ID pyrex glass tubing that is about 15 cm long. A small bend near the end of the tube allows for the drops to collect and drip onto the sample cups. The glass tube is heated, and a small capillary tube is placed inside. This is connected to an aspirator through a solenoid valve. This allows the deposition of the effluent onto the sample cups to be collected. There are four active positions on the carrousel (positions 1, 2, 3, and 5, in Fig. 2) used during the LC–DRIFT measurements, and one other (position 4) at which the sample cups reside immediately before the measurements of the DRIFT spectrum. Position 1 is where the sample is deposited on the KCl powder; position 2 and 3, where a slow

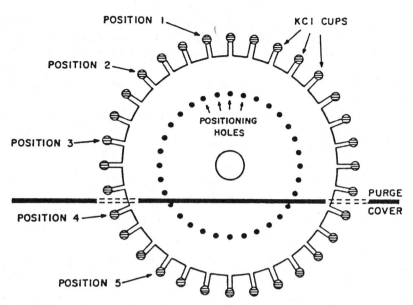

Figure 2. Schematic diagram of the carrousel used in the LC–DRIFT interface. A small light bulb is held above one of the positioning holes and a photodiode is held below it, to allow the position of the carrousel to be monitored. The sample is deposited at position 1, the solvent is eliminated at positions 2 and 3, and the spectrum is measured at position 5. Reproduced with permission from D. T. Kuehl and P. R. Griffiths, *Anal. Chem.*, **52**, 1394 (1980). Copyright 1980 American Chemical Society.

stream of air is drawn through the KCl for the final solvent elimination step; and position 5, where the IR spectrum of the deposited solute is measured.

The system is controlled by the microcomputer, which monitors the signal from the UV detector of the chromatograph. When the signal exceeds a certain threshold level, the computer sends the trigger signal to the interfacing device. Good spectra of samples separated by normal phase LC have been obtained for submicrogram quantities of all nonvolatile samples. However, some interferences were also observed. The source of those interferences may be any or all of the following: residual solvent remaining after the drying stage; contaminants in the solvent eluting from the chromatographic system; contaminants sorbed onto the KCl from the atmosphere; and contaminants present in small quantities in the KCl.

The first interference depends on the nature of the mobile phase—the more volatile the solvent, the smaller the problem. In addition, KCl is soluble in some solvents and this creates problems. The main contaminant sorbed onto the KCl from the atmosphere is water. This problem can usually be eliminated by measuring the KCl reference in the same way that the sample is measured, but water bands are often observed in the spectrum depending on the situation, such as during periods of high humidity or at very large ordinate expansions. DRIFT spectrometry yields much more intense absorption bands than conventional transmission measurements. Therefore, it is essential to keep the concentration of all contaminants to an absolute minimum in order to maintain low detection limits in this type of LC–FTIR interface.

Figure 3 shows the spectra measured from the chromatogram shown in Fig. 4. In commercial azobenzene, the trans isomer is the major component and the cis isomer represents about 1.7% of the mixture. The isomers were separated by LC with isocratic elution using a mixutre of *n*-hexane and ethyl acetate as the mobile phase at a flow rate of 2.0 mL/min. The trigger for data collection was set to be about 80% of the full height of the peak due to the cis isomer, because the peak height of the cis isomer is smaller than that of the trans isomer. There are four bands of azobenzene in the mid-IR region between wave numbers 800 and 600, which change in relative intensity depending on the particular isomer. This example is very demanding, because the minor component, that is, the cis isomer, is in the tail of the peak due to the major component, that is,

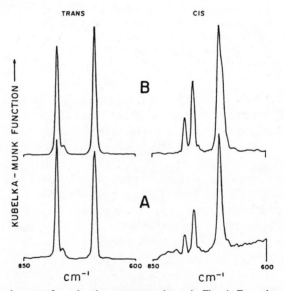

Figure 3. Infrared spectra from the chromatograms shown in Fig. 4. From the relative band intensities, the *trans*-isomer (98.3% abundance) is present at less than 10% of the concentration of the cis isomer (1.7% abundance) when the peaks are just resolved (resolution = 0.8). Reproduced with permission from D. T. Kuehl and P. R. Griffiths, *Anal. Chem.*, **52**, 1394 (1980). Copyright 1980 American Chemical Society.

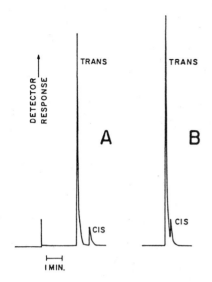

Figure 4. The chromatogram of *cis*- and *trans*-azobenzene, with the natural isomeric ratio, separated on Partisil 10 with a mobile phase of (*a*) 40% ethyl acetate, 60% hexane, and (*b*) 100% ethyl acetate. Reproduced with permission from D. T. Kuehl and P. R. Griffiths, *Anal. Chem.*, **52**, 1394 (1980). Copyright 1980 American Chemical Society.

the trans isomer, and is present in a quantity approximately 60 times less than the major component.

The conventional LC–DRIFT technique introduced here has largely solved all the problems associated with the flow-cell LC–FTIR combination, but still has a few limitations. The first is that solutes have to be significantly less volatile than the mobile phase solvent. This criterion is certainly satisfied for most samples separated by LC because volatile solutes are usually separated by GC. However, the second limitation is that water has not been successfully eliminated because of its high surface tension and latent heat of vaporization, thereby limiting separations to normal-phase absorption or size exclusion. In the flow-cell configuration, the situation is also the same because of the strength of the absorption bands of water. To overcome this limitation, alternatives have been tested by some authors (7,8).

One of them has been reported by Conroy et al. (7) recently. In that attempt, solutes present in aqueous eluents from reverse-phase LC columns are continuously extracted into dichloromethane, which evaporates easily and shows less interference from its IR absorption bands even if it remains in the KCl powder, and then deposited on the KCl for DRIFT measurement. Detection limits of less than 1 μg (injected) for many compounds separated by reverse-phase LC can be achieved routinely.

An alternative method of eliminating water from an aqueous mobile phase has been suggested by Kalasinsky et al. (8). 2,2-Dimethoxypropane (DMP) is continuously added to the column effluent, where DMP reacts with water to yield acetone and methanol, which are easily eliminated by evaporation. This technique is simpler than the Griffiths method (7), but has disadvantages: DMP is expensive, and fairly high flow rates are required. The liquid from the reaction cell is concentrated by differential evaporation, and the solutes are then deposited on KCl, after which DRIFT spectra are measured. Thus if the mobile phase contains an added buffer or ion-pairing regents, the IR absorption bands of these materials dominate the spectrum of each separated peak. In Griffiths' method, it is possible to retain these materials in the aqueous phase. Thus only the analytes are extracted from the aqueous phase into CH_2Cl_2.

The new attempts for reverse-phase conventional LC–FTIR described above are still in a very early development stage for practical uses, however. The most serious limitation in conventional LC–DRIFT is that solvent elimination and deposition must be triggered to function successfully by another LC detector, such as UV or RI. In the final and ideal system, the effluent from the LC column should almost certainly be continuously deposited on the KCl powder, thereby obviating the need for a trigger actuated by the signal from another LC detector. Such continuous LC–DRIFT interfaces have not yet been described in any presentation or publication, but one can expect that they will be available in the near future.

3.1.2. Flow-Cell Approach

Direct IR detection in a flow cell offers advantages in simplicity and universality. The only LC–FTIR detectors currently being manufactured use the flow-cell configuration. A flow cell was the basis for the first real-time IR detector designed for LC in 1977 by Foxboro/Wilks (9). This detector is a simple filter photometer capable only of monitoring a single IR band. In contrast, FTIR monitors the entire IR spectrum at higher sensitivity than dispersive or filter-type IR spectrometers. As a result, a straightforward single-beam flow-cell approach, similar to that used for GC–FTIR, has resulted in a practical LC–FTIR system designed by Vidrine and Mattson in 1978 (3).

In the direct flow-cell approach, there are many things to be considered in order to obtain successful interfacing between LC and FTIR. The first consideration is the solvent selection for LC separations. The most significant advantage of LC is to be able to choose a combination of stationary phase and mobile phase for a desired separation. However, selected solvents for separations generally have many strong absorption bands in the mid-IR region and make these ranges opaque. Vidrine (2) has discussed the relationships between optimum cell thickness and spectral windows of solvents. He concluded that, for any particular IR frequency, the measurement sensitivity is best if the cell thickness is chosen so that the solvent transmission is around 40% at the frequency of interest. In practice, a compromise giving maximum cell thickness consistent with solvent

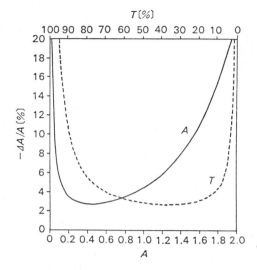

Figure 5. Relative error per 1% of measurement error for absorbance and/or transmittance.

transmission over all the bands of interest must be sought. When complete spectra are desired, excess sensitivity can be traded for more complete spectral coverage by selecting a shorter path length.

If we assume that Beer's law holds in absorption measurements, because the accuracy of the measurements in absorbance value, A, depends on the magnitude of the absorbance itself, the measurements should be performed at the optimum absorbance range by selecting the optimum cell thickness. Figure 5 shows the relationships between the relative error of each value and transmittance T or absorbance A. As seen, the relative error of absorbance A is minimized as follows:

$$T(\text{optimum}) = \frac{1}{e} = 36.79\% \tag{1}$$

$$A(\text{optimum}) = \log e = 0.4343 \tag{2}$$

Therefore the measurements should be done in the range from 0.25 to 0.70 absorbance in order to increase the accuracy. It can be deduced that the S/N ratio in an absorbance measurement is the highest near an absorbance of approximately 0.4. This means that, to optimize a flow-cell thickness so that maximum detector sensitivity may be obtained for a particular frequency, the solution should have an absorbance of approximately 0.4 at that frequency. Of course, solvents have many strong absorption bands and it is impossible to satisfy the above requirement at all the frequency ranges in the mid-IR region. In addition, the following points must be considered for flow-cell volume and path length to optimize the interface. Generally, the sensitivity increases with the increase of cell path length but chromatographic resolution decreases with it because of peak broadening owing to the increase of cell volume. Therefore, flow cell with smaller volume and larger path length is preferred for the purpose of IR monitoring. On the other hand, the path length must be decreased to reduce spectral interferences caused by the solvents when IR spectra of solutes are desired.

To compensate for the sensitivity loss due to the smaller path length, the amount of sample injected into the LC system must be increased, or more sensitive optical systems and detectors must be used. Because the former creates a problem of overloading the LC column, and the latter is also very hard to realize at present, the most transparent chromatographic solvents and the optimum flow-cell path length must be chosen. This is the most difficult problem to be solved and is the main reason why it is often difficult to use the flow-cell approach for conventional LC–FTIR interfacing.

The second consideration is the flow-cell material. For use with most organic solvents as mobile phases, the standard alkali salt window materials are optimum. The advantages and limitations of sodium chloride, potassium bromide, cesium bromide, and cesium iodide as cell-window materials in the mid-IR regions are widely discussed in many texts and catalogs (10). They are low-RI materials allowing high transmission, but are water soluble and somewhat fragile. Zinc selenide, silver chloride, and thallium bromoiodide (KRS-5) can be used with aqueous systems, but their higher RIs result in lower transmission and give rise to interference fringes. Barium fluoride, although fragile and slightly water soluble, is free of this problem. Polytetrafluoroethylene (Teflon) is also a good cell material that is compatible with most chromatographic samples and solvents.

Typical flow-cell designs are shown in Fig. 6. For small cell thickness, simple amalgam-sealed or demountable KBr or NaCl cells are suitable for many solvents if the cell holder has small-diameter tubing for inlet and outlet. This kind of cell is also suitable for medium path lengths if a spacer cut to minimize cell volume is used, as illustrated in Fig. 6a. These cells are easy to build and modify. An alternative cell, illustrated in Fig. 6b, uses cheap, undrilled windows. However, for this cell design, elastomer O-rings are required and they must be compatible with the solvent. For larger-path-length flow cells, the internal volume is a limiting factor in this design. It is a disadvantage to lengthen the optical path length further once a major part of the chromatographic peak can fit inside the cell at one time. Other design considerations are also required as follows: (1) dead volume in the cell should be minimized, (2) mixing volume in the cell should be smaller than the volume of expected chromatographic peaks, to retain chromatographic resolution, and (3) precautions should be taken to protect against leaks.

The third consideration, which is an important requirement for LC–FTIR, is

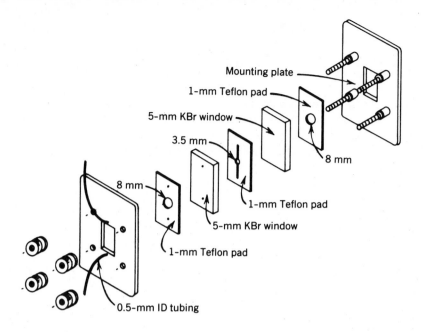

(a)

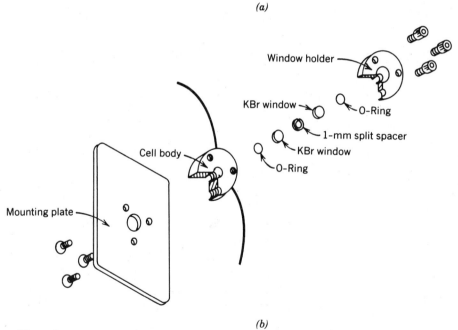

(b)

Figure 6. Flow-cell designs for LC–FTIR. (*a*) Common 19-mm × 38-mm demountable cell. (*b*) Using O-rings and undrilled windows. Reproduced with permission from Ref. 2. Copyright 1979 Academic Press.

a small beam diameter at the focus. If the normal beam is used, no special alignment is necessary and LC–FTIR can conveniently be done in the sample compartment, but sensitivity will not be high enough. Condensing optics can be used to obtain a smaller beam size to improve or maintain the sensitivity. However, this means performing critical alignments, using extra optics that entail optical losses, and causing further losses by very oblique rays in parts of the beam. Generally, commercially available beam condensers are suitable for that purpose.

The fourth consideration is the sensitivity. The wide variation in chromatographic peak volumes, optimum path length, and cell volumes make it impossible to quote a single detection limit for all LC work (10). The detection limit must be defined as detection of a specified amount of compound in the path length. The minimum detectable amount of a substance in any particular chromatogrpahic system can therefore be estimated from a knowledge of the peak volumes of solutes and the cell volume of the IR detector. For compounds that have strong absorption bands the detection limit is generally considered to be around a few hundred nanograms.

The most universal conventional LC–FTIR system based on a flow-cell configuration is commercially available in the United States from Nicholet.

Because LC–FTIR intrinsically produces three dimensions of information (time, frequency, and absorbance), plotting LC–FTIR data in three dimensions is the only way to represent all the available information from LC and FTIR. Figure 7 shows the chromatograms of a molecular size-exclusion separation of silicone oil, paraffin oil, and benzene. The 680–660 wave-number window is sensitive to the silicone (a phenylsiloxane) and benzene. The 1100–1040 wave-number window is sensitive to the silicone only, and the 2990–2850 window only to the paraffin oil. Plots of the three dimensions can be drawn automatically with the available system software as shown in Fig. 8. This system software also has the capability of monitoring a chromatogram on-the-fly with the use of the so-called Gram–Schmidt reconstruction procedure (11). Spectral subtraction allows this chromatographically incomplete separation to be resolved spectrometrically.

Direct flow-cell configuration for conventional LC–FTIR interfacing has resulted in an extremely useful detection technique for LC. Real-time chromatograms representing different IR bands allow functional-group analysis of the chromatographic peak, and complete spectra (except for regions of solvent opacity) can be retrieved for any point on the chromatogram. Still there are many limitations on the use of this interface, such as incompatibility for reverse-phase separations, which is the most popular separation mode in present LC. However, the application of LC–FTIR to some chromatographic problems can still be made. If the difficulties in reverse-phase separations can be solved, this approach will become a universal LC–FTIR system.

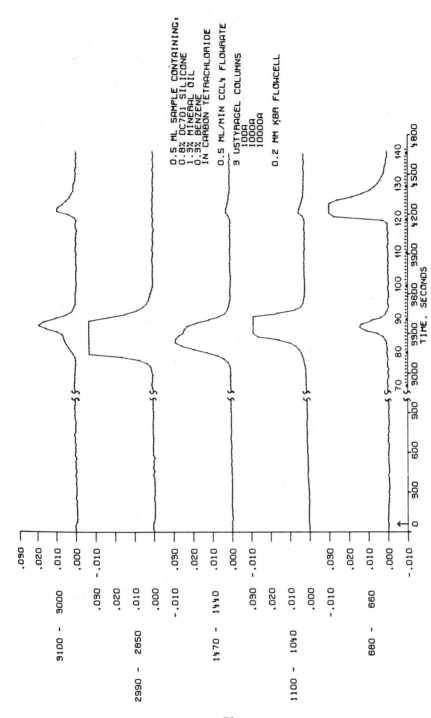

Figure 7. Molecular size separation of silicone oil, paraffin oil, and benzene. Reproduced with permission from Ref. 2. Copyright 1979 Academic Press.

79

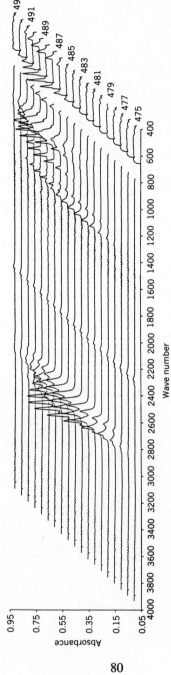

Figure 8. IR spectra from the chromatographic separation of Fig. 7 are plotted in order of acquisition. Each spectrum is slightly offset to represent a Z axis in time, demonstrating the three-dimensional nature of LC–FTIR. Reproduced with permission from Ref. 2. Copyright 1979 Academic Press.

80

3.2. Micro LC–FTIR

Recently, micro LC has been gaining popularity. Small column diameters and flow rates are characteristics of micro LC (12–14). The approximate 100-fold increase in solute concentration for microscale columns should improve the detectability of species as a function of sample amount injected. In particular, low solvent consumption of micro LC enables both IR-absorbing solvents to be easily eliminated and expensive solvents to be used for chromatography. These features offer the possibility of solving problems associated with conventional LC–FTIR techniques in both direct flow-cell configuration and solvent removal approaches.

In the following sections, alternative approaches such as solvent removal and flow cell in micro LC–FTIR interfaces are reviewed and discussed in detail.

3.2.1. Solvent Removal in Micro LC–FTIR Interfacing

As mentioned, micro LC requires only a small size of samples, as well as small amounts of the mobile phase and the packing materials, so that the associated components can be reduced more than those of conventional LC. Consequently, micro LC has an excellent potential for combination with any specific detectors such as electrochemical and mass spectrometric.

To overcome the problems associated with the conventional LC–FTIR techniques, a novel approach was proposed by Jinno et al. in 1981 (15–18). The approach is named the "buffer-memory" technique, in which micro LC is employed for the separation of the sample and a KBr crystal plate is used as the transport medium to accept the total column effluent.

An interface device developed for the KBr buffer-memory technique is shown in Fig. 9. The interface is fitted with a commercially available microfeeder that has been modified to be attached to the sample compartment of a FTIR spectrometer. The KBr crystal plate is set in the holder attached to the end of the moving rod of the interface. The eluent from the microcolumn is deposited on the crystal plate as a continuous narrow band. The crystal is transferred past the exit of the microcolumn for sample collection. The speed of collection can be controlled according to the micro LC separation conditions used. The buffer-memory crystal is then automatically brought into the IR beam as the eluent is collecting. A FTIR monitors the eluent as a simple LC detector or as a spectrometer. The interface device is equipped with an on–off twin timer that enables the buffer-memory plate to be stopped while IR spectra are being measured if conventional IR spectrometers are used.

The dead volume caused by the transfer line should be as small as possible. The transfer line used is PTFE tubing (40-cm by 0.1-mm ID), which caused no serious band broadening effect for micro LC peaks. The hook-shaped stainless

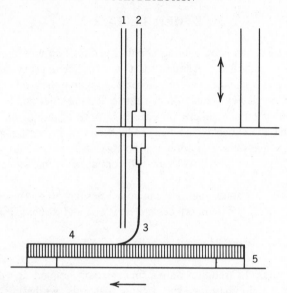

Figure 9. Interfacing device for micro LC–FTIR via buffer-memory technique: 1, nitrogen gas for eliminating solvents; 2, LC effluent from the microcolumn; 3, stainless steel capillary ; 4, KBr crystal plate, 5, plate support.

steel tubing attached to the PTFE serves as a spring coil to contract gently with the KBr plate. In order to facilitate evaporation of the carrier solvent, warmed nitrogen gas is passed over the sample on the crystal plate.

In LC–FTIR interfaces, it is very important to examine the ability for quantitation, that is, to know the applicability of Beer's law. For the purpose of confirmation, solutions containing measured amounts of di-*n*-propyl-2,4-dinitrophenylhydrazone (DNPH) (at 6, 4, 3, and 1 μg) were successively introduced and subsequently deposited onto a KBr crystal as a continuous band in normal-phase LC mode. The FTIR chromatograms derived from the buffer memory are shown in Fig. 10. Very high linearity is seen in this figure. As a rule, the present detection limit for DNPH seems to be less than 500 ng. In this instance, a small aperture (4 mm by 1 mm) was placed in front of the plate in order to minimize the noise from nondeposited areas of the memorized plate. The size of the beam focus of the FTIR spectrometer equipped with a 3X beam condenser was about 3.3 mm by 8 mm, whereas that of the solute-deposited band was spread over 2 mm with 1-mm width. Hence the measurement efficiency was reduced by a factor of about 20. It thus seems that matching some experimental parameters can result in an improvement of detection limit.

Another important aspect of quantitative analysis is the reproducibility of the deposition and subsequent absorbance measurements. The reproducibility of the

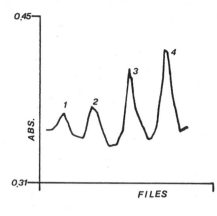

Figure 10. Peak height of the IR chromatogram at wave number 1255 as a function of the weight of sample injected. Solutions containing measured amounts of DNPH were successively injected into a microcolumn. Peaks: 1, 1μg, 2,3 μg, 3,4 μg, 4,6 μg. Reproduced with permission from Ref. 17. Copyright 1982 Elsevier.

chromatograms obtained by the buffer-memory technique is found to be about 6% (relative standard deviation) of the peak height of the chromatograms for five different deposits of 2 μg of DNPH.

One of the typical demonstrations of this approach has been in size-exclusion chromatography (SEC) of polystyrene. A mixture of four polystyrene standards was eluted through a microscale SEC column. The resulting UV chromatogram and FTIR chromatograms at characteristic absorption bands of the solutes are shown in Fig. 11a and b, respectively. It appears that the FTIR chromatograms obtained by the buffer-memory technique reflect the original chromatographic separation without much decrease in resolution. In this example, interferometric data were collected every 0.5 mm on the KBr plate. The flow rate of the mobile phase was 8 μL/min and the deposition rate of the effluent was 1.25 mm/min.

Lack of interference from the mobile phase solvents, which is the most significant advantage of the solvent elimination approach, also enables a wide wave-number range to be measured provided that absorption due to the background is kept constant in every part of the plate. Figure 12 shows such a chromatogram of standard polystyrenes measured at wave numbers 600–2000. It is important to note that the absorbance values obtained in this way are much higher than those in Fig. 11b. The FTIR chromatogram obtained with a wide wave-number range seems to be similar to those at fixed wave numbers. This result implies that IR can be used as a micro LC detector by measuring the total absorbance over a wide wave-number range even if the characteristic absorption locations of the sample are unknown. However, this is not the best way to obtain highly sensitive detection because of the low S/N ratio. IR can be useful as an universal detector in LC based on this nondispersive concept using the KBr buffer-memory technique, as in the Gram–Schmidt reconstruction method in the direct flow-cell approach.

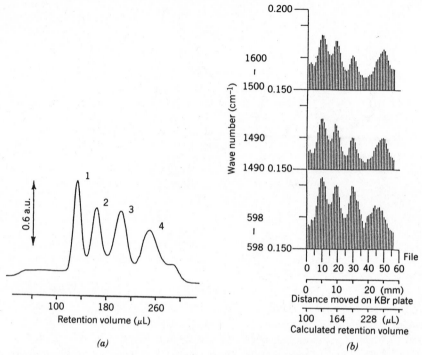

Figure 11. (*a*) SEC chromatogram of polystyrene standards measured with a UV detector at 254 nm. Micro-LC conditions: column, 1-mm ID × 22-cm PTFE tube; packing, TSKGel 3000H; mobile phase, THF 8 μL/min. Peaks: 1,MW 37000, 2,MW 10200, 3,MW 2800, 4,MW 500. (*b*) FTIR chromatograms at various wave numbers. Detector, TGS; accumulation, × 64; resolution, 8 wave numbers. Reproduced with permission from Ref. 17. Copyright 1982 Elsevier.

An additional, and more important, feature of the buffer-memory technique is that the effluent from the column is deposited on a piece of crystal plate in the form of a chromatogram. This means that it is fairly easy to preserve the solute deposited for subsequent detailed characterization, in comparison with the flow-cell techniques. Such examples have been demonstrated for sequential analysis of chromatographically separated organic and organic-bound metal species by some spectrometric methods, such as identification in situ by IR, X-ray fluorescence, and mass spectrometry to obtain information on the bonding between organic molecules and metal elements (18).

Initially the buffer-memory technique had been thought to be unsuitable for use in reverse-phase mode, because the collecting media are alkali salts and they are easily dissolved by aqueous solvents. However, measurements of micro

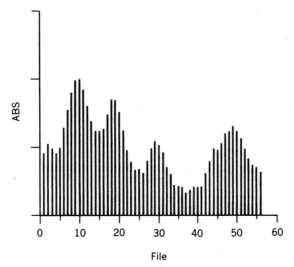

Figure 12. FTIR chromatogram of polystyrene standards for a wide wave-number range of 600–2000. Reproduced with permission from Ref. 17. Copyright 1982 Elsevier.

LC–FTIR in reverse-phase separations have become possible if a stainless-steel wire net is used as the medium instead of a KBr crystal (19).

An example for the separation of a mixture of caffeine, aspirin, and phenacetin is shown in Fig. 13. In this instance, an 846 mesh stainless-steel wire net which is 4 mm by 30 mm was used as the collecting medium, and the similar interfacing device as used in the KBr method was used to collect the effluent from the reverse-phase microcolumn. The chromatographic condition was a truly reverse-phase situation, which is described in the figure legend. No interference is observed from absorption of the mobile phase, methanol and water. The result is very impressive; it indicates that the buffer-memory technique can be a universal method for micro LC–FTIR interfacing in almost all separation modes.

The use of microcolumns does, of course, still present several disadvantages for LC–FTIR measurements, although they offer many merits. The principal problem is the capacity of the columns, which is often equal to, or even less than, the detection limits of the IR spectrometer. It is a common fact that column capacity decreases in proportion to cross-sectional area; thus the capacity of a 0.5-mm ID column is about 80 times less than that of a 4.6-mm ID column.

Therefore, Griffiths and Jinno considered that in view of the excellent microsampling capability of the DRIFT technique, the detection limits of the buffer-memory micro LC–FTIR technique could be reduced if the peaks could be efficiently deposited on a powdered alkali halide in analogous fashion to the

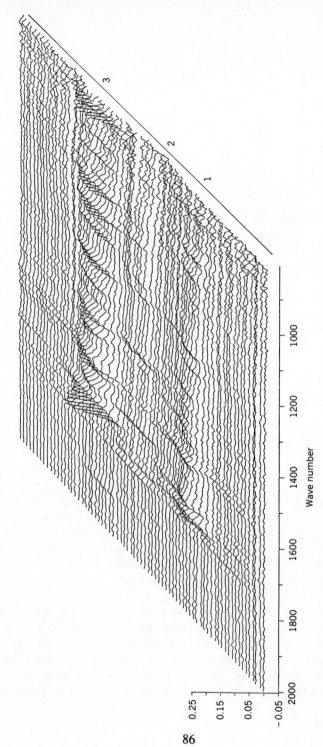

Figure 13. Reverse-phase separation of caffeine, aspirin, and phenacetin monitored by FTIR via buffer-memory technique. Column, 0.5-mm ID × 25 cm; packing, Develosil ODS-10; mobile phase, methanol–water 60:40; flow rate, 4 μL/min; accumultion, × 100. Peaks: 1 = caffeine, 2 = aspirin, 3 = phenacetin.

86

conventional LC–FTIR interface developed in 1979 (20). To illustrate the performance of the micro LC–DRIFT, a mixture of nitrobenzene and 4-chloronitrobenzene was separated using the 1- and 0.5-mm ID columns. The same mobile phase and linear velocity were employed for each separation, and the IR spectra of 4-chloronitrobenzene are shown in Fig. 14. In addition to the interesting fact that the higher flow rates in the 1-mm ID column improve the deposition characteristic, the S/N ratio of the spectrum of 10 ng of 4-chloronitrobenzene is sufficiently good that we can expect to obtain a recognizable spectrum for 1 ng of this material.

For reverse-phase separations with micro LC–DRIFT measurements, certainly the use of an alkali salt as a substrate is inappropriate for direct measurements. Substrates that are insoluble in water, in particular diamond powder, would seem to be more suitable (21). Analogous techniques that have been mentioned in discussions of conventional LC–DRIFT approaches should be applicable with

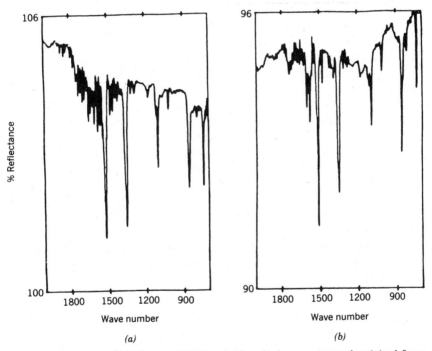

Figure 14. IR spectra of (*a*) 10-ng and (*b*) 15-ng 4-chloronitrobenzene separated on (*a*) a 1.0-mm ID and (*b*)0.5-mm ID column, respectively. The mobile phase was 2% methanol in hexane and the flow rates were 35 and 10 μL/min, respectively. Reproduced with permission from C. Conroy, P. R. Griffiths, and K. Jinno, *Anal. Chem.*, **57**, 822 (1985). Copyright 1985 American Chemical Society.

better efficiency for reverse-phase micro LC–DRIFT technique. Those studies are currently being studied in many laboratories.

3.3.2. Flow-Cell Approach in Micro LC–FTIR

Even if we use microcolumns instead of conventional columns for separation, the intrinsic problems in the direct flow-cell approach could not be solved entirely.

The merits in using microcolumns are as follows: lower flow rates in micro LC offer the possibility of using "exotic solvents" such as deuterated solvents or perfluorinated solvents, and higher mass concentration makes improved sensitivity in IR measurements. If we limit our interests to the detection of sample components having C—H and/or C=O fragments in their chemical structures, which are almost universal for organic compounds, some solvents such as deuterated solvents can be used without any interference, although they might otherwise be prohibitively costly in conventional LC separations. Because solvent consumption is quite low in microcolumn LC, these high-cost but better IR solvents can be used for LC separations.

We can discuss the usefulness of deuterated solvents as mobile phases in micro LC–FTIR prior to the demonstration of their performance via the flow-cell approach. As is well known, the C—H and O—H stretching vibrations are shifted to approximately $1/\sqrt{2}$ of their vibration frequencies by replacing the hydorgen with deuterium. Gore et al. (22) showed as early as 1949 that water used in conjunction with deuterium oxide enables one to obtain an IR absorption spectrum throughout almost the entire normal analytical region.

The discussion is restricted here to the solvents used in reverse-phase LC separations, because the very polar solvents used for this mode, such as water, methanol, and acetonitrile, absorb IR radiation much more strongly than any other organic solvents used in normal-phase and size-exlusion modes. These IR spectra are illustrated in Fig. 15, with their deuterated counterparts. The solvents were run in an airtight cell with KRS-5 windows, of which path length can be easily changed with lead spacers; the distance was defined in the usual way by an interference fringe pattern of the empty cell. No special precautions to prevent atmospheric exchange were taken during the measurements of the spectra. Water has intense absorption that covers the major part of the IR region, even at relatively short path length as indicated in Fig. 15. A much shorter path length (25 μm or less) is necessary to obtain reasonable information close to the most intense broad band (wave number 3400, O—H stretching), but working at such a very short path length is impractical in view of the lack of IR sensitivity.

Deuterium oxide is completely transparent in the region of X—X (X=C, O, N) stretching. Hence IR detection in this region is not precluded, although the molecules containing OH or NH groups undergo rapid exchange in some cases. At the same time, the shift of the wave number 1640 O—H band to wave number

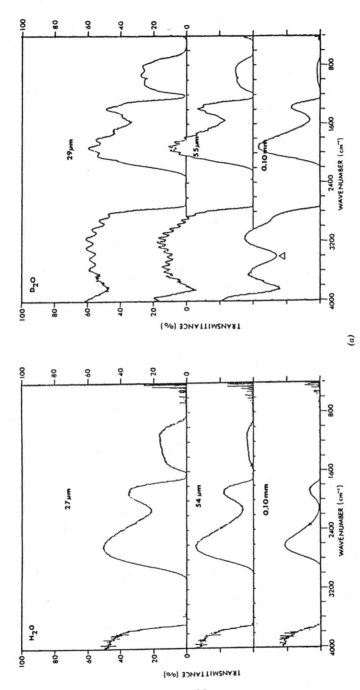

Figure 15. IR spectra of solvents used in reverse-phase chromatography. The bands marked can be assigned to O—H stretching vibrations that appeared as a result of D—H exchange between deuterium oxide and atmospheric moisture.

89

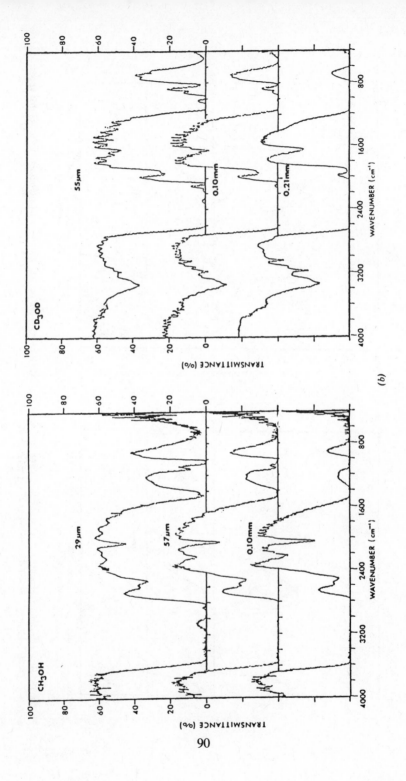

(b)

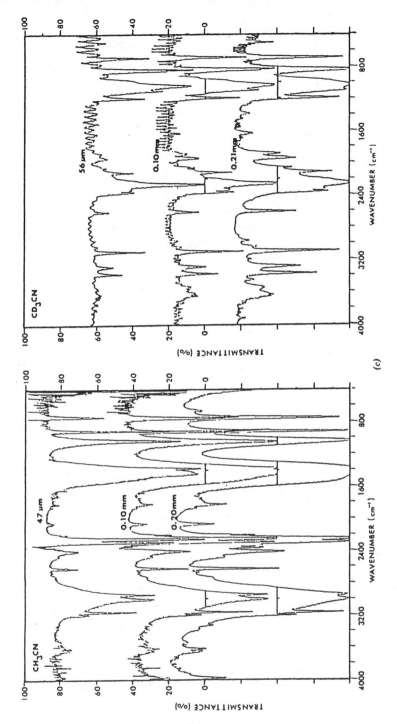

(c)

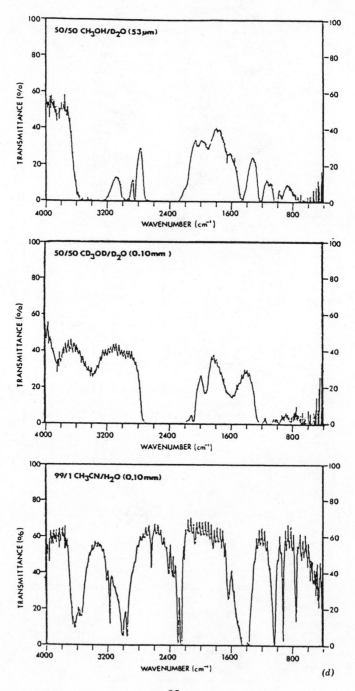

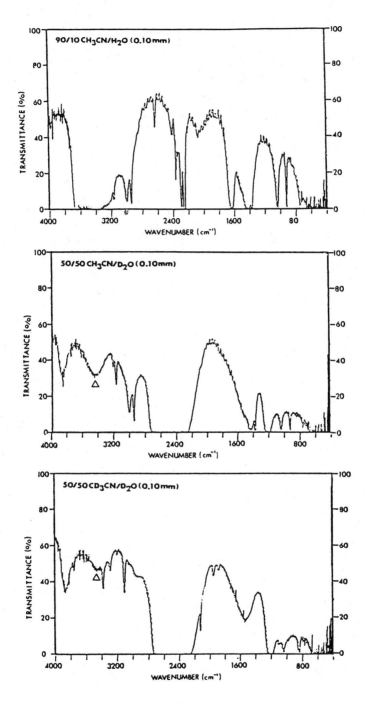

93

1210 clarifies the extremely interesting and useful X=Y (X, Y=C, O, N) region. Similarly both perdeuteromethanol and perdeuteroacetonitrile have a completely transparent window in a C—H stretching region. It is noted that acetonitrile is much more transparent than methanol, particularly in the C—H stretching region where methanol absorbs completely. Bearing in mind the above facts, we may expand this approach to LC employing FTIR detection with less cost by using a mixture of deuterium oxide and ordinary acetonitrile, which could cover most reverse-phase separations. Deuterium oxide is relatively less expensive than any other deuterated solvents. In addition, it is necessary to assess the chromatographic performance of deuterated solvents used as mobile phases. However, the discussion is beyond the scope of this chapter; readers seeking details about this subject can refer to other publications (24,25).

Some examples of FTIR spectrometric detection in conjunction with deuterated solvents are now given. In terms of solute detection (rather than solute identification), it is essential to find solvents that are transparent to IR radiation at the wave numbers of interest. There is no doubt that deuterated solvents satisfy this requirement. In Fig. 16, a comparison of the FTIR chromatograms of phthalates, separated by reverse-phase mode using deuterated and nondeuterated solvents, is shown. By the use of the deuterated solvent as the mobile phase, the chromatographic peaks monitored at wave number 2965 (C—H stretching), which were not clearly distinguished from the baseline fluctuation when protonated solvent was used, became clear. Further significant improvement because of the use of the deuterated solvent is observed at wave numbers 1300 and 1730, corresponding to C—O and C=O stretching vibrations, respectively. The reason is that the reduced absorbance at these wave numbers for the deuterated solvent offers high transparency.

As a demonstration of normal-phase separations, a mixture of 5α-cholestane, 5α-chorestan-3-one, and cholesterol was chromatographed on silica with a mobile phase of deuterocyclohexane–deuteromethanol–deuterochloroform. Figure 17 shows the ability of the system to select peaks corresponding to substances having specific molecular structure in addition to the general detection capability by monitoring the C—H absorption bands. Only 5α-cholestan-3-one with a carbonyl group can be detected at wave numbers 1710, and cholesterol with both alcoholic C—O and O—H bands can be selectively detected at wave numbers 3340 and 1050. Such an IR selectivity in conjunction with LC retention is the most promising way to identify the individual components of a complex sample.

As seen, the employment of deuterated solvents in place of protonated solvents enables one to see the C—H stretching region as well as the other regions where the solute absorbs, without affecting the separation. The selectivity or specificity of FTIR measurements allows us to discriminate even peaks that are insufficiently resolved chromatographically. A wide polarity range of deuterated compounds is readily available today from chemical suppliers. If eluents of weaker polarity

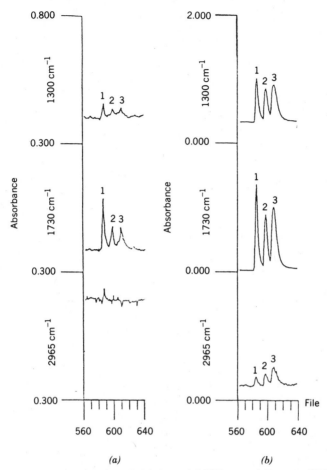

Figure 16. Reverse-phase separation of phthalates. (*a*) With acetonitrile–water 90:10: (*b*) with deuteroacetonitrile–deuterium oxide 90:10. Column, CHEMCOSORB ODS-H, 0.5-mm ID × 14 cm; flow rate, 4μL/min; accumulation, × 10; cell, AgCl (1.6 μL). Peaks: 1 = dimethylphthalate, 2,di-*n*-butylphthalate, 3,di-*n*-pentylphthalate. Reproduced with permission from Ref. 23. Copyright 1985 Friedr. Vieweg & Sohn.

than cyclohexane-d12 are required for an optimized separation, perhalogenated solvents such as FC-113 (1,1,2-trichloro-1,2,2-trifluoroethane) from Du Pont could be used because they also have favorable IR transparency.

It has been clearly shown that the direct flow-cell approach in micro LC–FTIR provides for universal detection and at the same time is a versatile technique for specific structural characterization.

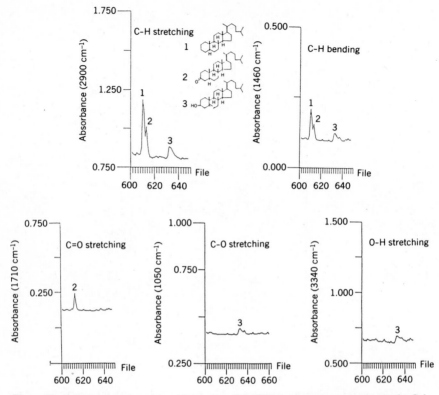

Figure 17. Selective detection of steroids by micro LC–FTIR in adsorption separation mode. Column, FineSIL-5, 0.5-mm ID × 11 cm; mobile phase, cyclohexane-*d*12–chloroform–*d*1–methanol–*d* 4(90.3:9.1:0.6); flow rate, 8 μL/min; accumulation, × 10; flow cell, AgCl(1.6 μL). Peaks: 1,5α-cholestane, 2,5α-flow cholestan-3-one, 3,cholesterol. Reproduced with permission from Ref. 23. Copyright 1985 Friedr. Vieweg & Sohn.

4. PRACTICAL USES OF LC–FTIR MEASUREMENTS

Because the only commercially available interfaces between LC and FTIR are flow-cell configurations, very few applications for solving practical analytical problems have appeared in the literature. Therefore, only two, very impressive examples of the practical applications with the LC–FTIR measurements are demonstrated in this section; one is by the flow-cell approach and the other is by the solvent elimination approach.

In the most impressive demonstration with the flow-cell LC–FTIR approach, described by Johnson and Taylor in 1983 (26), normal-phase chromatography on silica gel was carried out with freon-113 elution for the analysis of nonpolar

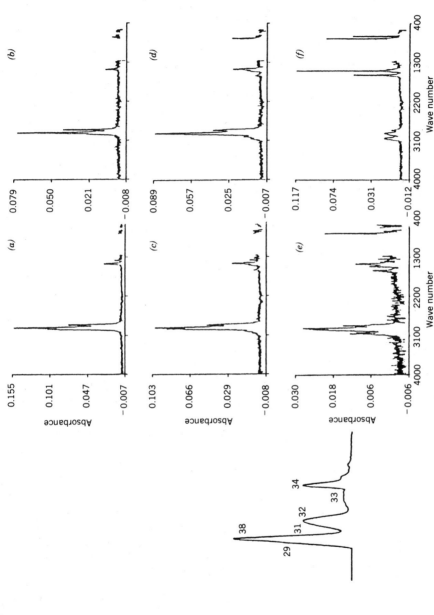

Figure 18. (*a*) Gram–Schmidt reconstructed chromatogram of semipreparative-scale separation of the nonpolar fraction of Mobil 92-026-019. (*b*) IR spectra obtained from the semipreparative-scale separation of the fraction of Mobil 92-026-019. (*a*) Peak 29; (*b*) peak 30; (*c*) peak 31; (*d*) peak 32; (*e*) peak 33; (*f*) peak 34. Reproduced with permission from C. C. Johnson and L. T. Taylor, *Anal. Chem.*, **55**, 436 (1983). Copyright 1983 American Chemical Society.

materials in coal-derived process solvent. In this work, separation on a semi-preparative scale of the nonpolar fraction of coal recycle process solvent was monitored by LC–FTIR measurements. The class separation observed in the previous model studies appears to hold in the process solvent based on examination of file spectra. This was confirmed upon observation of the IR spectra shown in Fig. 18. For example, peak 34 in Fig. 18 was assigned as an aliphatic-substituted biphenyl and has an elution volume corresponding to the model elution of 4-phenyltoluene, and so on. The class separation for jet fuel samples was also demonstrated.

A very interesting practical use of LC–FTIR (SEC–DRIFT) measurements in the study of reaction kinetics was reported by Miller and Ishida in early 1985 (27). For the analysis of the reactions of organofunctional trialkoxysilanes, which has required the use of many and varied techniques in order to obtain even a preliminary picture of the reactions involved, the SEC–DRIFT method using solvent elimination has been applied because of the low volatility of the solute and the high sensitivity of the technique. An example of the SEC–DRIFT for the reaction, in which 0.30 M phenyltrimethoxysilane in THF was catalyzed by water (pH 4.0) at room temperature, is illustrated in Fig. 19. Fig. 19a is the conventional digitized output from a 254-nm UV detector of the 7-h reaction

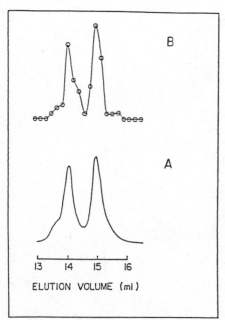

Figure 19. Size-exclusion chromatograms of a reacting solution of phenyltrimethoxysilane in acid-catalyzed aqueous THF at 25°C. Chromatogram A is the response of the fixed-wavelength (254 nm) UV detector and chromatogram B is constructed from the integrated intensity of a fundamental phenyl mode at wavenumber 1430, as monitored by diffuse reflectance FTIR. All peaks represent reaction products, and none of the starting material remains. Reproduced with permission from J. D. Miller and H. Ishida, *Anal. Chem.*, **57**, 283 (1985). Copyright 1985 American Chemical Society.

product, and Fig. 19*b* is the conventional SEC chromatogram resulting from DRIFT-integrated intensity data monitored at wave number 1430. The wave number band at 1430 represents the fundamental phenyl mode of the substituted benzene. The result of Fig. 19*b* indicates the use of the FTIR spectrophotometer as an additional fixed-wavelength detector and should correspond exactly to the output from the UV detector. Figure 20 shows the information produced by the SEC–DRIFT technique illustrated as a three-dimensional plot of elution volume, frequency, and intensity. By revealing the complete picture of the reactions involved in depicting the molecular differences between the chromatographically resolved reaction products, SEC–DRIFT has been successful in clarifying the reaction kinetics.

Both examples clearly show that LC–FTIR techniques can be a powerful and valuable tool in complex mixture analysis. Limitations on solvent windows in flow-cell analysis and limitations on volatility of mobile phases in solvent elimination will always be present in LC–FTIR techniques. However, those restrictions could be reduced with the use of micro LC–FTIR, which allows the chromatographer to use solvents with large IR windows (that were previously not considered because of their high cost) and solvents offer us easier elimination prior to IR measurements.

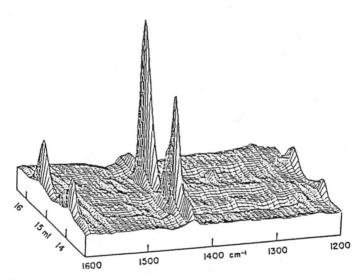

Figure 20. Three-dimensional reconstruction of SEC–DRIFT data in the mid-IR region, wave number 1500–1300 for the reacting solution of phenyltrimethoxysilane in acid-catalyzed aqueous THF at 25°C. Reproduced with permission from J. D. Miller and H. Ishida, *Anal. Chem.*, **57**, 283 (1985). Copyright 1985 American Chemical Society.

   With the developments of microcolumn technology and higher-sensitivity IR
detectors, LC–FTIR should become one of the most promising techniques for
solving many analytical problems.

5. FTIR DETECTION IN SUPERCRITICAL FLUID CHROMATOGRAPHY

In supercritical fluid chromatography (SFC), separations are performed by using
a supercritical fluid as the mobile phase with either a LC column or a GC column
(wall-coated open tubular capillary column), where the supercritical fluid is a
substance raised above its critical temperature and pressure. The physical properties of supercritical fluids such as the diffusion coefficients and viscosity are
intermediate between those of a gas and a liquid, which allow the use of GC or
LC columns without much problem. Those data are tabulated in Tables 1 and
2. Several detectors commonly used in either GC or LC can be used equally
well for SFC, such as UV–VIS, FID, and so on, but a need for identifying each
eluted component still exists. FTIR could have great potential to identify eluted
components by SFC.
   In combining SFC with FTIR, similar approaches for LC–FTIR as discussed
earlier, for example, direct flow cell and solvent elimination, can be realized.
However, more attractive possibilities exist for interfacing chromatographic techniques to FTIR, for some supercritical fluids such as carbon dioxide absorb
minimally in the mid-IR region, or have very high volatility.
   Two publications have recently appeared on SFC–FTIR interfaces. The first
is by Shafer and Griffiths (28) and the second is by Olesik et al. (29); both used
the flow-cell approach. In their work, they used carbon dioxide as the mobile
phase, because carbon dioxide is almost transparent in the mid-IR region. A

Table 1.   Properties of Supercritical Fluids for SFC

Compound	Critical Temperature (°C)	Critical Pressure (atm)	Critical Density (g/cm^3)
Carbon oxide	31.04	72.9	0.468
Sulfur dioxide	157.5	77.8	0.524
Ammonia	132.4	113.3	0.235
Nitrous oxide	36.4	71.5	0.452
n-Pentane	196.6	33.3	0.232
n-Hexane	234.2	29.6	0.234
n-Heptane	267.0	27.0	0.235
Xenon	16.6	38.0	1.105

Table 2.   Physical Properties of a Gas, a Liquid, and a Supercritical

Property	Units	Gas	Liquid	Supercritical Fluid
Density	g/mL	10^{-2}	1	0.3
Diffusivity	cm^3/s	10^{-1}	5×10^{-6}	10^{-3}
Dynamic viscosity	P	10^{-4}	10^{-3}	10^{-4}

typical example of SFC–FTIR is shown in Fig. 21. This presents a real-time reconstructed Gram–Schmidt chromatogram of a standard mixture of alkyl-phenols, aromatic ketones, and aromatic aldehydes. Figure 22 shows a spectrum of benzaldehyde obtained by averaging interferograms (across half-height points of the chromatographic peak), in which 2 μg of each component was injected. Those separations were pressure-programmed from 78.2 to 105.5 atm at a rate of 0.54 atm/min. The response of the Gram–Schmidt algorithm obviously does not depend solely on the concentration of each component, however.

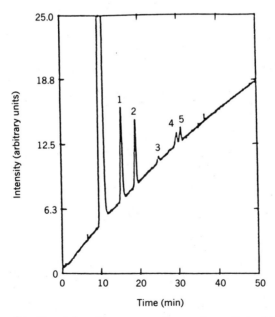

Figure 21. Gram–Schmidt real-time chromatogram of a test mixture. Peaks: 1,benzaldehyde, 2,*o*-chlorobenzaldehyde, 3,2,6-di-*t*butylphenol, 4,2-naphthol, 5,benzophenone. Reproduced with permission from Ref. 29. Copyright 1984 Friedr. Vieweg & Sohn.

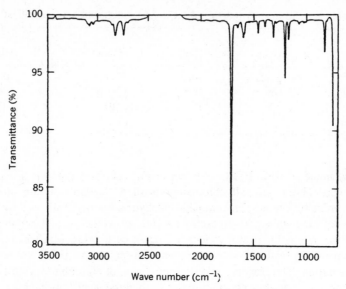

Figure 22. IR spectrum of benzaldehyde (corresponding to the chromatogram in Fig. 21). Reproduced with permission from Ref. 29. Copyright 1984 Friedr. Vieweg & Sohn.

Notwithstanding the above successes in SFC–FTIR via flow-cell approaches, there is an intrinsic problem with this approach. That is as follows: interference was observed in the measurements, by the decrease of carbon dioxide transparency when the pressure of supercritical carbon dioxide was increased. Because pressure-programming elution and adding organic modifier to supercritical fluid are common ways to optimize separations in SFC, Shafer et al. (30) has concluded that in SFC–FTIR the solvent elimination approach is to be preferred. Their demonstration of this SFC–DRIFT approach is shown in Fig. 23. In this example, a 0.5-μL injection of the quinone solution resulted in 350 ng per component on the column. The 5% methanol in carbon dioxide solution was heated to 50°C, pressurized to 200 atm, and passed through the microcolumn at a flow rate of 80 μL/min. The chromatogram observed is shown in Fig. 23a and the spectrum of acenaphthenequinone is shown in Fig. 23b.

Further improvements in sensitivity could be expected to be achieved by the redesign of the optics. It should be noted that no trace of methanol can be observed in the IR spectrum, so that pressure and temperature programming should not affect the quality of the spectrum. With the more recent investigations by Fujimoto et al. (31), which indicate that the slurry-packed microcapillary SFC–FTIR is possible via the KBr buffer-memory technique, the solvent elimination approach to SFC–FTIR interfaces should become the most promising.

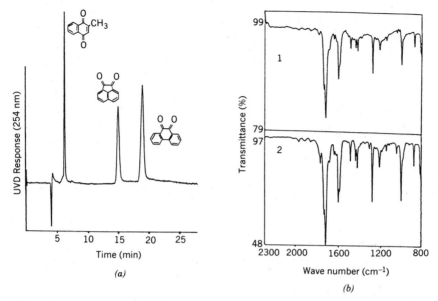

Figure 23. (*a*) Chromatogram of quinone mixtures separated by SFC (5% methanol–CO_2, 200 atm, 50°C); 350 ng per component injected. (*b*) IR spectra of (1) second peak in chromatogram shown in (*a*) and (2) authentic acenaphthenequinone. Reproduced with permission from K. H. Shafer, S. L. Pentoney and P. R. Griffiths, *JHRC&CC*, **7**, 707 (1984). Copyright 1984 Dr. Alfred Hüthig Verlag.

6. SUMMARY

FTIR techniques combined with LC have resulted in an extremely useful detector for LC. The problems encountered in LC–FTIR are discussed, and useful reviews on some approaches to solve them are also included. It is hoped that this chapter will serve as an introduction to a fairly new and interesting application of LC–FTIR. Spectroscopic theory and practice on FTIR are beyond the scope of this chapter, but it should be mentioned that many good sources of information on FTIR are available in books (2–4), as well as in the applications literature of the major FTIR systems manufacturers.

REFERENCES

1. P. R. Griffiths, J. A. deHaseth, and L. V. Azarraga, *Anal. Chem.*, **55**, 1361A (1983).
2. D. W. Vidrine, in J. R. Ferraro and L. J. Basile, Eds., *Fourier Transform Infrared Spectroscopy, Applications to Chemical Systems*, Vols 1–3, Academic, New York, 1979.

3. P. R. Griffiths, Ed., *Transform Techniques in Chemistry*, Plenum, New York, 1978.

4. R. J. Bell, *Introductory Fourier Transform Spectroscopy*, Academic, New York, 1972.

5. D. Kuehl and P. R. Griffiths, *J. Chromatogr., Sci.*, **17**, 471 (1979).

6. D. Kuehl and P. R. Griffiths, *Anal. Chem.*, **52**, 1394 (1980).

7. C. Conroy, P. R. Griffiths, P. J. Dutt, and L. V. Azarraga, *Anal. Chem.*, **56**, 2636 (1984).

8. K. S. Kalasinsky, J. T. McDonald, and V. F. Kalasinsky, paper presented at the Pittsburgh Conference on Analytical Chemistry and Applied Spectroscopy, Atlantic City, NJ, Mar. 1983, No. 357.

9. Foxboro/Wilks Corp., *Application Notes*, No. 2, Norwalk, CT, p. 1.

10. J. P. Coates, *Eur. Spectrosc. News*, **16**, 25 (1978).

11. J. A. deHaseth and T. L. Isenhour, *Anal. Chem.*, **49**, 1977 (1977).

12. R. P. Scott and P. Kucera, *J. Chromatogr.*, **125**, 251 (1976).

13. D. Ishii, K. Asai, K. Hibi, T. Jonokuchi, and M. Nagaya, *J. Chromatogr.*, **144**, 157 (1977).

14. T. Tsuda and M. Novotny, *Anal. Chem.*, **50**, 632 (1978).

15. K. Jinno and C. Fujimoto, *JHRC&CC* **4**, 466 (1981).

16. K. Jinno, C. Fujimoto, and Y. Hirata, *Appl. Spectrosc.*, **36**, 67 (1982).

17. K. Jinno, C. Fujimoto, and D. Ishii, *J. Chromatogr.*, **239**, 625 (1982).

18. C. Fujimoto, K. Jinno, and Y. Hirata, *J. Chromatogr.*, **258**, 81 (1983).

19. C. Fujimoto, T. Osuga, and K. Jinno, *Anal. Chim. Acta*,**178**, 159 (1985).

20. C. C. Conroy, P. R. Griffiths, and K. Jinno, *Anal. Chem.*, **57**, 822 (1985).

21. J. M. Brackett, L. V. Azarraga, M. A. Castles, and L. B. Rogers, *Anal. Chem.*, **56**, 2007 (1984).

22. R. C. Gore, R. B. Barnes and E. Peterson, *Anal. Chem.*, **21**, 382 (1949).

23. C.Fujimoto, G. Uematsu, and K. Jinno, *Chromatographia*, **20**, 112 (1985).

24. K. Jinno, *JHRC&CC* **5**, 366 (1982).

25. K. Jinno and C. Fujimoto, *J. Liq. Chromatogr.*, **7**, 2059 (1984).

26. C. C. Johnson and L. T. Taylor, *Anal. Chem.*, **55**, 436 (1983).

27. J. D. Miller and H. Ishida, *Anal. Chem.*, **57**, 283 (1985).

28. K. Shafer and P. R. Griffiths, *Anal. Chem.*, **55**, 1939 (1983).

29. S. V. Olesik, S. B. French, and M. Novotny, *Chromatographia*, **18**, 489 (1984).

30. K. H. Shafer, S. L. Pentoney, and P. R. Griffiths, *JHRC&CC*, **7**, 707 (1984).

31. C. Fujimoto, Y. Hirata, and K. Jinno, *J. Chromatogr.*, **332**, 47 (1985).

CHAPTER

4

INDIRECT ABSORBANCE DETECTORS

MICHAEL D. MORRIS

Department of Chemistry, University of Michigan, Ann Arbor, Michigan

1. INTRODUCTION

It is possible to make light absorption measurements indirectly, by measuring the heat evolved as excited molecules relax back to the ground state. Such measurements, in principle, provide for nonfluorescent molecules, the advantages normally associated with fluorimetry. In particular, as the concentration of absorbing molecules approaches zero, the heat evolved also goes to zero. By contrast, the zero concentration limit of a direct absorption measurement is 100% transmission. It is easier to measure small signals against a nominally zero background than small differences in large signals, and the indirect measurement should represent a more favorable measurement case.

Less obviously, indirect absorption measurements are well-suited for measurements in systems with small volumes and/or short light paths. Indirect absorption measurements are usually made with laser light sources. The temperature rise is a function of power density, and the effects are strongest close to the focus of a tightly focused beam. Long paths provide only modest increases in signals. Lasers, which can be focused to nearly diffraction-limited diameters, are the obvious light sources for such measurements. The volumes probed are often no more than a few to a few tens of nanoliters.

The major techniques used for indirect absorbance measurements are photoacoustic spectroscopy and thermal lens spectroscopy, as well as some variants on these. The principles and some applications of these techniques have been recently reviewed (1). In addition, the photoacoustic effect has been the topic of major monographs (2,3) and reviews (4).

The photoacoustic effect was observed in the nineteenth century, but was not widely investigated as an analytical tool until the early 1970s. The thermal lens effect, on the other hand, was discovered in the early years of laser experimentation. It was immediately recognized as a technique with potential for low absorbance measurements. However, the technique escaped the notice of the chemical community until the early 1970s (5). Exploitation of thermal lens spectroscopy by analytical chemists did not really begin until the end of the decade (6). Application of both techniques to liquid chromatographic detection began in the early 1980s and continues as an active research area today.

2. A BRIEF INTRODUCTION TO LASERS

The great majority of indirect absorbance detectors utilize laser light sources. Laser beams have high spectral brightness and spatial characteristics that allow tight focusing and other manipulations with simple optical systems. These properties suggest that lasers will continue to be the dominant sources for this type of detector in the future. For the novice, we present a brief overview of laser principles and a description of the lasers used for indirect absorbance measurements and of ancillary techniques for wavelength shifting. The interested reader can find discussions of laser theory and operations at any desired level of detail. The references contain a short list of monographs with introductory discussions of laser operation and related topics (7–9).

Figure 1 shows the energy level scheme in a four-level laser. The system is excited from the ground state, level 0, to state 3, by some external energy source, which need not be optical. If there is a very rapid transition into long-lived metastable level 2, it is possible to build up a population inversion in level 2. That is, the number of atoms or molecules, N_2, in state 2, can be greater than the number in state 1, N_1.

A population inversion is thermodynamically impossible in a two-level·sys-

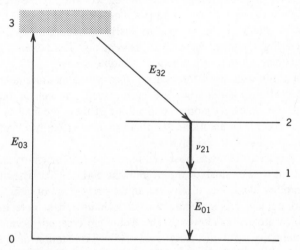

Figure 1. Energy levels in a four-level laser. Laser emission occurs at $\nu_{21} = \Delta E_{21}$.

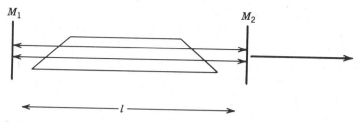

Figure 2. A generalized laser system. M_1 is a totally reflecting mirror; M_2 is a partially reflecting mirror. The mirror separation is l.

tem, because for $E_a < E_b$, $N_a < N_b$ for any finite temperature. This relation, the Boltzmann condition, does not hold if there is an indirect route to populating state b. Thus a laser must involve a minimum of three energy levels. A four-level system is more efficient, and most practical lasers involve four energy levels.

If $N_2 > N_1$, then incoming radiation at ν_{12} does not increase the population in level 2, but stimulates a downward transition from level 2 to level 1. This process is called stimulated emission. It follows that a beam of light at frequency ν_{12} passing through the inverted system will gain photons rather than be attenuated. In fact, the beam intensity will build exponentially, rather than decay exponentially, if the population in state 2 can be kept high enough to overcome losses from spontaneous emission or radiationless transitions. That means that there is some threshold excitation, or pumping, rate that must be maintained.

The stimulated emission system is usually enclosed in a cavity, as shown in Fig. 2. The laser cavity is based on the Fabry–Perot interferometer. M_1 is a total reflector, and M_2 is made partially reflective. Light waves circulate between the mirrors, building up a high intensity. The laser output is extracted from mirror M_2, leading to a steady-state circulating intensity between the mirrors.

This process occurs only if a standing wave can be set up between the mirrors, as described below. The reflectivity of M_2 is chosen to maximize the power that can be extracted from the system. If the gain of the system is high, the reflectivity can be low. If the gain of the system is small, then only a small fraction of the light can be removed from the laser on each pass.

A standing wave can be generated in the laser cavity only at frequencies closely related to the frequencies at which the constructive interference conditions are satisfied in a Fabry–Perot interferometer of the same mirror spacing, d, as given by

$$\nu = n\left(\frac{c}{2d}\right) \tag{1}$$

Equation 1 defines the longitudinal modes of the laser cavity. These are the discrete frequencies at which the laser operates. For most laser systems, $1/v \ll d$, so that n is a large number. The laser can operate at more than one frequency. The number depends upon the mirror spacing and the bandwidth of the transition on which stimulated emission occurs.

Equation 1 neglects the spatial distribution of light across the resonator. Resonant conditions must be satisfied everywhere. That can occur only for certain spatial distributions. These are called the transverse modes of the laser. For a cavity composed of square mirrors, with width a, $a \ll d$, Eq. 2 is a good approximation to the complete resonance conditions:

$$v = \frac{c}{2}\left[\left(\frac{n}{2}\right)^2 + \left(\frac{m}{2a}\right)^2 + \left(\frac{l}{2a}\right)^2\right]^{1/2} \tag{2}$$

The transverse modes are defined by the integers m and l. The modes are called transverse electric and magnetic modes, abbreviated TEM_{ml}.

Many lasers are based on a confocal resonator. In this configuration curved mirrors are used. The radius of curvature of each mirror is equal to the distance between the mirrors. Confocal resonators have lower diffraction losses than plane mirror resonators. For the confocal resonator, Eq. 3 describes the longitudinal and transverse modes:

$$v = \frac{c[2n + (1 + m + l)]}{4d} \tag{3}$$

A laser is usually designed to operate in TEM_{00}. This mode has the lowest diffraction losses and the lowest divergence of any transverse mode. For a cavity with circular mirrors, TEM_{00} provides an intensity distribution which has a maximum in the center of the beam and a Gaussian falloff of intensity with radial distance from the center.

Equation 4 defines the spatial distribution of electric field intensity, $A_{00}(x,y)$ across the laser beam in the TEM_{00} mode:

$$A_{00}(x,y) = \exp\left[\frac{-\pi(x^2 + y^2)}{d\lambda}\right] \tag{4}$$

The radial distance at which the field amplitude drops to $1/e$ of its maximum value is called the beam radius, w_s. The beam radius is given by

$$w_s = \left(\frac{\lambda d}{\pi}\right)^{1/2} \tag{5}$$

The characteristic low divergence of a laser beam is a direct result of the use of this resonator configuration. A Gaussian beam propagates through a distance z with a beam radius $w(z)$ related to w_s by

$$[w(z)]^2 = w_s^2 \left[1 + \left(\frac{z}{z_0} \right)^2 \right]$$ (6)

The parameter z_0, called the confocal parameter, is defined by

$$z_0 = \frac{\pi w_s^2 n}{\lambda}$$ (7)

From these equations it follows that at sufficiently large distances the half-angle divergence, ϑ, is given by

$$\vartheta = \frac{\lambda}{\pi w_s}$$ (8)

For most lasers, the divergence angle is 1–5 mrad.

The earliest visible lasers used narrow atomic or molecular electronic transitions. Consequently, their outputs consisted of emission over one or more narrow wavelength bands. Argon ion and He–Ne are typical examples. In the mid 1960s it was discovered that laser action could be obtained from many luminescent organic molecules. The spectrum of the laser output roughly matched the emission spectrum of the molecule. Replacement of M_1 with a wavelength-selective reflector, such as a diffraction grating, allows tuning the wavelength of the laser over most of the fluorescence band of the molecule.

These tunable lasers are called dye lasers. Typically, they are tunable over a range of 20–50 nm. The energy source for pumping the dye is most commonly another laser, but flash lamps are also used. Although CW lasers can be used to pump dye lasers, most employ pulsed lasers with pulse durations of 1–20 ns. Simple configurations have output bandwidths of 0.01–0.1 nm. Dyes are available for tuning in the wavelength range from roughly 320 to 1000 nm.

Nonlinear optical techniques can be used to generate coherent radiation at wavelengths not generated by the laser itself. Many birefringent solids, when exposed to intense laser light, are distorted sufficiently that their response is no longer linear in incident power. The nonlinear response causes generation of new emissions at the second harmonic of the incident light frequency. This effect, called frequency doubling or second harmonic generation, is *not* a violation of the law of conservation of energy. Two incident photons are required for each photon generated at the harmonic frequency.

The materials used for second harmonic generation are salts of common oxyanions. They include potassium and ammonium dihydrogen phosphate and arsenate, and their deuterated analogs. Carefully grown and cut crystals are required. The phosphates and arsenates work well for generating wavelengths down to about 250 nm. Below that other materials, such as potassium pentaborate, are used.

The same phenomenon can be used to generate frequencies that are the sum and difference of two laser frequencies, if two beams are incident on the crystal. Frequency mixing is used to cover frequency ranges not conveniently accessible to doubling, or to extend the range available from one laser dye.

Frequency conversion efficiency by these techniques is typically 1-10% if high peak power pulsed lasers are used. Several crystals are required to cover the entire UV range. Frequency conversion efficiency at any wavelength depends on the angle the incident beam makes with respect to the optic axis of the crystal. It is necessary to choose an angle such that the fundamental propagating as the ordinary ray and the harmonic as the extraordinary ray travel in the same direction. Because RI is a strong function of wavelength, the angle is also wavelength dependent. Consequently, the crystal must be rotated as the incident wavelength is changed. Fully automated systems are available.

Raman shifting provides an alternative route to some wavelength tunability in the ultraviolet. Hydrogen, methane, and deuterium all produce strong stimulated Raman scattering when illuminated with pulsed laser radiation. This results in production of new laserlike radiation at several wavelengths separated from the incident laser frequency by exact multiples of the vibrational frequency of the molecule.

Raman shifting requires a very simple apparatus. A Raman shifter is just a tube filled with gas to a pressure of 3-6 atm. A focusing lens is placed before one end of the tube. At the exit end a prism is used to separate the wavelengths, and a second lens is used to recollimate the radiation. Both Stokes (red) and anti-Stokes (blue) shifted radiation can be generated. Hydrogen is the most popular fill gas, because it provides the largest Raman shift (4155 cm^{-1}, about 35-40 nm in the UV).

Indirect absorbance detectors usually employ one of a small number of types of lasers which are listed in Table 1. The laser type is given, along with its output wavelength range, typical power levels, and approximate price range.

The most popular CW laser has been the argon ion laser, usually operated on one of the blue or blue-green lines. Moderate power (50-500 mW) is available on several lines in the 458-528-nm range in lasers costing $10,000-15,000. These lasers require water-cooling and consume substantial amounts of electricity. They are reliable, stable, and easy to operate. Small air-cooled argon ion lasers intended mostly for graphic arts applications are available from several vendors. They provide 10-50 mW at 488 or 514 nm. In general, the operating

Table 1. Lasers Used for Indirect Absorbance

A. CW Lasers

Laser	Wavelength (nm)	Power (mW)	Price Range ($)
Argon ion	351, 353 458, 476, 488 496, 502, 514, 529	10–2000	4000–25,000
He–Cd	325 442	2–15 3–50	2000–13,000
He–Ne	633	0.5–50	500–13,000

B. Pulsed Lasers

Laser	Wavelength (nm)	Pulse Duration (ns)	Pulse Energy (mJ)	Repetition Rate (pulses/s)	Price Range ($)
Nitrogen	337	1–15	0.1–10	1–100	3000–20,000
Excimer	143 249 308 351	10–20	5–500	1–200	20,000–50,000
Nd–YAG	1064 532^a 355^a 266^a		100–1000 10–100 5–50 1–10	10–20 10–20 10–20 10–20	25,000–50,000
Pulsed dye lasers[b]					
N$_2$ laser pumped	350–800	1–15	0.01–1	1–50	6000–30,000
Excimer laser pumped	330–800	10–20	1–30	1–50	40,000–80,000
Nd–YAG laser pumped	400–800	10–20	1–30	10–20	40,000–80,000

[a]Harmonics.
[b]Harmonic generation extends wavelength range down to 217 nm.

lifetime of the laser tube decreases with increasing power output. Water-cooled laser tubes have operating lifetimes of 3000–5000 h. Air-cooled tubes have lifetimes of 5000–10,000 h. Tube replacement costs are substantial, 30–50% of the cost of a complete laser. However, tube rebuilding services are now available from several vendors.

Large argon ion lasers operate on two UV lines, 351 and 353 nm. These outputs are derived from Ar^{2+}. It is difficult to obtain high enough populations of doubly ionized argon ions. Consequently UV outputs are weak relative to the visible emissions possible from the same laser. In addition, emission is at too high a wavelength to be generally useful for chromatography.

The UV line of the He–Cd laser has been used in fluorimetric detectors for several years. Recently, He–Cd lasers providing 10–15 mW of UV output have become available. This is sufficient power for indirect absorbance detectors. Where a 325-nm source is useful, the He–Cd laser is a logical candidate. The laser is air-cooled, requires little or no operator adjustment, and has adequate pointing stability for chromatographic purposes. Tube lifetime is about 4000–5000 operating hours.

The He–Cd laser also emits at 442 nm. It is necessary to use different mirrors for blue and UV emission. Note that most efficient UV operation is obtained from lasers with mirrors permanently sealed to the ends of the laser cavity, to avoid the losses due to separate windows.

The standard probe laser is a small He–Ne laser, usually with a power of 1–2 mW. Small He–Ne lasers have no operator adjustments. They provide good pointing stability and a working lifetime of 15,000–20,000 h. He–Ne lasers are available at powers as high as 50 mW. The larger lasers are bulky and, because they are not widely used, expensive.

Semiconductor lasers based on Ga–As–Al operate at the red end of the visible and the near infrared. Most commercial devices operate at 800–900 nm. Power conversion efficiency is high, so that the devices are quite compact. Semiconductor lasers operating in the red, near 650 nm, have been demonstrated, but are not commercially available.

Because of their small size, semiconductor lasers have divergences of 5–20°, rather than the milliradians characteristic of other lasers. Pulsed and CW versions are available. Typically, the average power is 3–10 mW, but CW arrays are available with powers to 1 W. With a suitable power supply, a CW laser can be modulated electrically. The devices are easily damaged by electrical overload. With properly designed power supplies, they have operating lifetimes that are too long to measure accurately.

These devices may merit serious consideration for special applications because they provide an extremely attractive combination of power, price, size, and ease of operation.

Almost all lasers for UV operation are pulsed systems. The most important devices are the Nd–YAG and excimer lasers.

Excimer lasers are discharges operated with rare gas–halogen mixtures. Laser action takes place in states of rare gas–halogen compounds, which exist only in the excited state. Wavelengths from 193- (argon fluoride) to 351-nm (xenon fluoride) are conveniently available. In addition to argon fluoride and xenon fluoride, two other excimer lasers are widely used. These are krypton fluoride (249 nm) and xenon chloride (308 nm).

The laser consists of a discharge chamber containing the desired rare gas and halogen at low pressures and a mixture of other rare gases, primarily helium, to provide a total pressure of 2–3 atm. The pulse width depends on the characteristics of the cavity and discharge circuitry. It is usually in the range 5–25 ns. The maximum pulse repetition rate is determined largely by the power supply, and is typically 25–200 pulses/s. Pulse energies are typically 10–500 mJ in laboratory-size lasers.

The beam quality from a conventional excimer laser, with flat mirrors, is poor. However, use of unstable resonator (curved mirror) optics produces a beam of acceptable spatial characteristics for some indirect absorbance detection.

Although excimer lasers produce convenient UV wavelengths, they have several disadvantages. First, their use requires handling of halogen gases, notably fluorine. Second, changing wavelengths means changing the gas fill mixture, by evacuating and refilling the system. Finally, the available excimer lasers are all too powerful for chromatographic use. The peak powers available are high enough that a focused beam can damage quartz windows. There is no reason, however, that a low-power excimer laser could not be manufactured. Such a device would also be less expensive than lasers now on the market, which cost $20,000–45,000.

The other important UV source is a frequency tripled or quadrupled Nd–YAG laser. The Nd–YAG laser operates at 1.064 μm. Its second harmonic is at 532 nm, and its third and four harmonics are at 355 and 266 nm, respectively. The fourth harmonic is at a useful wavelength. Nd–YAG lasers generate beams with excellent spatial characteristics. As with excimer lasers, however, the peak powers are sufficiently high that damage to optics or samples is a serious problem. An Nd–YAG laser with accessories for harmonic generation costs $25,000–50,000.

Tunable laser radiation is available from dye lasers. These use fluorescent organic molecules, typically coumarins or rhodamines, as the active medium. They can be pumped with a pulsed laser, such as an Nd–YAG or excimer laser, a CW laser, such as an argon ion laser, or a flash lamp. The most popular devices employ pulsed lasers as pump radiation. Most commonly, the second or third harmonic or Nd–YAG is employed, but excimer laser pumped dye lasers are also frequently encountered. Small dye lasers pumped by compact nitrogen lasers (337 nm) are also available, at very low prices.

Excimer laser pumped dye lasers can produce wavelengths as low as 320 nm. Nd–YAG or nitrogen laser pumped lasers can produce wavelengths down to about 380 nm. Ultraviolet wavelength coverage down to about 210 nm is available by frequency doubling the dye laser output and, if an Nd–YAG laser is used, by combinations of frequency doubling and mixing. Although UV systems are easy to use, they are expensive. A complete Nd–YAG system with accessories for UV operation can cost more than $80,000.

At present, an excimer laser or Nd–YAG laser with a Raman shifter provides the chromatographer with about the best available compromise between cost and versatility.

3. THEORY OF THE THERMAL LENS

3.1. General Aspects

A thermal lens (Fig. 3) is simply the result of the RI gradient that forms in a heated medium. Most materials expand when heated. Therefore dn/dT, the RI temperature coefficient, is usually negative. If the material is heated with a laser beam or other source with cylindrical symmetry, the thermal lens itself will be negative.

A rigorous treatment of the thermal lens leads to intractable equations (10,11). Consequently, some simplifying assumptions are necessary, and actual lens behavior usually deviates slightly from the approximate governing equations.

Gordon and co-workers (10) presented the first theoretical treatment of thermal lens theory. Their treatment was used by Hu and Whinnery (11) to describe the simplest one-laser thermal lens measurement systems. Later workers have extended this theory to other experimental configurations. Occasionally, a different set of simplifying assumptions is used.

Gordon et al. assume that the sample is a liquid that is heated by a TEM_{00} Gaussian source, whose radius, w, is constant throughout the length of the sample cell. If conduction is the only mode of heat transfer in the liquid, the temperature distribution as a function of radial distance from the beam center and time from the start of the laser irradiation, $\Delta(r,t)$, is given by

$$\Delta T(r,t) = \frac{A\pi w_0^2}{8k}\left[Ei\left(\frac{2r^2}{w^2}\right) - Ei\left(\frac{2r^2}{8Dt + w^2}\right)\right] \qquad (9)$$

In Eq. 9, A is the heat density of the laser beam; that is, $A = 0.24P$, where P is the power density in watts per square centimeter. The thermal properties of the liquid are lumped into D, the thermal diffusivity. D is defined by $D = k/\rho C_p$, where k is the thermal conductivity, ρ is the density, and C_p is the specific heat of the liquid.

Equation 9 is complicated if one assumes that the laser beam is brought to a focus and diverges, so that its radius changes through the length of the sample. Consequently, a constant power density is assumed over the entire length, b, of the sample.

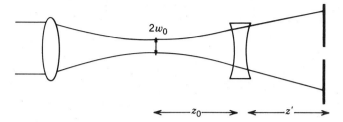

Figure 3. The thermal lens effect. Light source from a laser is focused to a radius w_o. The thermal lens is formed at z_0, approximately one confocal parameter beyond the beam waist. A limiting aperture is located a distance z' beyond the thermal lens.

There are two other problems with Eq. 9. The assumption of an infinite radius for the medium generates a steady-state infinite temperature gradient. Fixing a finite boundary much larger than the laser beam generates removes this mathematical artifact. Second, the exponential integral is a transcendental function that is not easily employed in equations relating temperature gradient to the focal length of the thermal lens (12).

Gordon et al. solve the second problem by approximating the exponential integral as the series defined by

$$Ei(-x) = \gamma nrx - x + \frac{x^2}{4} \cdots$$

$$\gamma = 1.781 \cdots$$

(9a)

They retain only terms as high as quadratic. This approximation is valid for small values of r. Thus near the axis of the laser beam Eq. 10 is a good approximation to the temperature distribution:

$$\Delta T(r,t) \cong \frac{0.06bP}{\pi k} \left[\ln\left(1 + \frac{8dT}{w^2} \right) - \frac{16Dt}{w^2 + 8Dt} \frac{r^2}{w^2} \right]$$

(10)

where b is the length of the illuminated region of liquid.

The focal length of the thermal lens can be calculated by substitution of Eq. 10 into Eq. 11 for RI distribution, $n(r,t)$, where n_0 is the RI at the reference temperature:

$$n(r,t) = n_0 + \left(\frac{dn}{dT} \right) \Delta T(r,t)$$

(11)

$$n(r,t) = n_0 + \left(\frac{dn}{dT}\right)$$
$$\times \frac{0.06bP}{\pi k} \left[\ln \left(1 + \frac{8Dt}{w^2} - \left(\frac{16Dt}{w^2 + 8Dt}\right) \frac{r^2}{w^2}\right]\right. \tag{12}$$

It is usually adequate to neglect the logarithmic term. Equation 12 can then be recast in the form of

$$n = n_0 \left[1 + \delta\left(\frac{r}{w}\right)^2 \right] \tag{13}$$

Here, δ is given by

$$\delta = -\frac{0.12bP}{n_0 K \pi} \left(\frac{dn}{dT}\right) \frac{8Dt}{w^2 + 8Dt} \tag{14}$$

A parabolic RI distribution described by Eq. 13 functions as a lens whose focal length is given by

$$F = -\left(\frac{1}{2\sigma}\right) \frac{w_0^2}{l} \tag{15}$$

Combining Eqs. 14 and 15 gives the standard thermal lens equation (11):

$$F(t) = \frac{\pi n_0 K w^2 (w^2 + 8Dt)}{0.24bPl\left(\frac{dn}{dT}\right) 8Dt} = F_\infty \left(1 + \frac{t_c}{2t}\right) \tag{16}$$

The steady-state focal length, F_∞, is given by

$$F_\infty = \left(\frac{k\pi n_0 w^2}{0.24bPl}\right) \bigg/ \left(\frac{dn}{dT}\right) \tag{17}$$

The thermal lens is detected by measuring its effect on a laser beam. The probe laser may be the one used to generate the lens, or it may be an independent laser. If an independent probe is used it must have much lower power than the heating laser, so that it has a negligible effect on the thermal lens properties of the solution.

The assumption that w is constant throughout the length of the sample is a thin lens assumption. It is not necessary that the heating laser be collimated. However, its divergence angle must be small enough to maintain a constant radius through the length of the sample. In fact, in the simplest experimental configuration maximum sensitivity is obtained if the sample is not placed exactly at the beam focus, but about one confocal parameter beyond it.

Figure 3 shows the basic thermal lens experimental configuration. The maximum sensitivity is obtained if the thermal lens is generated at a distance z from the focal length, which generates the maximum fractional change $\Delta w/w$ in beam radius observed at some distance z in the far field.

The effect of a thermal lens, in the thin lens approximation, is to change the radius of curvature of the Gaussian laser beam at the lens. If the radius of curvature of the beam before the lens is R_0, then after the lens it is R. Equation 18 describes the thermal lens system, or any thin lens system:

$$\frac{1}{F(t)} = \frac{1}{R_0} - \frac{1}{R} \tag{18}$$

Because the defocused laser beam is a Gaussian beam, it is described by a beam waist w_0', and its radius of curvature, according to Eqs. 19 and 20:

$$w_0'^2 = \frac{w^2}{1 + (\pi w^2/\lambda R)^2} \tag{19}$$

$$z' = \frac{R}{1 + (\lambda R/\pi w^2)^2} \tag{20}$$

We want to choose w and R to maximize the fractional change in the beam radius in the far field, where $w^2 \gg w_0'^2$. Equation 21 applies:

$$w = \frac{\lambda(z + z')}{\pi w_0'} \tag{21}$$

By substitution of Eqs. 19 and 20 into 21, and using Eqs. 16 and 17, we find that Eq. 22 describes the fractional change in beam radius, $\Delta w/w$:

$$\frac{\Delta w}{w} = \frac{1}{R} w^2 \left\{ \frac{\pi w_0'}{\lambda w} \left[\left(2\frac{w_0^2}{w^2} \right) - 1 \right] \pm \frac{\pi}{\lambda} \left(\frac{w_0}{w} \right) \left[1 - \left(\frac{w_0^2}{w^2} \right) \right]^{1/2} \right\} \tag{22}$$

Because $1/\Delta R = -1/F$, Eq. 23 follows:

$$\frac{\Delta w}{w} = \left(-\frac{\theta}{1 + t_c/2t}\right) \left\{\left(\frac{b'}{w}\right) \left[2\left(\frac{w_0^2}{w^2}\right) - 1\right] \pm \frac{w_0}{w} \left[1 - \left(\frac{w_0^2}{w^2}\right)\right]^{1/2}\right\} \quad (23)$$

$$\theta \equiv \frac{P_{abs} \left(\frac{dn}{dT}\right)}{\lambda k} \quad (24)$$

P_{abs} is the absorbed laser power, which is proportional to the indcident power, P, and the absorption coefficient, α.

The first term in curly brackets in Eq. 22 or 23 is negligible compared to the second, since $w \gg w_0'$. Equation 25 is an adequate approximation:

$$\frac{\Delta w}{w} = \pm\left(\frac{-\theta}{1 + t_c/2t}\right) \left(\frac{w_0}{w}\right) \left[1 - \left(\frac{w_0^2}{w^2}\right)\right]^{1/2} \quad (25)$$

Equation 25 maximizes where $w = \sqrt{2}\, w_0$, one confocal distance beyond the waist of the input beam. Equation 25 minimizes if the lens is generated one confocal parameter before the focus of the beam. Curiously, Eq. 25 predicts that the thermal lens can not be observed by its effect on the input laser beam, if the sample is placed exactly at the focus of the input beam.

The thermal lens is measured by placing an aperture in the beam at some distance from the sample cell in which the lens is generated. The change in the laser intensity passed through the beam is a measure of the change in the square of the beam radius. It can be shown that Eq. 26, essentially the square of Eq. 25, applies:

$$\frac{I(t)}{I(0)} = \left(1 - \frac{\theta}{1 + (t_c/2t)} + \frac{\theta^2}{2[1 + (t_c/2t)^2]}\right)^{-1} \quad (26)$$

Equation 26 relates the thermal lens behavior to the sample absorbance and to its thermal and optical properties. At sufficiently low absorbance and laser power the linear term dominates and the time-dependent intensity through the aperture is directly proportional to absorbance. However, if either the laser power or absorbance is sufficiently large, the full quadratic dependence becomes important.

Many thermal lens experiments assume a linear concentration/signal relation, or measure a component of the signal that contains only the linear response. The quadratic response is inherent in any experiment that depends on heat conduction.

Table 2.   **Enhancement Factors for Common Solvents**

Solvent	k (mW/cm·K)	$10^4 dn/dT$ (K^{-1})	F/P (mW^{-1})
CCl$_4$	1.02	-5.8	3.88
C$_6$H$_6$	1.44	-6.4	3.05
(CH$_3$)$_2$CO	1.60	-5.0	2.16
CH$_3$OH	2.01	-3.9	1.33
H$_2$O	6.11	-0.8	0.09

Reprinted with permission J. M. Harris and N. J. Dovichi, *Anal. Chem.*, **52**, 695A (1980). Copyright 1980 American Chemical Society.

The quantity $\theta/\alpha = (dn/dT)/\lambda k$ is often called E, the enhancement factor in the analytical literature (6). The enhancement factor is the relative thermal lens response per unit of laser power. Table 2 (6) lists enhancement factors for several common solvents. From the table it is clear that water is a poor choice of solvent for a thermal lens experiment, whereas nonpolar solvents such as carbon tetra-chloride are quite good. The relatively poor sensitivity of water is a result of both a large thermal conductivity and a small RI temperature coefficient. The methanol–water mixtures used as reverse-phase LC solvents typically have enhancement factors in the range 0.5–1/mW of laser power.

The thermal lens model proposed by Gordon et al. (10) assumes that the RI distribution in the laser beam is parabolic. This assumption leads, ultimately, to simple equations such as Eq. 26 and analogous equations for other experiments. A laser beam has a Gaussian intensity profile, not a parabolic profile. The parabolic expression is used for mathematical convenience and, ultimately, be-cause it works adequately except at very high absorbances. The Gaussian intensity profile can be retained, and approximations introduced elsewhere in the model (13), describing the aberrations in the thermal lens more fully. The somewhat more complicated equations describe a strong thermal lens better than the basic parabolic lens approximation (13). An empirical equation that incorporates some of the features of the aberrant lens into the parabolic model equations describes strong thermal lens behavior better than either one alone (14). For the purposes of analytical chemistry, the simpler parabolic approximation would appear to be adequate.

3.2.   The Repetitively Chopped Thermal Lens

The theory of the repetitively chopped dual-beam thermal lens has been derived by Swofford and Morrell (15) as an extension to the theory of the time-dependent single-beam lens. For the chopped beam case, it is necessary to account for the "memory" effect of previous chopper cycles. That is, the lens from earlier cycles has not completely died away during a given cycle. Thermal lens effects caused by the probe laser are assumed to be negligible. This condition can be satisfied by judicious choice of probe wavelength and by keeping probe laser power small compared to pump laser power.

The thermal lens equations have no closed-form solution for the conditions of the pump and probe experiment. However, the system achieves a steady state after a few chopper cycles, and simple limiting equations are quite adequate.

Equation 27 describes the focal length, $F(t_m;m)$, of the thermal lens formed during the chopper-open portion of its cycle:

$$\frac{1}{F(t_m;m)} = \frac{\partial n}{\partial t} \frac{\alpha P_1}{\pi w_1^2 Jk} \left[\frac{1}{1 + t_c/2t} + \right. \tag{27}$$
$$\left. \sum_{n=1}^{m-1} \left(\frac{1}{1 + 2A_n/t_c} - \frac{1}{1 + 2(A_n + \tau/t_c)} \right) \right]$$

In Eq. 27 A_n is defined by Fig. 4 and Eq. 28:

$$A_n = (t_m + \tau') + (n - 1)(\tau + \tau') \tag{28}$$

The on-time of the chopper is τ. A_n is the on-time from the present cycle back through n cycles, as defined by Fig. 4.

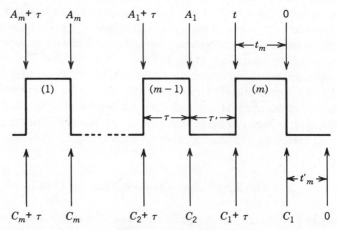

Figure 4. Timing sequences for calculation of the strength of the repetitively chopped thermal lens. The chopper is open for time τ and closed for time τ^1.

Similarly, the focal length of the thermal lens on the chopper-closed portion of the cycle is defined by

$$\frac{1}{F'(t'_m;m)} = \frac{\partial n}{\partial t}\frac{\alpha P_1}{\pi w_1^2 Jk}\sum_{n=1}^{m}\left(\frac{1}{1 + 2C_n/t_c} - \frac{1}{1 + 2(C_n + \tau)/t_c}\right) \quad (29)$$

C_n is defined as the off-time from the present cycle back through n cycles, as given by Eq. 30 and Fig. 4:

$$C_n = t'_m + (n - 1)(\tau + \tau') \quad (30)$$

These equations lead to a cumbersome expression for the response of the system. However, if the distance from beam focus to the probe, z', is approximately one confocal parameter, and if $F(t;m) \gg 2z'$, the equations simplify to

$$\frac{\Delta I(0,\tau;m)}{0(0;0)} = 2z'\left(\frac{1}{F(0;m)} - \frac{1}{F(\tau;m)}\right)$$

$$= 2z'\frac{\partial n}{\partial t}\frac{\alpha P_1}{\pi w_1^2 Jk}\left\{\frac{-1}{1 + t_c/2\tau}\right.$$

$$\left. + \sum_{n=1}^{m-1}\frac{2[1 + 2n(\tau + \tau')/t_c]}{[1 + 2n(\tau + \tau')/t_c^2 - (2\tau/t_c)^2]}\right\}$$

Even Eq. 31 is not expressible in closed form. The fundamental (chopper frequency) component can be extracted numerically, and accounts for about 85% of the signal. However, it is generally adequate to include only the leading term in parentheses, as in Eq. 32. Equation 32 is a good steady-state approximation to the behavior of the lens:

$$\frac{\Delta I(0,\tau;m)}{\Delta I(0,0)} = 2z'\frac{\partial n}{\partial T}\frac{\alpha P_1}{\pi w_1^2 Jk}\left(\frac{-1}{1 + t_c/2\tau}\right) \quad (32)$$

For most choppers, $\tau = \tau'$, and $1/2\tau = f$, the chopper frequency. Equation 32 can be reformulated as

$$\frac{\Delta I(0,\tau;m)}{\Delta I(0,0)} = 2z'\frac{\partial n}{\partial T}\frac{\alpha P_1}{\pi w_1^2 Jk}\left(\frac{-1}{1 + Ft_c}\right) \quad (33)$$

Equation 33 predicts that the lock-in amplifier signal will be proportional to the absorbance of the sample and to the laser power. At low frequency a limiting signal is expected, whereas at frequencies greater than $1/t_c$ an inverse dependence on chopper frequency is predicted.

Carter and Harris (16) point out that the above derivation neglects the quadratic component of the concentration dependence of the thermal lens. This dependence generates a signal at the second harmonic of the chopper frequence. The non-linearity is less than 3% below absorbance 0.001 and increases to 31% at absorbance 0.05. Consequently, working curves are nonlinear above absorbance 0.001 and actually go through a maximum at sufficiently high absorbance. The linear dynamic range of the pump/probe techinque can be increased by decreasing the pump laser power. However, the dynamic range is obtained at the cost of decreased sensitivity.

Equations 27–33 describe only the thermal lens formed by the heating laser. No assumptions are made about the character of the probe beam. The probe should be well-matched to the pump beam. Maximum sensitivity will occur if the probe beam has the same diameter and divergence, or equivalently, the same focused diameter and confocal parameter, as the heating beam. This condition is rarely completely satisfied.

Carter and Harris (16) have derived a similar approximate description of the pump and probe system, explicitly including the effects of the probe laser. They have shown that sensitivity, relative to a single laser experiment, is given by Eq. 34. The subscripts s and p refer to the heating and probe lasers, respectively. Because $\partial n/\partial T$ is a slowly varying function of wavelength, the magnitude of the observed signal is inversely proportional to probe laser wavelength.

$$\frac{E_p}{E_s} = \frac{(\partial n/\partial T)_p \, \lambda_s}{(\partial n/\partial T)_s \, \lambda_p} \tag{34}$$

Short-wavelength probe lasers are preferred, in principle. In practice, small He–Ne lasers are almost universally used as probe lasers, because they are compact, stable, and inexpensive. Only if the sample itself absorbs at 633 nm is another probe laser, usually argon ion, employed.

Fang and Swofford (17) point out that the strongest thermal lens is formed if the heating beam is focused into the sample. The maximum sensitivity to change occurs if the probe is focused one confocal distance before the heating laser. This condition can be met only by independent focusing of probe and heating lasers. However, the sensitivity increase is only a factor of 2. Few workers appear to have bothered with the added complication of independent beam focusing to achieve this relatively small increase.

A larger thermal lens signal can be obtained in longer cells (17). A limiting signal is achieved in a cell longer than about two confocal distances. The reason is that the lens becomes weaker as the heating beam becomes larger before and after focus. This point is discussed in detail in Section 3.4. Note, however, that the existence of a distance-independent signal is useful. This behavior means that strong signals can be achieved in short path cells if tightly focused lasers are used.

3.3. The Pulsed Thermal Lens

The thermal lens can also be formed with a pulsed laser. In this case it is necessary to use an independent probe laser. The lasers are assumed to have identical waists and divergences. Absorption by the probe beam is assumed to be negligible.

The thermal lens can be described by very simple equations (18) if one further assumes that the formation time of the lens is negligible compared to its decay time and that the laser pulse itself is infinitely short. The lasers used in thermal lens measurements have pulse durations of 1–20 ns. The rise time, t_r, of the thermal lens is the time it takes an acoustic pulse to travel the radius of the focused laser at the speed of sound. The rise time is given by Eq. 35, where v_s is the speed of sound in the medium:

$$t_r = \frac{v_s}{w_0} \tag{35}$$

Because the lasers are focused to a radius of 10–100 μm, and the speed of sound in most liquids is in the range of 900–1700 m/s, the rise time is typically 10–100 ns. The decay time is t_c, about 1–10 ms. Thus the assumptions are justified.

From these assumptions Twaroski and Kliger (18) derive the focal length of the pulsed thermal lens:

$$\frac{1}{f(t)} = \frac{1}{F_0}\left[\left(1 + \frac{2t}{t_c}\right)^{-2}\right] \tag{36}$$

$$\frac{1}{F_0} = \frac{\alpha(Q/0.5t_c)1(\partial n/\partial t)}{\pi J k w_1^2\,(1 + 2t/t_c)^2} \tag{37}$$

In Eqs. 36 and 37, Q is the energy in the laser pulse and F_0 is the initial focal length of the thermal lens, at the time the laser pulse heats the sample.

The initially strong lens decays with a time dependence that is ultimately t^{-2}. Note that the initial focal length is the same as that produced in a steady-state thermal lens experiment by a CW beam of power $P = Q/0.5t_c$.

For the thin lens approximation the optimum position for the cell is approximately 0.707 confocal parameter beyond the focal point of the focusing lens, at a distance $z' \simeq 0.707b$. In this case, the ratio of the change in beam intensity observed through a limiting aperture, relative to the unperturbed ($t \to \infty$) intensity, is given by

$$\frac{\Delta I_{bc}}{I_{bc,t \to \infty}} \simeq \frac{-2z'}{F(t)} \tag{38}$$

The value of t_c is generally a few milliseconds. This means that if the pulse repetition rate is more than 100–200 pulses/s, the system is effectively a CW system. Small thyratron-triggered excimer or nitrogen lasers can be pulsed that rapidly, and some metal vapor lasers operate at pulse rates of 10^3–10^4/s. However, most pulsed lasers actually employed in thermal lens systems do not behave as pseudo-CW devices.

Twaroski and Kliger (18) consider the more general case of n-photon absorption. The general form of the equations is similar to the one-photon case.

3.4. Path-Length Dependence of the Thermal Lens

In almost all photothermal measurements, the signal is generated by using a focused laser beam. Because the signal depends on the power density in the laser beam, photothermal signals do not increase linearly with cell length (19). This behavior is generally observed for any phenomenon that depends upon power density. Analogous behavior has been reported for inverse Raman spectroscopy (e.g., 20).

The diameter $w(z)$ at a distance z from the focus of a Gaussian laser beam is given (21) by

$$[w(z)]^2 = w_0^2[1 + (z/z_c)^2] \tag{39}$$

The beam radius at focus is w_0. The confocal parameter, z_c, is defined by

$$z_c = \frac{w_0^2 n}{\lambda} \tag{40}$$

The position dependence can be treated (19) by substituting the basic thermal lens response Eq. 16 into Eq. 39 and integrating over cell length, b, from z_A to z_B. Equation 41 results:

$$\frac{\Delta I}{I} = \frac{2\theta z_c}{b} \left[\int_{z_A}^{z_B} \frac{z}{z_c^2 + z^2} \, dz + \theta^2 z_c^2 \int_{z_A}^{z_B} \frac{dz}{(z_c^2 + z^2)} \right] \tag{41}$$

The thermooptical parameters have been lumped together as $\theta = -2.303P(dn/dT)A/\lambda k$.

Evaluation of Eq. 41 yields Eq. 42, the explicit length dependence of the thermal lens signal:

$$\frac{\Delta I}{I} = \frac{\theta z_c}{b} \ln\left[\frac{z_c^2 + (7_A + b)^2}{z_c^2 + z_a^2}\right] + \frac{\theta^2 z_c}{b} \arctan\left[\frac{bz_c}{z_c^2 + z_a^2 + bz_a}\right] \quad (42)$$

Equation 42 is valid only in the limit of a weak thermal lens. The derivation assumes that the beam at any point is unperturbed by the presence of the lenses formed at earlier points along its path. This limitation can be removed by treating the system as a series of thin lenses, with the lens at each point calculated from the state of the beam after perturbation by all lenses formed earlier. The generalized equations have not yielded a closed-form solution, although a numerical approximation provides a good fit to experiment (19).

Figure 5 shows the predicted response as a function of cell length. A limiting response is obtained for a cell whose length is many confocal parameters long. Maximum sensitivity is obtained when the experiment is operated under thin

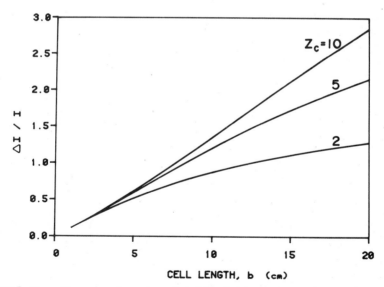

Figure 5. Thermal lens response versus sample path length, $\theta/b = 2.303; E_a = 0.1\ \text{cm}^{-2}; n = 1.0$, $\lambda = 514.4$ nm. Reproduced with permission from Ref. 19.

lens conditions, with a focusing lens that gives a confocal parameter that is long compared to the cell path length.

Carter and Harris (22) have considered the implications of focused beams for small volume measurement. They suggest that a reasonable maximum path length is $b_{max} < z_c/3$. The critical time for the system depends on w^2. For this value of b_{max}, the beam diameter remains close to the minimum and the critical time varies over a range of no more than $\pm 15\%$. This condition provides a single time constant for time-resolved measurements and a constant relative contribution from linear and quadratic components of the thermal lens response.

Assuming that the beam remains truly cylindrical, the boundaries of the region generating the thermal lens can be calculated. A cell length of $z_c/3$ can be accommodated in a volume defined by a cylinder of radius $w = 2w_0$. The minimum sample volume is given by

$$V_{min} = \pi(4w_0)^2 b = 16z_c\lambda b \geq 48\lambda b^2 \tag{43}$$

The implication of Eq. 43 is that the tighter focusing improves the mass sensitivity in a tightly focused short-path cell system, relative to a loosely focused long-path cell system, because the volume increases quadratically with path length.

Absorbance detection limits should be independent of volume, because absorbance detection limits are really heat detection limits. In practice Carter and Harris find that absorbance detection limits decrease by a factor of 2 when the cell volume is reduced from 2.6 μL to 29 nL. They suggest that the improvement results from two factors. First, the tightly focused beam diverges more rapidly. The detector can be placed close to the sample and is less sensitive to pointing errors and misalignment. Second, the tightly focused beam should be less sensitive to thermal or refractive index gradients in the sample.

These properties of photothermal detectors are the basis of their utility for LC detectors. Strong signals can be obtained for tightly focused beams in short path-length, low-volume systems.

4. THE PHOTOACOUSTIC EFFECT

The photoacoustic effect is the pressure change generated by the heat released following light absorption. If the absorbing medium is a gas, the photoacoustic effect generated by a source modulated at frequencies from 10^2 to $\sim 10^4$ Hz is a sound wave, and can be detected with a microphone. The term photoacoustic spectroscopy refers to detection of the pressure change in any medium, by any pressure sensor.

Photoacoustic spectroscopy has been extensively reviewed (1–4). Here we describe only those aspects of theory immediately pertinent to chromatographic detectors.

4.1. The Photoacoustic Effect with a Repetitively Chopped Source

The simplest theoretical treatment of photoacoustic spectroscopy with a modulated CW laser assumes that a sample is a liquid in a cylinder of finite radius heated by a laser focused tightly enough to be considered a line source (23). The walls of the cylinder are assumed to be a cylindrical piezoelectric transducer. Thermal effects are unimportant, if the cylinder radius, r, is much larger than the beam radius, w. In this case, Eq. 44 describes the pressure, $p(r)$, produced by the laser beam of power P_0, modulated at angular frequency ω:

$$p(r) = \frac{\alpha P_0 \omega \beta}{4C_p} H_0^{(2)} \left(\frac{\omega r}{s} \right)$$ (44)

In Eq. 44 α is the absorption coefficient, β is the volume expansion coefficient and C_p is the specific heat at constant pressure, and s is the speed of sound in the liquid. $H_0^{(2)}$ is a Hankel function of the second kind, which represents a wave propagating radially outward.

The radial velocity, $v(r)$, is given by

$$v(r) = \frac{j\omega v P_0 \beta}{4\rho c_p s} H_1^{(2)} \left(\frac{\omega r}{s} \right)$$ (45)

Here $H_1^{(2)}$ is a Hankel function of the second kind.

The total time-average acoustic power, P, is given by Eq. 46, where z is the direction of laser propagation:

$$P = \frac{2\pi r}{2} \int_0^\infty e^{-2\alpha z} \operatorname{Re}(pv_r^*)dz = \frac{\pi r}{2\alpha} \operatorname{Re}(pv_r^*)$$ (46)

The right side of Eq. 46 ultimately simplifies to

$$P = \frac{\omega \beta^2 \alpha P_0^2}{16\rho c_p^2}$$ (47)

Curiously, Eq. 47 predicts that the acoustic power is quadratic in incident optical power. In fact, most of the delivered optical power, as noted by Tam (24), is converted to heat. The conversion efficiency of a pulsed photoacoustic experiment is much higher.

4.2. The Pulsed Photoacoustic Effect

Patel and Tam (4) have outlined a rigorous theory for the photoacoustic signal expected from a pulsed laser source. They assume that the pulse is deposited in a cylinder of Gaussian cross section and narrow diameter relative to the distance, r, at which the signal is observed. They assume that the pulse is detected piezoelectrically and that the laser pulse width 2τ is long compared to the response time of the detector, or the nonradiative relaxation time of the sample. The laser pulse is assumed to have a Gaussian time dependence and a total pulse energy E_0. The pressure changes induced by the pulses are taken to be small compared to the ambient pressure. The system is assumed to be adiabatic.

For these conditions, a good approximation to the pressure change $p(r,t)$, observed at a distance r from the laser beam, is given by

$$p(r,t) = \frac{v_s \beta E_0}{\pi^{3/2} C_p \tau^3} \int_{-\infty}^{t - r/v_0} \frac{t' e^{(t'/\tau)^2} dt'}{[v_s^2(t - t')^2 - \sigma^2]^{1/2}} \tag{48}$$

In Eq. 48 v_s is the speed of sound in the medium, E_0 is the pulse energy, and β is the volume temperature coefficient of expansion of the sample.

The signal maximizes at $t \cong r/v_s \pm 0.7/\tau$. The integral can be evaluated for large r, for which the integrand is most important at the start of the experiment, when $t - t' \pm r/va$. In this case we can make the following approximation:

$$v_s^2(t - t')^2 - r^2 \cong 2r[v_s(t^- - \tau') - r] \tag{49}$$

For this case, the pressure is given by

$$p(r,t\pm) \cong \pm \frac{\beta \alpha E_0}{\pi C_p \tau^2} \left(\frac{v_s E}{2\pi r} \right)^{1/2} \equiv \pm P_0(r) \tag{50}$$

Equation 50 describes a peak positive and negative pressure (rarefaction) change. The pulse is delayed from the start of the laser pulse by time τ_d, as given in

$$\tau_d = \frac{r}{v_s} \tag{51}$$

Time-gating of the photoacoustic signal is important in rejection of signals because of scattered light falling on the detector or cell walls and in distinguishing the primary signal from the echoes that accompany it.

5. EXPERIMENTAL PROBLEMS

Convective mass transport is potentially a serious error source in any photo-thermal or photoacoustic spectroscopic measurement. In their classic early paper Gordon et al. (10) observed convective upward motion of dust parti-cles in the beam. A dust particle required about 5 s to travel one beam width, in agreement with a very approximate calculation of convection velocity, approxi-mately 0.03 cm/s. They concluded that convection, while observable, was a minor phenomenon.

Whinnery and co-workers (25) estimated convective velocities by assuming an unperturbed thermal lens temperature distribution, and calculating the motion in a gravitational field for a liquid of known velocity. The boundary problem could not be solved exactly, but, for a liquid of viscosity μ, Eq. 52 is an upper bound for the upward convective velocity v_z, if the maximum temperature rise is ΔT_m:

$$V_z = \frac{\rho \alpha g \Delta T_m \pi w_0^2}{16\mu} \tag{52}$$

In weakly absorbing pure solvents, only small temperature changes are ob-served, and convective velocities of 1–3 mm/min are expected. In a steady-state thermal lens experiment, convection causes a blurring of the top of the thermal lens on the ordinary time scale.

At higher absorbances, convection can cause oscillations in the thermal lens. Oscillatory behavior has been reported by Buffett and Morris (26). Using cylin-drical spectrophotometer cells filled with iodine in carbon tetrachloride, they were able to induce pronounced oscillations, as shown in Fig. 6. Dye tracing experiments demonstrated that these oscillations result from circulation of the liquid to the top of the cell, around the edges and back to the axis of the laser beam, as shown schematically in Fig. 7. Complete circulation causes stirring of the solution, generating the oscillatory behavior. The periodicity could be readily observed in the power spectra of thermal lens measurements taken on systems for periods of 10–30 min.

Similar convective oscillation has been observed in other systems. The effect is probably quite general in any experiment in which a fluid is heated directly or indirectly for an extended period.

For example, Epstein and co-workers (27) have demonstrated that convective

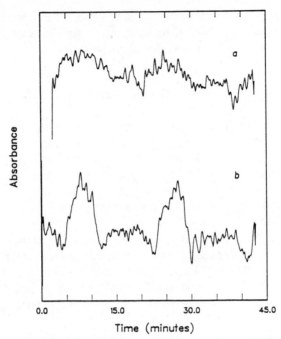

Figure 6. Thermal lens signal of iodine in carbon tetrachloride, 2-cm cylindrical cell. (*a*) Lasers positioned at center of cell. (*b*) Lasers positioned near edge of cell. Reproduced with permission from Ref. 26.

oscillations are common in systems monitored by laser-induced fluorescence or phosphorescence. They comment that much "oscillatory" behavior is due to convection, not to chemical effects. Their observations have been verified by LaPlante and co-workers (28,29), who have obtained power spectra with very sharply defined convective frequencies including overtones, in some geometries.

A laser is not necessary as the heating source. Similar effects can be induced in liquids in which an electrically heated wire is immersed (30,31).

In LC systems, the mobile phase linear velocity is usually high enough so that convective oscillations are not observed during an actual experiment. However, convective oscillations are readily observed in some chromatographic cells without solvent flow (26). A 1-mm-diameter cylindrical cell that is illuminated along the cylinder axis can show oscillations with a period of a few seconds. These oscillations can be a nuisance in the initial alignment of photothermal detector systems.

Photoacoustic and photothermal detectors are all subject to noise associated with mobile phase flow itself and, with some laser sources, with flow-induced

Figure 7. Convective flow patterns in a cylindrical thermal lens cell. Reproduced with permission from Ref. 26.

noise in the laser gain medium. The evidence at hand suggests that flow-related noise is the limiting noise source in LC systems. The noise becomes proportional to laser power at relatively low powers.

Flow noise limits the effective heating laser power that can be profitably used in indirect absorbance detectors, in some cases to just a few milliwatts. Alternatively, the flow noise limitation implies that the benefits of indirect absorbance detectors can be obtained with compact, inexpensive lasers.

Flow cells for LC deliberately include provision for turbulent flow. The reason is to avoid formation of regions of stagnant liquid. These regions would decrease resolution of the system. The problem is eliminated by introducing the mobile phase into the flow cell with a right angle turn. This technique results in flow cells that have the familiar "T", "U", or "Z" flow geometries.

The pump itself adds flow noise to the system. Standard dual piston pumps, even with good pulse dampening, generate some irregularities in mobile phase velocity. Mobile phase velocity irregularities are minimized with syringe pumps, but at the cost of some versatility.

The requirement for turbulent flow and the existence of pump flow fluctuation can not be completely reconciled with the requirements of indirect absorbance detectors. Turbulent flow or any flow fluctuation disrupts the spatial distribution of heated sample, generating acoustic noise and noise in refractive index gradients. These phenomena, then, degrade S/N ratios in indirect absorbance detectors.

Quite similar problems are encountered in the design of CW dye lasers, where a high linear velocity is needed in the active region of the dye cell to prevent buildup of degradation products or dye molecules in the triplet state. Here high velocity is induced by constricting the flow path, but at the cost of high acoustic noise.

Flow fluctuation problems were encountered early in the development of indirect absorbance detectors. It is not surprising, then, that most early reports of photoacoustic and thermal lens detectors contain extensive discussions of flow fluctuation and methods to overcome it (32–35).

Dovichi and Harris point out that mobile phase flow is equivalent to an increase in the thermal conductivity of the system (33). Consequently, increasing flow rate is expected to decrease thermal lens or photoacoustic response. Over a limited range of flow rates, the rate of change of enhancement factor with flow velocity, dE/dv, is constant.

The absolute change in sensitivity is not especially great at chromatographic flow rates, but flow-rate fluctuations are equivalent to sensitivity fluctuations, that is, to noise. For a simple flow system, noise in a time-resolved thermal lens experiment can be described quite accurately as Gaussian noise consisting of a constant static component and an independent and constant flow-rate fluctuation term:

$$\sigma^2 = \sigma_0^2 + A^2 \left(\frac{dE}{dv}\right)^2 \sigma_v^2 \tag{53}$$

The standard deviation in the flow system is σ, the component from static thermal lensing is σ_0, and from flow-rate fluctuations is σ_v. This approximation is valid, because the time-resolved measurements emphasize low frequencies, down to 1–5 Hz.

The behavior of a photoacoustic or thermal lens system based on a chopped CW laser is somewhat different. What is observed depends on whether or not the modulation frequency is in a region where the pump/column/cell combination is acoustically noisy. The acoustic noise power is typically high below 100–200 Hz, and falls off rapidly at higher frequencies. The details of acoustic noise falloff depend strongly on the mechanical design of the system and the mobile phase flow rate.

Oda and Sawada studied the modulation frequency dependence of photoacoustic signals and S/N ratio. Their cell (Fig. 8), when pumped by a Japan Spectroscopic Co. TRI ROTAR piston pump, is very noisy at low frequencies, but quiet at frequencies above 2 kHz. Because the photoacoustic response shows a resonant peak at 4 kHz, the S/N ratio also maximizes at this frequency. This behavior is summarized as Fig. 9.

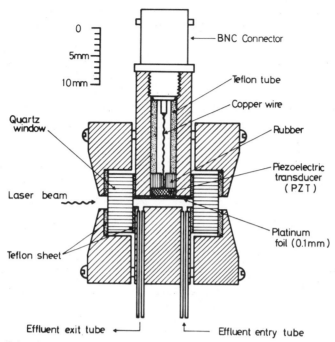

Figure 8. Schematic of photoacoustic flow cell and detector assembly. Reproduced with permission from J. Oda and T. Sawada, *Anal. Chem.*, **53**, 471 (1981). Copyright 1981 American Chemical Society.

Pulseless pumping minimizes turbulence problems in indirect absorbance detectors. Buffett and Morris (35) found that the combination of a Kratos SFA-234 flow cell, which employs a modified "Z" flow geometry and an LDC dual piston Mini-Pump with pulse dampener, could not be employed in thermal lens detectors. However, the same cell performed well when used with a syringe pump (ISCO 314), which produces pulseless flow.

More recently, Pang (36) has defined the conditions for satisfactory operation of a thermal lens detector with a fairly noisy piston pump. Using an LDC dual-piston mini-pump and a straight-through capillary flow cell, he studied the frequency dependence of S/N ratio in a thermal lens system. A capillary provides minimum turbulence, but does introduce some optical complications, because it functions as a strong cylinder lens. The results are summarized as Fig. 10.

Surprisingly, in this system, S/N ratio is not a monotonic function of frequency. Thermal lens systems do not show resonances, unlike photoacoustic detectors. Thus the thermal lens signal decreases monotonically with increasing

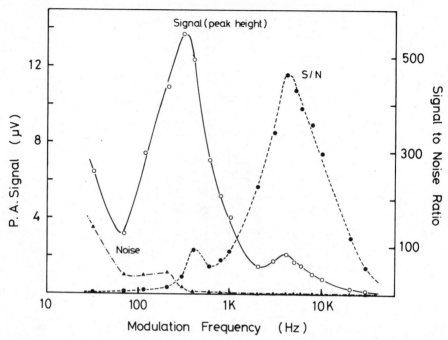

Figure 9. Modulation frequency dependence of photoacoustic signal in flow cell of Fig. 8. Test sample 1×10^{-5} M 2'Cl-DAAB. 1-mL/min flow rate, 488 nm, 500-mW laser irradiation. Reproduced with permission from J. Oda and T. Sawada, *Anal. Chem.*, **53**, 471 (1981). Copyright 1981 American Chemical Society.

modulation frequency. The noise, however, is not monotonic, but shows peaks at frequencies related to 60 Hz. The source of this noise is the 60 Hz and harmonic noise on the probe laser intensity, generated by imperfect power supply regulation. Probe laser intensity fluctuations are strobed at the high chopping frequencies. The aliased signals appear as low-frequency noise whenever the modulation frequency is near a line-frequency harmonic.

Perhaps the most extreme example of flow-related noise can be found in supercritical fluid measurements (37). In carbon dioxide near its critical point ($P = 87$ atm, $T = 28°C$), noise in thermal lens measurements in a flow system has been reported to be two orders of magnitude worse than the noise in a static system under the same temperature/pressure conditions. This extraordinary flow sensitivity is a direct result of the rapid change of both density and thermal and optical properties of a fluid near its critical point.

Betz and Nikelly (38) have shown that flow-related noise can be reduced in dual-beam absorbance detectors by using carefully matched flow conditions in sample and reference cell. Pang and Morris (39) have observed flow noise

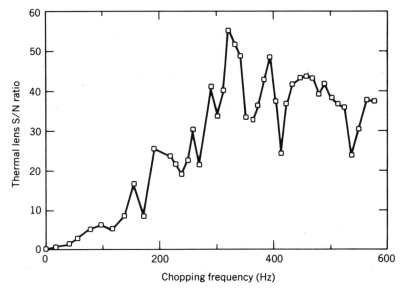

Figure 10. Flow-rate dependence of thermal lens enhancement in aqueous cobalt chloride. Data from T.-K. J. Pang, unpublished observations.

reduction using matched flows in a series differential thermal lens detector. To the extent that turbulence patterns are matched, turbulence-induced aberrations in a thermal lens can be made to cancel. Imprecise measures of flow matching were used in those experiments, but noise reduction by a factor of 2.5–4 was observed. Improvements to their system could easily lead to a factor of 10 reduction in turbulence-generated noise.

It is clear that the best operating conditions are likely to be specific to a given pump/cell combination and the type of experiment carried out. Any new apparatus should be tested to find the signal acquisition conditions that maximize the S/N ratio.

6. REALIZED INDIRECT ABSORBANCE DETECTORS

Indirect absorbance detectors are most valuable for chromatographic systems where small working volume is the major requirement. Noise levels are about 1–5 microabsorbance. This noise level can be obtained in working volumes of a few nanoliters. Although most workers have employed high power lasers, the evidence suggests that low power (5–25 mW) CW or low pulse energy (5–15 mW) pulsed lasers are adequate for most applications.

Detectors based on photoacoustic spectroscopy and detectors based on the thermal lens effect and related phenomena have been proposed. The performance of the two classes of detectors has been quite similar. It is too early to say which class of detector will ultimately prove more practical. However, the decision will probably be based on criteria of design simplicity, rather than fundamental performance limitations.

6.1. Photoacoustic Detectors

Oda and Sawada (32) have shown absorbance detection limits (S/N = 2) of 7.9×10^{-6} in an LC detector with CW laser excitation. They employed 500 mW argon ion 488 nm excitation and modulated at 4.035 kHz, using an acoustooptic modulator. They used a piezoelectric detector incorporated in the home-made flow cell shown in Fig. 8. The cell employs conventional U-tube flow geometry. The piezoelectric ceramic is coupled to the sample by a piece of polished platinum foil. Quartz end windows enclose the cell. The choice of modulation frequencies was determined largely by excess noise caused by mobile phase flow fluctuations.

The cell is designed for operation with a conventional HPLC system, employing a 20μL sampling loop and a flow rate of approximately 1 mL/min. The active volume, about 19 μL, is quite large.

The cell has been tested with isomers of chloro-4-(dimethylamino)azobenzene (Cl-DAAB). These azobenzenes have similar molar absorptivities at 488 nm (approx. 7700 L/m · cm) and 254 nm (9600 L/m · cm). This circumstance facilitates comparisons between photoacoustic detectors and fixed-wavelength UV absorbance detectors. The photoacoustic detector was shown to have linear response over more than four decades of absorbance. Its linearity is quite similar to that of an absorbance detector. The precision of replicate measurements is almost identical to the precision of measurements made with an absorbance detector in the same system.

Chromatograms taken near the detection limits are shown as Fig. 11. Oda and Sawada (32) report a 25-fold lower detector limit (approximately 7.9×10^{-6} absorbance) for photoacoustic detection than for UV detection (2×10^{-4} absorbance). They suggest that the detection limits are determined by noise in the lock-in amplifier, about 5 nV, and that a different lock-in amplifier could reduce detection limits to the value limited by noise in the piezoelectric transducer itself, about 1 nV.

Winefordner and co-workers (34,40) have adapted the windowless flow cell principle of Diebold and Zare (41) to LC. The windowless cell (Fig. 12) is intended for measurement of any or all of photoacoustic, fluorescence, and photoionization signals simultaneously. The three phenomena of heat evolution,

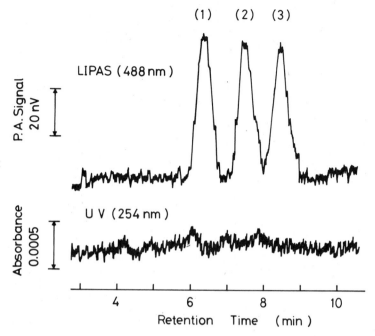

Figure 11. Chromatograms of DAAB isomers. (1) 2'Cl-DAAB, (2) 3'Cl-DAAB, (3) 4'Cl-DAAB, all 1.56 ng/20 μL. Upper curves, photoacoustic detector; lower curve, UV absorbance detector. Reproduced with permission from J. Oda and T. Sawada, *Anal. Chem.*, **53**, 471 (1981). Copyright 1981 American Chemical Society.

luminescence, and ionization are the most probable outcomes of laser excitation of any sample.

Originally (41) the windowless cell was used to minimize background fluorescence from cell windows. It can also serve to minimize contamination and memory problems. However, it is less certain that a windowless cell is optimum for a photoacoustic measurement. The design does not appear to provide efficient coupling of the photoacoustic signal to the transducer. For example, Winefordner and co-workers (42) have compared the performance of their windowless cell to the performance of the more conventional piezoelectric transducer design of Fig. 13. Detection limits with the conventional design are approximately 10× below those obtained with the windowless cell.

Winefordner and co-workers have used pulsed nitrogen and excimer lasers as light sources (34,40,42) for photoacoustic detectors. The nitrogen laser energy used had low (0.3 mJ) pulse energy. Thus high detection limits were obtained with it. Excimer (XeCl 308 nm) laser pulse energies as high as 11 mJ were

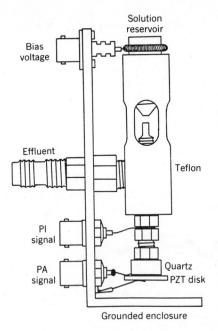

Bias voltage

Solution reservoir

Effluent

Teflon

PI signal

PA signal

Quartz

PZT disk

Grounded enclosure

Figure 12. Windowless flow cell for simultaneous photoacoustic, fluorescence, and photoionization measurements. Reproduced with permission from E. Voigtman, A. Jurgensen, and J. D. Winefordner, *Anal. Chem.*, **53**, 1921 (1981). Copyright 1981 American Chemical Society.

employed, and the high pulse energy laser allowed lower detection limits. Above 2-mJ Winefordner and co-workers observed a decrease in sensitivity. Around 2 mJ, the signal becomes large enough that proportional noise due to pulse–pulse fluctuations in laser energy becomes important. Detection limits no longer decrease in inverse proportion to pulse energy, but approach a limiting value.

The use of pulsed lasers in photoacoustic detectors poses certain problems. In addition to the desired signal, photoacoustic signals generated by the cell windows and walls are observed, as are echoes from the original signals. The observed transducer response is a series of oscillations. The desired signal is captured with a gated integrator, which is opened only during the time when the appropriate signal pulse is incident on the detector.

The arrival time of the signal pulse at the transducer is quite reproducible. With the cell of Fig. 13, the arrival time is 20–50 μs, depending on the exact laser–transducer distance and the mobile phase composition. Because the pulse travels through the mobile phase, the arrival time depends on the speed of sound in that phase. If a solvent gradient is used an aperture set for a fixed time after each laser pulse would give systematic errors because the speed of sound would be changing with the mobile phase composition. In principle, the aperture delay could be programmed to track the changing delay time. How successful this tactic would be is uncertain.

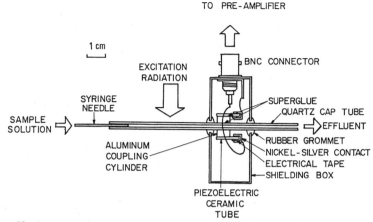

Figure 13. Quartz capillary tube flow cell and piezoelectric transducer for photoacoustic detection. Reproduced with permission from Ref. 42.

All of the pulsed laser photoacoustic detector work reported has used polynuclear aromatic hydrocarbons as test substrates. These compounds typically have high quantum yields for fluorescence and therefore correspondingly low quantum yields for nonradiative processes. As Table 3 shows, the detection limits for photoacoustic and UV absorbance detection are similar. This suggests that photoacoustic spectroscopy detection limits would be lower than absorbance detection limits by about an order of magnitude for nonfluorescent or weakly fluorescent compounds.

6.2. Thermal Lens Detectors

Several groups have reported use of the thermal lens effect and related thermooptic phenomena for liquid chromatography detection. Thermal lens spectroscopy is less well-developed than photoacoustic spectroscopy. Consequently, some of the chromatographic work is intertwined with development of thermal lens technology itself.

Leach and Harris (43) were the first to employ the thermal lens effect for LC detection. They used a reverse-phase system (50:50 methanol–water) and a 250 × 4.6-mm column packed with ODS on 5-μm silica and 18-μL 10-mm path-length flow cell. They employed an argon ion laser (190 mW) in a single-beam experiment, using a modified time-resolved measurement.

Leach and Harris demonstrated operation of this system with separation of the three *o*-nitroaniline isomers. With a 5 s time constant, they obtained ab-

Table 3. Limits of Detection (LOD) of Several Polynuclear Aromatic Hydrocarbons for UV Absorbance, N_2, and XeCl Excimer PAS Detection of HLPC Effluents[a]

Compound	UV Absorbance LOD (μg/mL) $\lambda = 254$ nm	N_2-laser PAS LOD (μg/mL) $\lambda = 337.2$ nm Pulse energy $= 0.3$ mJ	XeCl laser PAS LOD (μg/mL) $\lambda = 308$ nm Pulse energy $= 11$ mJ
Azulene	0.03 (0.2)[b]	0.4 (0.4)	0.3 (0.4)
Biphenyl	0.02 (0.1)	—[c]	—
7,8-Benzoflavone	0.08 (0.7)	0.7 (0.8)	0.2 (0.2)
Phenanthrene	0.01 (0.09)	—	0.6 (0.7)
Fluoranthene	0.06 (0.5)	0.5 (0.6)	0.2 (0.2)
Pyrene	0.08 (0.7)	0.4 (0.5)	0.3 (0.3)
Chrysene	0.02 (0.2)	—	0.3 (0.3)
3,4-Benzopyrene	0.05 (0.4)	1.0 (1.0)	0.4 (0.5)

[a]Data from Ref. 34.
[b]In parentheses, absolute limits of detection in nanograms, calculated by multiplying the concentration LOD (μg/mL) by the illuminated volume of sample solution in the cell (mL). The UV absorbance detector has an 8 μL flow cell (or illuminated volume), as compared to an illuminated volume of ~1.2 μL for the PAS flow cell.
[c]No detection.

sorbance detection limits of 1.5×10^{-5} absorbance per centimeter. These initial thermal lens results demonstrated detection limits about 10 times lower than absorbance detectors of the period.

Subsequently, Buffett and Morris (44) adapted the two-laser pump/probe configuration to chromatography. They also used a conventional column (250 $\times$ 4.6-mm, 10-μm LiChrosorb RP-18) and flow rates and employed a commercial LC flow cell. This group also used argon ion 458-nm radiation, but reduced the power to 90 mW. They were able to obtain absorbance detection limits of about 2×10^{-6} with a 1-s time constant. In these preliminary experiments no attempt was made to assess the absorbance range over which the technique could operate. Typical chromatograms are shown as Figure 14.

Buffett and Morris noted the effects of flow fluctuations on their thermal lens measurements. They used a modulation frequency of 125 Hz in order to avoid breakup of the thermal lens. They observed linear response over about a 500-fold absorbance range, but did not test the performance of their system at absorbances higher than 6×10^{-4}. However, nonlinear performance is expected if the absorbance is much above 0.001 and the quadratic terms in the thermal lens response become increasingly important (45).

The performance improvement of the Buffett/Morris detector over the Leach/

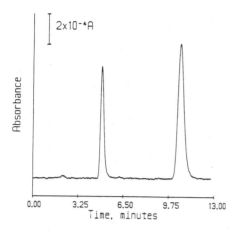

Figure 14. Thermal lens detection of 22-ng *o*-nitroaninline (ONA) and 136-ng *N,N*-dimethyl-3-nitroaniline (NND). Experimental conditions (1-mL/min flow rate, 80% methanol, 90-mW Ar⁺ 458 nm). Reproduced with permission from C. E. Buffett and M. D. Morris, *Anal. Chem.*, **54**, 1821 (1982). Copyright 1982 American Chemical Society.

Harris system can be attributed to two factors. First, the use of an 80:20 methanol mobile phase, rather than a 50:50 composition, increased the sensitivity of the experiment. Second, Leach and Harris were forced to simplify the data extraction algorithm used with their time-resolved system. The effect was to exchange S/N ratio for computation time.

The computational problems in the Leach/Harris technique arise directly from the nonlinear behavior of a time-resolved thermal lens experiment. Equation 54 is the thermal lens governing expression with terms lumped into one parameter, θ, which is defined in Eq. 55:

$$I(t) = I(0) \left[1 + \frac{\theta}{(1 + t_c/t)} + \frac{0.5\,\theta^2}{(1 + t_c/t)^2} \right] \quad (54)$$

$$\theta = \frac{-2.303\,P(dn/dt)\,A}{1.91\,\lambda k} \quad (55)$$

A real-time fit to equation to solve for θ, that is, A, is not feasible. Leach and Harris linearized the equations by assuming the value of t_c to be the independently determined value for the mobile phase employed and by taking $I(0)$ to be the first measured point of the transient. Equation 54 can be rearranged to

$$\theta(t) = \left(1 + \frac{t_c}{t} \right) \left[\left(\frac{2I(0)}{I(t)} - 1 \right)^{1/2} - 1 \right] \quad (56)$$

Equation 56 was used to calculate $\theta(t)$. The average value of $\theta(t)$ is used to define the absorbance for each transient.

This procedure emphasizes errors introduced by noise in the measurement of $I(0)$. Replacement (46) of the simple average value of θ with a weighted average improves the precision of the measurement by about 40%.

A two-laser configuration is advantageous for chromatographic experiments based on CW lasers. Lock-in amplifier signal extraction is experimentally convenient and powerful. The two-laser configuration is required for experiments with pulsed lasers. Only pulsed lasers currently provide complete UV wavelength coverage.

As the required cell volume decreases, it becomes increasingly difficult to implement a two-laser configuration. Low-volume flow cells have short path lengths and require tight focusing of the laser beams. The effects of small random changes in laser beam pointing direction become serious under these conditions. If beam wander is bad enough two-laser thermal lens systems become impractical, if not actually impossible to use.

The problem was first encountered by Buffett and Morris (35) in a study of the application of a thermal lens detector to microbore LC. They found that a short focal length lens (32 mm) was required to operate their system with a 1 mm path length cell. The tight focusing accentuated beam pointing stability problems. They reported that daily realignment of the system was necessary even if the laser was allowed to warm up for close to 1 h before use. They observed that the noise level in the system depends quite critically on the overlap of the heating and probe beams. Small changes in alignment could increase the noise from 9×10^{-7} to 7×10^{-6} absorbance.

A CW laser can be made to function as both pump and probe laser. Pang and Morris (47) have shown that a CW laser modulated with a square wave generates a thermal lens response at the harmonic of the modulation frequency used, about 315 Hz. This second harmonic signal can be extracted with a lock-in amplifier. Because only one laser is used, the system is unconditionally stable. The only alignment needed is centering of the flow cell and limiting aperture on the laser beam and adjustment of the positions of focusing and relay lenses.

The second harmonic technique minimizes optical problems, but generates very small signals. The second harmonic signal is typically about 0.1% of the signal at the modulation frequency. The signal at the fundamental is just the intensity change as the chopper alternately passes and blocks the laser light. Extraction of the small harmonic signal from the much larger fundamental is a difficult task, even for a sophisticated lock-in amplifier. Consequently, second harmonic detection is achieved at the cost of a three- to fourfold increase in noise over the best two-laser results.

Alternatively, pump and probe beams can be obtained as fractions of the same laser with a beam splitter and encoded by plane of polarization (48,49). Because both beams come from the same laser, the optical path can be arranged so that the two beams will track each other if the pointing direction changes.

Once initially aligned, this configuration requires little or no further adjustment. In effect, the design transfers the burden from the lock-in amplifier to the polarization optics. High-quality polarizing prisms and beam combiners are inexpensive.

Buffett and Morris also demonstrated that good detector performance does not require high laser power, but only an adequately high heating laser/probe laser power ratio. They were able to reduce the heating laser power from 100 to 7.5 mW with no degradation in S/N ratio if they maintained a power ratio of 30:1 or more in the two beams. Therefore, thermal lens detectors can be based on compact and inexpensive lasers. If pulsed lasers are used, pulse energies can be low enough to eliminate completely the problems of thermal damage to the chromatographic cell.

Sepaniak and co-workers (50) have applied thermal lens detection to open tubular LC. They employ a modified version of a time-resolved one-laser configuration. In this experiment the laser is chopped at 500 Hz and the thermal lens measured as the overflow over an optical fiber using a gated integrator with a 100-μs aperture. Capillary tubes with internal diameters of 100 or 200 μm are used as sample cells. An argon ion laser (458 nm, 500–800 mW) is used as a light source.

These experiments demonstrate the straightforward applicability of thermal lens measurements in small-diameter cells. Sepaniak and co-workers obtain baseline noise levels of about 3×10^{-5} absorbance, as shown in Fig. 15. The very high modulation frequency and rather peculiar detection scheme does not provide

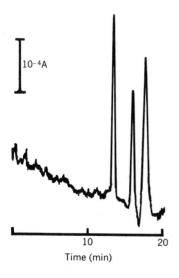

Figure 15. Thermal lens detection of o-nitroaniline, 4,5-dimethylaniline, and N,N-dimethyl-3-nitroaniline. Reproduced with permission from M. J. Sepaniak, J. D. Vargo, C. N. Kettler, and M. P. Muskarinec, *Anal. Chem.*, **56**, 1252 (1984). Copyright 1984 American Chemical Society.

optimum detection of a thermal lens from generated with a CW laser. Despite these problems, the Sepaniak experiments provide noise levels competitive with the best conventional absorbance detectors. As the authors realize, a better experimental design could easily drop absorbance detection limits to the microabsorbance range.

6.3. Related Detector Systems

Yeung and co-workers have developed novel systems for detecting refractive index changes formed by radiationless relaxation (51,52). Figure 16 shows the Woodruff and Yeung (51) interferometric detector. In this system the flow cell is placed in the cavity of a commercial Fabry–Perot interferometer. The interferometer is illuminated by a low-power HeNe laser. It is scanned under computer control when mobile phase only is passing through the flow cell. This measurement establishes the refractive index of the mobile phase. When the mobile phase is heated by eluate absorption the RI changes. A scan of the interferometer now establishes the new refractive index. Repeated scans can be converted to a plot of the time dependence of the refractive index and, through that, to an indirect absorption chromatogram. The scan requirements of the interferometer limit the effective time constant of the instrument to 15 s or longer. The detectability (S/N = 3) is about 2.6×10^{-6} absorbance.

Although the response speed of the device was inadequate, it illustrated an important property. Indirect absorption measurements based on refractive index

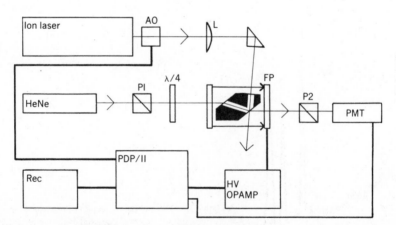

Figure 16. Fabry–Perot interferometer as indirect absorption detector. AO, acoustooptic modulator; P1, P2, polarizing prisms; λ/4, quarter-wave plate; FP, Fabry–Perot interferometer; HeNe, single-frequency HeNe laser; PMT, photomultiplier tube. Reproduced with permission from S. D. Woodruff and E. S. Yeung, *Anal. Chem.*, **54**, 271 (1985). Copyright 1985 American Chemical Society.

changes do not require a Gaussian heating beam. In fact, the interferometer was arranged so that the RI across the illuminated area was constant. The Yeung experiments provide evidence that indirect absorption detectors that use laser beams with poorly defined spatial qualities can be successful. Certainly, inexpensive excimer lasers would be likely candidates for these systems.

Bobbitt and Yeung (52) have developed an indirect absorption detector based on change in optical rotation of an optically active eluent. Heat evolution causes a change in the sample density and a corresponding change in rotation of the eluent. Detectability is about 1.4×10^{-6} absorbance.

6.4. Applications to Related Techniques

Leach and Harris (53) have demonstrated application of their one-laser technique to flow injection analysis, using carbon tetrachloride as a solvent. Absorbance detection limits of about 8.5×10^{-7} were obtained. Yang and Hairrell (54) have used a crossed-beam system with flow injection analysis, using bromophenol blue in ethanol as a test system.

Thermal lens detection has been applied by Leach and Harris (37) to flow injection analysis in supercritical carbon dioxide. The sensitivity of the measurement is about 150 times greater than in carbon tetrachloride. However, flow-induced noise problems are much worse than in the usual liquid solvents. Near the critical point the thermooptic properties of carbon dioxide are changing rapidly with temperature. In effect, the sensitivity of the measurement is too great, and any small change in properties becomes a large change in measured signal. Thus the noise in a flow system measurement is about two orders of magnitude greater than the noise in a static system measurement. With well-controlled flows thermal lens spectroscopy may become the technique of choice for simple absorbance measurements in supercritical fluid chromatography.

REFERENCES

1. D. S. Kliger, Ed., *Ultrasensitive Laser Spectroscopy*, Academic, New York, 1983.
2. Y.-H. Pao, Ed., *Opto-Acoustic Spectroscopy and Detection*, Academic, New York, 1977.
3. A. Rosenwaig, *Photoacoustics and Photoacoustic Spectroscopy*, Wiley-Interscience, New York, 1980.
4. C. K. N. Patel and A. C. Tam, *Rev. Mod. Phys.*, **53**, 517 (1981).
5. J. R. Whinnery, *Acc. Chem. Res.*, **7**, 225 (1974).
6. J. M. Harris and N. T. Dovichi, *Anal. Chem.*, **52**, 695a (1980).
7. W. Demtröder, *Laser Spectroscopy*, Springer-Verlag, New York, 1982.

8. K. Thayagarajan and A. K. Ghatak, *Lasers: Theory and Applications,* Pergamon, New York, 1981.

9. A. Yariv, *Introduction to Optical Electronics,* 2nd ed., Holt, Rinehart and Winston, New York, 1976.

10. J. P. Gordon, R. C. C. Leite, R. S. Moore, S. P. S. Porto, and J. R. Whinnery, *J. Appl. Phys.,* **36,** 3 (1965).

11. C. Hu and J. R. Whinnery, *Appl. Opt.,* **12,** 72 (1973).

12. M. Abramowitz and I. A. Stegun, Ed. *Handbook of Mathematical Functions,* U.S. Government Printing Office, Washington, 1964, p. 228.

13. S. J. Sheldon, L. V. Knight, and J. M. Thorne, *Appl. Opt.,* **21,** 1663 (1982).

14. C. A. Carter and J. M. Harris, *Appl. Opt.,* **23,** 476 (1984).

15. R. L. Swofford and J. A. Morrell, *J. Appl. Phys.,* **49,** 3667 (1978).

16. C. A. Carter and J. M. Harris, *Anal. Chem.,* **55,** 1256 (1983).

17. H. L. Fang and R. L. Swofford, in D. S. Kliger, Ed., *Ultrasensitive Laser Spectroscopy,* Academic, New York, 1983, p. 176.

18. A. J. Twaroski and D. S. Kliger, *Chem. Phys.,* **20,** 253 (1977).

19. C. A. Carter and J. M. Harris, *Appl. Spectrosc.,* **37,** 166 (1983).

20. G. R. Daigneault and M. D. Morris, *Appl. Spectrosc.,* **35,** 591 (1981).

21. A. Yariv, *Quantum Electronics,* 2nd ed., Wiley, New York, 1975.

22. C. A. Carter and J. M. Harris, *Anal. Chem.,* **56,** 922 (1984).

23. Y. Kohanzadeh, J. R. Whinnery, and M. M. Carroll, *J. Acoust. Soc. Am.,* **57,** 67 (1975).

24. A. C. Tam, in D. S. Kliger, Ed., *Ultrasensitive Laser Spectroscopy,* Academic, New York, 1983, pp. 1–108.

25. J. R. Whinnery, D. T. Miller, and F. Dabby, *IEEE J. Quantum Electron.,* **QE-3,** 382 (1967).

26. C. E. Buffett and M. D. Morris, *Appl. Spectrosc.,* **32,** 455 (1983).

27. I. R. Epstein, M. Morgan, C. Steel, and O. Valdes-Aguilera, *J. Phys. Chem.,* **87,** 3955 (1983).

28. J. P. LaPlante, J. C. Micheau, and M. Giminez, *J. Phys. Chem.,* **88,** 4135 (1984).

29. M. Gimenez, J. C. Micheau, D. Lavabre, and J. P. LaPlante, *J. Phys. Chem.,* **89,** 1 (1985).

30. R. Anthore, P. Flament, G. Gouesbet, M. Rhazi, and M. E. Weill, *Appl. Opt.,* **21,** 2 (1982).

31. M. E. Weill, M. B. Rhazi, and G. Gouesbet, *C. R. Acad. Sci. Paris,* Ser. II, **294,** 567 (1982).

32. J. Oda and T. Sawada, *Anal. Chem.,* **53,** 471 (1981).

33. N. J. Dovichi and J. M. Harris, *Anal. Chem.,* **53,** 689 (1981).

34. E. Voigtman, A. Jurgensen, and J. D. Winefordner, *Anal. Chem.,* **53,** 1921 (1981).

35. C. E. Buffett and M. D. Morris, *Anal. Chem.,* **55,** 376 (1983).

36. T.-K. J. Pang, unpublished observations.

37. R. A. Leach and J. M. Harris, *Anal. Chem.*, **56**, 2801 (1984).

38. J. M. Betz and J. G. Nikelly, *J. Chromatogr. Sci.*, **21**, 478 (1983).

39. T.-K. J. Pang and M. D. Morris, *Anal. Chem.*, **57**, 2153 (1985).

40. E. Voigtman and J. D. Winefordner, *J. Liq. Chromatogr.*, **5**, 2113 (1982).

41. E. J. Diebold and R. N. Zare, *Science*, **196**, 1439 (1977).

42. E. P. C. Lai, S. Y. Su, E. Voigtman, and J. D. Winefordner, *Chromatographia*, **15**, 645 (1982).

43. R. A. Leach and J. M. Harris, *J. Chromatogr.*, **218**, 15 (1981).

44. C. E. Buffett and M. D. Morris, *Anal. Chem.*, **54**, 1821 (1982).

45. C. A. Carter and J. M. Harris, *Anal. Chem.*, **55**, 1256 (1981).

46. R. A. Leach and J. M. Harris, *Anal. Chim. Acta*, **164**, 91 (1984).

47. T.-K. J. Pang and M. D. Morris, *Anal. Chem.*, **56**, 1467 (1984).

48. T.-K. J. Pang and M. D. Morris, *Appl. Spectrosc.*, **39**, 90 (1985).

49. Y. Yang, *Anal. Chem.*, **56**, 2336 (1984).

50. M. J. Sepaniak, J. D. Vargo, C. N. Kettler, and M. P. Muskarinec, *Anal. Chem.*, **56**, 1252 (1984).

51. S. D. Woodruff and E. S. Yeung, *Anal. Chem.*, **54**, 1174 (1982).

52. D. R. Bobbitt and E. S. Yeung, *Anal. Chem.*, **57**, 271 (1985).

53. R. A. Leach and J. M. Harris, *Anal. Chim. Acta*, **164**, 91 (1984).

54. Y. Yang and R. E. Hairrell, *Anal. Chem.*, **56**, 3002 (1984).

CHAPTER

5

FLUOROMETRIC DETECTION

MICHAEL J. SEPANIAK and CHARLES N. KETTLER

Department of Chemistry, University of Tennessee, Knoxville, Tennessee

1. INTRODUCTION

The versatility and exceptional efficiencies afforded by modern LC columns have made LC indispensable for routine chemical analyses and for many specialized separations. Despite rapid advances in column technology (e.g., columns packed with very small diameter, bonded-phase particles, and open capillary columns), the use of LC has been slowed by the absence of tunably selective and highly sensitive detectors. If the sample being analyzed contains naturally fluorescent compounds, or compounds that can be derivatized to a fluorescent form, fluorescence detection can provide the selectivity and excellent sensitivity needed to complement the separation power of modern LC columns.

We begin this chapter by briefly discussing some theoretical concepts behind fluorometry. More complete theoretical and practical descriptions of the technique can be found in several monographs on the subject (1–4). A basic understanding of these concepts is not essential to the routine use of commercial fluorometric LC detectors for established fluorometric analyses. However, the chromatographer often is placed in unique situations, such as building a fluorometric LC detector from instrumentation available in the laboratory or improving the analytical characteristics of fluorometry by employing state-of-the-art equipment, for which an understanding of these fundamental concepts is important.

Subsequently, we describe the basic instrumentation and analytical methodologies that are important in fluorometric LC detection. Sections describing routine and modern research uses of fluorometry in LC are provided to illustrate the scope of applications of this technique.

Throughout these discussions we emphasize topics in fluorometry that require special attention because of their use in LC detection. These topics include the effect of solvent flow on fluorescence signal recovery, the compatibility of typical solvents used in LC and fluorometry, and the design of sample (flow) cells that maintain chromatographic efficiency while providing high detection sensitivity.

148

2. BASIC THEORETICAL CONCEPTS

2.1. Electronic States

The fluorescence process involves the spin-allowed emission of electromagnetic radiation by a chemical species following promotion to an excited electronic state. Fluorescence excitation in molecules is accomplished through the absorption of a quantum of radiation and involves changes in electronic structure. A basic understanding of electronic structure and electronic states of organic molecules is important in predicting what sorts of molecules will fluoresce efficiently and what experimentally controllable parameters in LC influence the analytical characteristics of fluorometric detection.

The atoms that make up a molecule form chemical bonds when electrons occupy bonding molecular orbitals that are formed via the overlap of atomic orbitals. The geometry of the molecular orbital is often used for classification purposes. A σ bond is formed when a pair of electrons occupies a molecular orbital that is symmetrically arranged around an axis joining the nuclei of the atoms forming the bond. Because the electronic charge is localized between the positively charged nuclei, the electrons are very tightly bound by the molecule. The energy required to promote σ electrons to higher-energy, vacant orbitals is generally too large for them to be involved in the fluorescence excitation process.

π Bonds are formed when electrons occupy bonding molecular orbitals that are perpendicularly arranged about the internuclear axis. The electrons involved in these bonds are less tightly bound, and in the case of aromatic molecules are thought to be delocalized. These π electrons can be promoted to vacant orbitals through the absorption of near UV–VIS radiation and are very important in the fluorescence excitation and emission processes. Some molecules contain atoms (e.g., nitrogen) that have paired valence shell electrons not directly involved in the formation of chemical bonds. These so-called nonbonding, n, electrons can also be promoted to higher orbitals through the absorption of near UV–VIS radiation.

In addition to these bonding and nonbonding orbitals, each molecule contains a group of higher-energy, generally unoccupied, antibonding orbitals. If all the electrons in a molecule are in their lowest available energy orbitals the molecule is said to be in its ground electronic state. When an electron is promoted to an unoccupied orbital the molecule exists in an unstable transient condition referred to as an excited electronic state. Excited electronic states can be classified as singlet or triplet states according to the spin pairing of the electrons of the molecule. A singlet state is one in which all the electrons in a molecule have a paired electron with opposite spin; a triplet state exists when two unpaired electrons have the same spin.

2.2. Electron Transitions

The ground and important excited electronic states of a typical organic molecule are graphically depicted in Figure 1. An asterisk by an orbital indicates that it is antibonding. The excited states are represented using the symbols for the orbitals that are occupied by a single electron (the electrons are depicted with an arrow, the direction of which indicates its spin) and the distinction between singlet and triplet states is made using a superscript. It is evident that for a typical organic molecule there is one ground electronic state, which is usually a singlet, and several excited states corresponding to $n \rightarrow \pi^*$ and $\pi \rightarrow \pi^*$ one electron orbital transitions. The probabilities for radiative transitions depend on factors including orbital overlap, molecular symmetry, and multiplicity (5). These factors result in typical molar absorptivity ranges for transitions to $^1(n\pi^*)$, $^1(\pi\pi^*)$, and $^3(\pi\pi^*)$ excited states of 10^1–10^3, 10^3–10^5, and 10^{-3}–10^{-1}, respectively. Because of poor orbital overlap between highly localized nonbonding (n) orbitals and π^* orbitals the values for molar absorptivities for $n \rightarrow \pi^*$ transitions are not as large as for $\pi \rightarrow \pi^*$ transitions, which have better orbital overlaps. Radiative transitions involving electron spin (multiplicity) changes are quantum mechanically forbidden and hence have very small molar absorptivities. Because

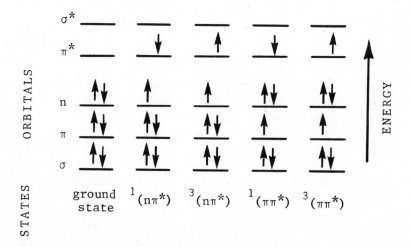

Figure 1. Electronic configuration of the various transitions involved in the formation of excited states.

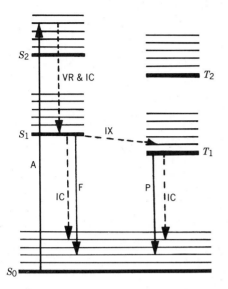

Figure 2. Diagram of luminescence pathways; heavy lines denote ground vibrational levels in an electronic state and lighter lines indicate higher vibrational levels.

the first step in the overall fluorescence process is excitation via the absorption of radiation, it is clear that the nature of the excited state(s) involved plays an important role in determining the magnitude of the fluorescence signal.

Actually in an organic molecule there are many excited states with energies above the ground state. These states are often represented as in Figure 2 using S_0, S_1, S_2, and so on for the ground and excited singlet states and T_1, T_2, and so forth for the excited triplet states which, by Hund's rule, are lower in energy than the corresponding singlet states. Associated with each of these electronic states is a manifold of vibrational levels. At room temperature absorption (A) originates in the lowest vibrational level of the ground electronic state and terminates in some vibrational level of an excited state. For example, the absorption process shown in Figure 2 results in the promotion of a molecule to the fourth excited vibrational level of its second excited singlet state (S_2). Spectral congestion, which is caused by the existence of many closely spaced vibrational (and rotational) levels and many energy-altering, solvent-related microenvironments, produces the broad featureless absorption spectra characteristic of condensed phase spectroscopy.

There are several sequential and/or competing processes that serve to bring the excited molecule rapidly back to its ground state. The first of these processes, vibrational relaxation (VR), brings the molecule from an excited vibrational level in the excited electronic state to the lowest vibrational level of that excited state. The time frame for this relaxation is roughly equivalent to a few molecular vibrations or 10^{-14}–10^{-12}s. Once in the lowest vibrational level of the excited electronic state the molecule is in a more stable condition and further relaxation generally proceeds more slowly. If this state is not the lowest excited electronic state then the proximity of lower-energy excited electronic states, and the overlap in energy of upper vibrational levels of these lower excited states with the lowest vibrational level of the higher excited electronic state, results in efficient relaxation via a nonradiative process termed internal conversion (IC). Once vibrational relaxation and internal conversion have brought the excited molecule to the lowest vibrational level of the lowest excited electronic singlet state (S_1), generally within picoseconds, slower (less efficient) radiative processes became competitive modes of relaxation.

Three competing processes, radiative fluorescence emission (F), nonradiative intersystem crossing (IX) to a triplet state, and nonradiative internal conversion, are most important in the relaxation of molecules in their S_1 state. Once a molecule has intersystem crossed to the triplet manifold it can emit radiation, phosphorescence (P), but this involves a multiplicity change that is quantum mechanically forbidden, and consequently phosphorescence seldom competes well with internal conversion as a means of depopulating the triplet state. Fluorescence emission is the dominant radiative mode of relaxation in organic molecules. Furthermore, because of very rapid nonradiative relaxation, fluorescence is generally observed only from the lowest vibrational level of the lowest excited singlet state, even if highly excited electronic states are initially populated via the absorption of radiation.

The quantum efficiency Φ_F of fluorescence emission is expressed

$$\Phi_F = \frac{k_F}{k_F + \Sigma\, k_d} \tag{1}$$

where k_F is the rate constant for fluorescence emission and $\Sigma\, k_d$ is the sum of the rate constants for all the nonradiative processes that can depopulate S_1. These rate constants include those for internal conversion (k_{IC}) and intersystem crossing (k_{IX}) as well as rate constants for quenching and photochemical reactions.

If k_F is much larger than the other rate constants appearing in Eq. 1, Φ_F approaches its upper limit of 1. The values of k_F for $\pi\pi^*$ and $n\pi^*$ lowest excited singlet states are in the ranges of 10^7–10^9 and 10^5–10^7, respectively. The relatively small values of k_F for molecules with $n\pi^*$ lowest excited singlet states and the small singlet–triplet energy differences for these molecules, which result in relatively large k_{IX} values, account for the poor fluorescence of molecules with $n\pi^*$ lowest excited singlet states.

2.3 Structural and Environmental Effects

The molecular structure and environment of an analyte determine the values of the rate constants discussed above, and hence determine if fluorometric detection is fascile. Except when fluorescence derivatization techniques (see Section 5.2) are employed, the chromatographer has little control over the molecular structures of the analytes in a sample. However, it is important in choosing a suitable detector for a particular analysis to be able to predict, based on molecular structure, if the analytes will have reasonably large values of Φ_F. It is not always possible or expedient to consult the literature when making such predictions. Consequently, a few general guidelines are provided to aid in deciding on the feasibility of fluorometric detection.

The low-energy $\pi\pi^*$ states that result in large values of Φ_F are generally found in highly conjugated organic molecules. Most unsubstituted aromatic molecules are at least moderately fluorescent and their fluorescence is enhanced by having a rigid, planar structure. The restricted motion or lack of flexibility in rigid molecules reduces the probability of nonradiative relaxation. The effect of increasing the number of fused rings in an unsaturated aromatic molecule is to increase fluorescence intensity and to shift the excitation and emission spectral bands to longer wavelengths.

The intensity of fluorescence in heterocyclic aromatic molecules is determined largely by whether S_1 is a $n\pi^*$ or $\pi\pi^*$ state. For example, pyridine, a six-member ring with one nitrogen, is not fluorescent but the addition of a fused benzene ring in the formation of quinoline results in a weakly fluorescent molecule. Presumably the lowest excited singlet state in pyridine is a $n\pi^*$ state, whereas in quinoline the extensive π electron system lowers the energy of the $\pi\pi^*$ state and, depending on the solvent employed (see below), the lowest or nearly the lowest excited state in the molecule is a $\pi\pi^*$ state.

For substituted aromatic molecules, substituents that are electron donating (e.g., $-NH_2$ and $-OH$) enhance fluorescence, whereas electron-withdrawing substituents (e.g., $-NO_2$ and $-CO_2H$) decrease fluorescence. Groups such as $-SO_3H$, which do not interact strongly with the π bonding electrons of an aromatic molecule, have relatively little effect on fluorescence intensities. Heavy atom substituents such as $-Br$ and $-I$ tend to diminish fluorescence by increasing k_{IX}. The fluorescence of heterocyclic aromatic molecules is especially sensitive to substitution. For instance, although pyridine does not fluoresce, 3-hydroxypyridine in its nonionized form is moderately fluorescent.

Unlike molecular structure, the chromatographer has considerable control over the molecular environment of a fluorescing analyte. Temperature and, much more importantly, mobile phase composition are parameters that determine the molecular environment of the analyte and should be considered in terms of their effects on the separation and on the analytical characteristics of fluorometric detection. Enthalpy changes associated with phase transfer in LC are much smaller than in GC and this makes temperature adjustment a relatively ineffective means of adjusting distribution coefficients in LC. Nevertheless, separations are occasionally performed at elevated temperatures in order to effect a small change in column selectivity or to improve the mass transfer characteristics of the chromatography. Temperatures are generally not maintained above 80°C because of the thermal instability of commonly used bonded phase packings. Nonradiative relaxation, which competes with fluorescence, depends on molecular collisional processes. As the temperature is increased molecular motion and collisional frequency increase. A rough general rule is that fluorescence intensity decreases approximately 1% for every 1°C increase in temperature.

The viscosity, polarity, and pH of the mobile phase are parameters that can influence fluorescence. Control of these parameters is often a compromise between optimum separation conditions and optimum detection conditions. For example, high viscosity solvents are generally desirable in fluorometry because they restrict molecular motion and decrease the rate constants for nonradiative relaxation processes. In contrast, high-viscosity mobile phases generally have poor mass transfer characteristics and lead to large column pressure drops.

Molecules in excited electronic states generally possess molecular geometries and electronic charge distributions that are substantially different than in the ground state. In most instances the excited state is more polar than the ground state and an increase in solvent polarity produces a greater stabilization of the excited state. The result is a shift in both excitation and fluorescence emission spectra to longer wavelengths. This "red shift" is general and its magnitude can depend on the electronic states involved. The difference in ground state–excited state stabilization is generally greater for $\pi\pi^*$ excited states than for $n\pi^*$ excited states and, in fact, it is sometimes possible to exchange lowest excited singlet states by changing solvents. This effect is observed in quinoline, which has a fluorescence quantum efficiency that is 1000 times greater in water than in benzene, owing to the existence of a $\pi\pi^*$ lowest excited singlet state in the highly polar water. Thus it would probably be desirable to use reverse-phase LC, which employs highly polar mobile phases, in the separation and fluorometric detection of heterocyclic aromatic molecules such as quinoline. However, the chromatographer should be careful not to overgeneralize because there are many exceptions to the trends cited in this section.

The fluorescence of many aromatic molecules that contain acid and base functionalities is sensitive to the pH and hydrogen-bonding ability of the mobile phase. For example, there are significant shifts in the fluorescence spectral bands and fluorescence intensities of vitamin B_6 (pyridoxine) and some of its metabolites as pH is varied. The ionic forms of these compounds exhibit the largest fluorescence intensities. In our work, postchromatographic column pH adjustment proved an effective means of maximizing the fluorescence signals of these compounds when fixed-wavelength laser excitation was employed (6). With regard to the hydrogen-bonding ability of the mobile phase, it is generally observed that fluorescence intensities decrease when solvents that can hydrogen bond with the analyte are employed. For example, aqueous solvents have been reported to quench the fluorescence of certain hydrogen bonding compounds including indoles and catecholamines (7,8).

The presence of solutes in the mobile phase can also dramatically influence fluorescence intensity. The fluorescence quenching of an analyte by other solutes is quite common. The proposed mechanisms for this quenching are varied and depend on the nature of the analyte, solute, and solvent. Two types of quenchers that are frequently present in LC mobile phases are oxygen and heavy metal or halogen ions. Dissolved oxygen tends to quench the fluorescence of many organic molecules by promoting intersystem crossing. The dissolution of dissolved gases at the detector is a common occurrence in LC and has prompted many chromatographers to "degas" their mobile phases before use. This practice is especially important when fluorometric detection is to be employed for compounds that have been reported to exhibit oxygen quenching. The buffer ions and counter ions often incorporated in mobile phases to influence the retention of a solute can also result in fluorescence quenching. For example, we have employed cetyltrimethylammonium salts in mobile phases to modify capillary columns dynamically and to act as ion pairing reagents (9). It was observed that the fluorescence peak heights for certain NBD–Cl derivatives of alkylamines were as much as 70 times larger for the chloride salt than for the bromide salt. Presumably the heavier bromide ions in the mobile phase more efficiently promote intersystem crossing. Thus it is not necessary for the heavy atom to be a substituent of an organic molecule in order to exhibit quenching effects.

Because the fluorescence of an analyte in solution depends on many dynamic molecular processes that can be influenced by the environment of the analyte, it is clear that the choice of mobile phase is particularly important when fluorometric detection is employed in LC.

2.4. Signal Intensity

The signal intensity in fluorometry (I_f) is given by the product of the intensity of the radiation incident on the sample (I_0), the fraction of I_0 that is absorbed, the quantum efficiency for fluorescence, and the instrumental efficiency for collecting the fluorescence emission (k), which is radiated in all directions. Using Beer's law to determine the fraction of light absorbed, the working expression given in Eq. 2 is obtained:

$$I_f = I_0 (1 - e^{\varepsilon bC})\Phi_F k \tag{2}$$

where ε, b, and c are molar absorptivity, path length, and concentration, respectively. When sample absorbance is small this expression reduces to

$$I_f = I_0 2.3 \varepsilon b C \Phi_F k \tag{3}$$

which shows a linear relationship between the measured signal intensity and the concentration of the sample. The term "fluorescence advantage" is often used to describe the direct proportionality between signal intensity and the intensity of the incident radiation.

3. BASIC FLUORESCENCE INSTRUMENTATION

The most effective use of fluorescence detection in LC requires the working chromatographer to be familiar with the basic design of a fluorescence detector. A block diagram in Fig. 3 illustrates the basic components of a fluorescence detector. A discussion of each component is provided in order to clarify the parameters of importance for each element in the figure.

3.1. Components

3.1.1. Excitation Sources

When considering an excitation source for fluorescence detection there are three important factors to be evaluated: source intensity, wavelength of the emitted radiation, and lamp stability. Fluorescence depends linearly on the intensity of the exciting source and therefore it is usually advisable to use as intense a source as possible. This is especially true when sensitivity is a major concern. Deuterium and xenon gas discharge lamps are often employed as continuum sources for LC fluorescence detection.

The desired excitation wavelengths are selected by a monochromator or filter. The xenon arc lamp is employed in many of the commercially available LC fluorescence detectors. This lamp provides continuous emission in the 250–1300-nm range, with the most intense emission in the range of approximately 300–500 nm. Xenon lamps do not exhibit the stability of deuterium lamps and require well regulated, externally cooled power supplies for stable operation. Typical power output is 75–150 W with the 150 W lamp being employed in many commercial LC detectors.

Deuterium lamps generally provide less output power than xenon lamps; however, the emission range is smaller and greater intensity is available in the UV region. Because most analytes absorb most strongly in the region below 300 nm the deuterium lamp is often the excitation source of choice. The concomitant

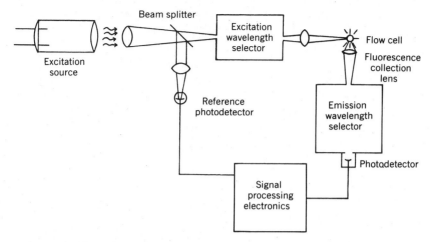

Figure 3. Block diagram of a basic fluorescence detector for LC.

disadvantage of the deuterium lamp is that its restricted output range limits its range of applications.

Mercury, cadmium, and zinc lamps are also employed for excitation in fluorescence detection. These sources offer high-intensity narrow-band outputs in the UV and near UV. The wavelengths available from these lamps are mercury (254, 365, 405 nm), cadmium (229, 326 nm) and zinc (214, 308, 335 nm). These sources are ideal for fluorescence detection of analytes that have an absorption maximum corresponding to the output of the lamp, thus providing some measure of selectivity and good sensitivity for the compound of interest. Lasers offer several advantages over conventional excitation sources for fluorometric detection. Their specialized use in LC detection is covered in Section 6.1.

3.1.2. Excitation Wavelength Selection

The two common methods to isolate specific wavelengths emitted by the source for fluorescence excitation involve the use of filters or monochromators. Filters are low in cost and are simple to use; however, there is generally a reduction in selectivity relative to wavelength selection using a monochromator because the filters usually pass a wider range of wavelengths. Short-wavelength cutoff filters pass light at wavelengths lower than their rated value. The rated value is the wavelength at which 50% of peak transmission occurs. The transmission of short-wavelength cutoff filters is typically >65%; however, this drops off rapidly as the wavelength decreases. Narrow band-pass interference filters can be employed to isolate specific bands from sources such as mercury, cadmium, or zinc lamps; however, the transmission of these filters in the UV region is 20% or less. Although filters are simple to use, if a change in excitation wavelength is desired, it is necessary to change filters, which is not as convenient as simply changing a monochromator setting.

Grating monochromators are often employed in order to provide a greater range of wavelength selection, smaller spectral band pass, and the ability to scan the excitation spectrum. The dispersive element in the monochromator is usually a ruled grating, although some monochromators employ holographic gratings. Gratings obey the Bragg equation (10) and therefore pass higher orders of radiation in addition to the first order. These higher orders are simply integer fractions of the wavelength of the first order. For example, if the monochromator is set to pass 500-nm radiation in the first order, 250-nm radiation will also be passed by the monochromator in the second order.

A concern for both filters and monochromators used for fluorescence excitation is their ability to transmit the incident radiation as defined by their band pass and reject light outside the band pass. In each case the transmittance is wavelength dependent. If narrow-band interference filters with band passes of approximately 10 nm are employed at wavelengths greater than 400 nm, the maximum transmission at the center of the band pass is typically about 50% and transmission outside the stated band pass (stray light rejection value) is typically about 0.01%. For wavelengths less than 400 nm the transmission is 20% or less whereas transmission outside the band pass remains the same at approximately 0.01%.

In the case of a monochromator the transmission efficiency maximum can be as high as 80%. The distribution of light into various orders can be changed by modifying the angle of the facets of the grating. Gratings are often "blazed" for greater efficiency in certain spectral regions (11). The dispersion efficiency is highest at the blaze wavelength and falls off on either side of this wavelength. The stray light rejection capabilities of most small monochromators is equal to or better than that observed with narrow-band interference filters.

3.1.3. Emission Wavelength Selection

The isolation of fluorescence emission involves a trade-off between selectivity and sensitivity of the fluorescence measurement. Monochromators provide good wavelength selection with the possibility of very narrow band pass; however, narrow band passes reduce the total fluorescence recovered. The emission is often very broad band in nature and the narrow band pass samples only a small part of the total emission. The band pass of the monochromator is determined by the product of the reciprocal linear dispersion and the slit width. The reciprocal linear dispersion of the monochromator, usually given in units of nanometer per millimeter, is a set value, but the slit width can be easily changed by the operator. If sensitivity is not a problem and greater selectivity is desired, then it is sometimes possible to improve selectivity by reducing the slit width. This can be advantageous when it is necessary to eliminate or decrease fluorescence from coeluting compounds or impurities. A typical band pass for the small monochromators used in LC detection is less than 20 nm and, although there are filters available with small band passes, their employment is prohibitively expensive because it requires many filters to cover the same spectral range as a monochromator.

Long-wavelength cutoff filters pass wavelengths greater than their rated value. Typical transmission for long-wavelength cutoff filters is 80%. These filters are often used to remove specular, Raman, and Rayleigh scatter and isolate the fluorescence of analytes over a very large spectral region. Consequently selectivity is poor, but one filter can be used for many analytes. Although these filters are very effective at blocking scattered excitation radiation, they are usually made from absorbing colored glasses that tend to fluoresce and therefore can contribute to background.

Long-wavelength filters can also be used to reject unwanted radiation from higher grating orders, which can contribute to the background. For example, if the excitation wavelength is at 250 nm and the emission monochromator is set to isolate fluorescence emission at 500 nm, then scattered excitation radiation will pass the emission monochromator in the second order. A long wavelength cutoff filter can be used in conjunction with the emission monochromator to eliminate this higher-order radiation problem.

Narrow-band interference filters can be employed to isolate regions of the spectrum with bandwidths of 10–70 nm. The stated bandwidths are usually for the points where transmission has fallen to one-half the peak transmission. Although these filters can eliminate both long- and short-wavelength intereferences, the transmission is usually less than 50% and the small band pass makes these filters generally less convenient than monochromators for isolating fluorescence emission.

3.1.4. Flow Cells

The point at which LC and fluorescence detection truly come together is at the flow cell. The peaks eluting from the chromatographic column must pass through a suitable cell that will allow appropriate illumination for fluorescence excitation and subsequent fluorescence signal recovery. The cell must be designed so that there is a minimum loss of chromatographic efficiency and a maximum fluorescence signal.

The general rule is that for negligible solute band dispersion the volume of the flow cell must be no more than 20% of the volume of the solute band as it elutes from the column. Actually, the shape of the flow cell and the flow cell–column connections are equally important in minimizing dispersion. The contributions of the column, injector, and detector to the total solute band variance (σ^2_{total}) are additive as shown in Eq. 4:

$$\sigma^2_{total} = \sigma^2_{column} + \sigma^2_{injector} + \sigma^2_{detector} \tag{4}$$

If the flow cell is a simple round capillary tube, the detector's contribution to the solute band variance ($\sigma^2_{detector}$) is given by

$$\sigma^2_{detector} = \frac{2D_m L}{v} + \frac{d^2 v L}{96 D_m} \tag{5}$$

Where D_m is the solute diffusion coefficient in the mobile phase, v is the mobile phase linear velocity in the flow cell, and d and L are the flow cell diameter and length, respectively. Equation 5 generally underestimates the detector's contribution to band variances because of the dispersive effects of poorly swept volumes at connections to the flow cell or sharp bends within the flow cell.

The geometric design of the flow cell must provide proper "washout." Washout can be considered to be the number of flow cell volumes needed to pass a solute band through the flow cell completely. An ideal washout would be one cell volume, which could be achieved by using a straight piece of tubing having no bends such that laminar flow is maintained. Often it is necessary to construct flow cells that have constraints put upon them by the fluorescence instrument. This is especially true when a flow cell is "added on" to a commercial spectro-fluorometer. The chromatographer must be careful to connect the column to the flow cell with tubing and fittings that do not provide areas that are poorly swept by the flow of the mobile phase. These poorly swept areas are potential mixing areas and can cause solute band dispersion and increase the washout. Most commercial instruments provide specifications for the illuminated volume of the flow cell. For these types of specifications adjacent solute bands within the flow cell must be separated by at least this volume in order to preserve resolution.

The importance of illuminated volume and washout requirements is illustrated in Fig. 4. The geometric design of a particular fluorescence flow cell is shown here. The exciting light enters perpendicular to the plane of the paper and the fluorescence is collected at a right angle to this. Problems can occur at any point of the cell where there is a breakdown of the flow profile. The cross-hatched areas of this figure represent two solute bands. As seen in Fig. 4a, the eluting solute bands are moving toward the illuminated volume and are chromatographically resolved. Figure 4b illustrates the band broadening that has occurred within the flow cell. This broadening is due to poor mobile phase mass transfer (14) and, more importantly, poor washout of the poorly swept areas at the sharp

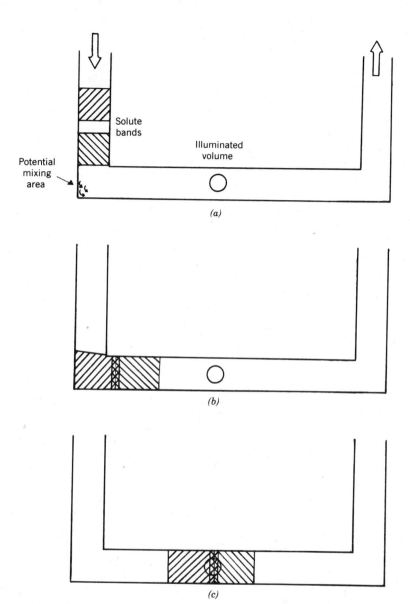

Figure 4. Flow-cell diagram illustrating importance of proper geometric design and illumination volume.

corners of the flow cell. This has caused the bands to overlap and has destroyed the separation. Figure 4c shows the overlapped bands within the flow cell illuminated volume. It is obvious that the band overlap due to broadening will cause the bands to be simultaneously detected within the illuminated volume. If band dispersion due to cell geometry were not present there is still the concern that the illuminated volume would be too large and that there would not be sufficient separation of the bands to be selectively illuminated. When this is the case, a reduction of illuminated volume can minimize this problem.

Lyons and Faulkner have addressed the problem of optimizing detectability for fluorometric LC flow cells (12). They evaluated the contribution to background via scattering of excitation light by reflection and refraction at surfaces within the flow cell (specular scatter) and Rayleigh scatter from the flow cell materials and flow stream. They concluded that employment of a square flow cell with a square bore can reduce the reflected and refracted scatter to an insignificant level. Washout can be a major concern when flow moves from a round to a square tube. This problem is particularly significant in microscale LC where the elution volumes are very small and column efficiency is often vulnerable to small changes in the flow profile. The microscale techniques require miniaturization of current instruments or new ideas in flow cell design (see Section 6.2) to employ fluorescence detection effectively (13).

The problem of Rayleigh scatter from the flow-cell material and flow stream is significant only when the reflected and refracted scatter are truly minimized. Contributions to background due to fluorescence of the flow cell material are best minimized by constructing the flow cell of quartz or fused silica.

3.1.5. Fluorescence Emission Collection

The collection of fluorescence from flow cells generally occurs at right angles to the excitation source so that the collection of scattered light is minimized. Collection in this manner with $f/1$ optics limits the amount of fluorescence collected to approximately 1/16 of that which is available. Collection optics can be aligned such that the fluorescence and Rayleigh and Raman scatter from the solution under illumination can be imaged onto the entrance slit of the emission monochromator, and the specular scatter from the walls of the flow cell can be rejected. This is accomplished by spatially selecting that portion of the image which does not contain the specular scatter from the flow-cell walls. Rejecting the specular scatter will reduce the background and enhance the S/N ratio. Unfortunately the Rayleigh and Raman scatter emanate from the same region of the flow cell as the fluorescence and cannot be spatially rejected and therefore must be dealt with in a different manner.

The collection optics generally do not image the fluorescence emission emitted from various points along the path of the excitation beam within the flow cell with equal efficiencies. If, as an example, the collection optics most efficiently image the fluorescence emitted from the center of the cell, then absorption of excitation radiation by concentrated analyte solutions within the cell can result in a reduction of the radiation reaching the center of the cell. A signal is then observed that is less than that predicted by Eq. 3. This is referred to as the inner filter effect. The influence of the inner filter effect on the linearity of calibration plots are discussed below.

The flow cell employed in the Kratos SF 980 LC fluorometer (Fig. 5) addresses

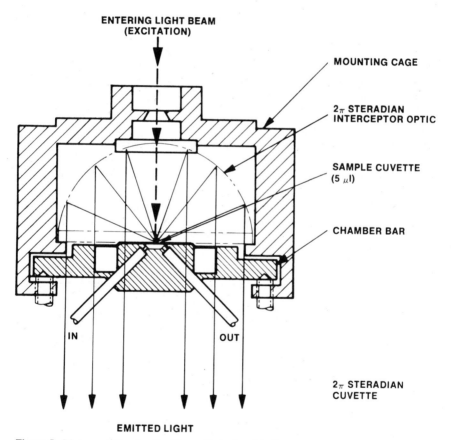

ENTERING LIGHT BEAM
(EXCITATION)

MOUNTING CAGE

2π STERADIAN
INTERCEPTOR OPTIC

SAMPLE CUVETTE
(5 μl)

CHAMBER BAR

IN   OUT

2π STERADIAN
CUVETTE

EMITTED LIGHT

Figure 5. Diagram of 2π steradian flow cell employed in Kratos FS 970 fluorescence detector. Courtesy of Robert Weinberger, Kratos Analytical.

the problems of inner filter effect and low fluorescence collection efficiency. The inner filter effect is minimized by using a flow cell that is rectangular with the wide side of the cell being exposed to the exciting radiation and thus the depth of penetration required to reach the back of the cell is reduced. Fluorescence is collected by employing a concave mirror around the flow cell and making the back of the flow cell reflective. This design increases the collection of fluorescence and scattered radiation. The increased amount of scatter necessitates the use of filters to reduce the amount of this scattered radiation which reaches the photomultiplier tube, thus realizing the advantage of increasing the amount of fluorescence collected. This detector has shown a limit of detection of 0.8 pg of anthracene injected onto a 4.6-mm $\times$ 25-cm C_{18} column measured at a S/N ratio of 3 (15).

3.1.6. Detectors

The intensity of fluorescence emission is typically very low, especially when trace level analyses are being performed and when proper measures have been taken to minimize background. This low light level requires the use of a sensitive detection device. Photomultiplier tubes (PMT) convert radiant energy in the UV, visible, and IR regions of the electromagnetic spectrum into an electrical current (16). The PMT is the most practical and sensitive photodetector available and is used in most commercial fluorescence detectors. Photomultiplier tubes are devices consisting of a photocathode, a series of dynodes, and an anode all contained within a vacuum tube. The photocathode is normally operated at a high negative voltage and each succeeding dynode held at a slightly lower negative voltage. The anode is usually held at ground. When a photon of appropriate energy strikes the photocathode it can liberate a photoelectron that is accelerated toward the first dynode owing to its positive charge with respect to the photocathode. When the photoelectron strikes the dynode it generally releases two or more secondary electrons, which are accelerated down the dynode chain in an analogous manner. After propagating down the dynode chain the pulse initiated by a single photon consists of 10^5–10^7 electrons arriving at the anode. It is this photocurrent that is employed in measuring fluorescence. It is intuitive that excessively large incident light fluxes can damage the PMT, whereas at very low light levels the limiting noise factor is often the dark current. Dark current is the current arising while the PMT is operated without any radiation striking the photocathode and is the result of thermionic emission from the photocathode and dynodes.

The ideal characteristics of a fluorescence photodetector are high sensitivity with good S/N ratio, linearity of response, low dark current, and rapid response times. Photomultiplier tubes, as the name implies, provide an amplification of the incoming signal. This signal amplification provides the sensitivity necessary for acceptable measurements of very low fluorescence intensities. The selection of the proper PMT for each given application is very important. The spectral response of a PMT is largely determined by the composition of the photocathode. Photomultiplier tubes can be obtained with good spectral sensitivity in the 200–900 nm range; however broad-range PMTs are very expensive in comparison to PMTs with limited spectral range. If the wavelength of maximum fluorescence emission of the analyte of interest is known,then it is best to select a PMT with good spectral sensitivity in that region.

When the application requires emission spectra to be obtained over a large region, then it is important to realize that the difference in response of the PMT over different regions of the spectrum will be reflected in the emission spectrum. Correction for the wavelength variance of the detector response, as well as for excitation source intensity variations, is necessary in order to obtain "corrected" spectra. Comparison of fluorescence spectra of the same compound obtained with different instruments is invalid unless the spectra are corrected for both source and detector wavelength variance (17).

The photocurrent from the PMT is often passed through electronic low-pass noise filters. These filters attenuate high-frequency signals and pass the low-frequency signals. They are capable of rejecting a great deal of noise with minimal distortion of the analytical information.

A simple noise filter is characterized by its time constant. The time constant is the time required for the recorder to reach 63.2% of its final value. A noise filter will reduce the amplitude of the noise inversely with the square root of the time constant. For example, to reduce the noise by a factor of 2 the time constant must be increased fourfold. The chromatographer must be aware that excessively long time constants can have a serious effect on the recorded chromatographic peak shapes. Peaks become asymmetric when excessively long time constants are used. The peak maximum is shifted approximately by the same amount as the time constant and the peak height is reduced. However, the integrated peak area is not affected. In order to reduce base-line noise to a minimum with negligible peak distortion the time constant should be on the order of 10% of the peak width (in time units) of the narrowest peak in the chromatogram.

The photocurrent resulting from the detection of an eluting solute band is usually processed in one of three ways. The most conventional method is to measure the DC signal current. This can be done by measuring the voltage drop across a load resistor with a voltmeter or by employing an operational amplifier circuit.

When scanning of the fluorescence emission spectrum is desired, often the ratio method is employed. A portion of the exciting light is split off and measured by a second photodetector. Any source fluctuations arising from wavelength changes are monitored by this detector and the resulting signal is fed into a photomultiplier dynode voltage control circuit where it is used to control the dynode voltage of the detectors. As the excitation energy fluctuates, the resulting fluorescence signal does the same. Taking the ratio of the reference and fluorescence signals provides compensation for source drift and fluctuation.

For situations where the fluorescence emission level is very low, photon counting is often employed (18). If the rate of PMT photoelectron ejection at the cathode and the frequency response of the system is such that individual current pulses arriving at the anode can be resolved, then the number of pulses per unit time can be counted. Because the count rate is directly related to photons arriving at the photocathode, we are "counting" photons.

Photon-counting circuits employ a discriminator circuit that can selectively discriminate between photocathode pulses and pulses that are a result of thermally ejected electron emission of the dynodes, which contributes to the dark current, and signal pulses. The ability to decrease the dark current contributions to background by selecting the pulses that are most probably due to fluorescence photons enhances the S/N ratio, intensifies the sensitivity to low light levels, and provides the opportunity to employ long-term signal integration. Other advantages to photon counting include direct digital processing of the spectral information and greater analytical precision (18). When scattered radiation levels are low, the photon count resulting from the thermal ejection of cathode electrons can be significant. In this case the PMT can be cooled to reduce the background photon count.

Photon counting is often employed when light levels are less than roughly 10^6 photon/s arriving at the photocathode. At fluxes greater than this the typical frequency response of the measurement system is inadequate to resolve incoming pulses and thus conventional DC current measurement is used.

Recently electronic array detectors have been employed for fluorescence spectrometry detection (19,20). Television tube type detectors such as the vidicon and solid-state devices such as the photodiode array have been used. These detectors are typically mounted so that a portion of the emission spectrum falls upon the array after being dispersed by a grating. Because each individual sensing element of the array is monitoring a certain portion of the emission spectrum these devices are often referred to as multichannel detectors.

These multichannel detectors offer some advantages and disadvantages in comparison with PMTs (21). The multichannel detector allows simultaneous measurements over an extended wavelength range whereas a PMT is able to make only a single measurement per unit time. PMTs are somewhat more sensitive although modifications of the arrays do provide near-photon-counting sensitivity. PMTs can be chosen for a particular spectral response whereas multichannel detectors provide good response across the spectrum from 300 to 800 nm with only minimal response in the UV. PMTs are easier to use and lower in cost, but do not offer the convenience of accessing the large amounts of data obtainable with a multichannel detector. The multichannel array detectors provide the opportunity to rapidly scan (full spectrum scan in less than 1 s) the emission spectrum of fluorescent compounds. This is advantageous in LC in that the emission spectrum of eluting solutes can be obtained without stopping flow in order to scan the spectrum mechanically.

Jadmec and co-workers (22) employed a silicon-intensified target vidicon tube to record fluorescence spectra of polycyclic aromatic hydrocarbons separated from a marine diesel fuel oil sample. The vidicon detection allowed them to obtain spectra of compounds not detected with UV absorption detection.

Hershberger and co-workers (23) employed a video detector for the determination of fluorescent compounds in a complex shale oil mixture separated by LC. By simultaneously monitoring selected wavelengths they were able to separate spectrally and quantify both benzo[a]pyrene and benzo[e]pyrene, which were chromatographically unresolved. This capability can enhance the chromatographic process when extremely complex samples are being analyzed.

Phase sensitive and gated detection are two additional detection modes that are employed to enhance S/N ratio (24). A lock-in amplifier can be used to extract a synchronous AC signal from an otherwise noisy detector output. The lock-in accomplishes this by comparing the input signal wave form of interest with a reference wave form. The signal and reference wave forms are typically derived from the same source. In a fluorescence experiment the excitation source is mechanically chopped at a fixed frequency and therefore the fluorescence is modulated at the same frequency. By first amplifying only the signals of the same frequency as the chopper and then comparing the phase of all of the signal components with the chopper reference wave form (most choppers put out a wave form synchronized with the chopping) the resulting signal is demodulated (25). By passing the demodulated signal through a low-pass electronic filter, noise contributions that are of random amplitude, phase, and frequency are significantly reduced and only signals that are synchronous with the chopping are passed. It is important to note that the most significant contribution to background in fluorometric LC detection is scattered light. Because this scattered light is also modulated by the chopping of the excitation source the reduction in this noise is minimal. Diebold and Zare (26) employed a lock-in amplifier to obtain very sensitive fluorescence detection of aflatoxins separated by LC. Using the 325-nm radiation of a He–Cd laser for excitation they were able to obtain a limit of detection of 750 fg. The fluorescence response was linear over three orders of magnitude.

Gated detection usually employs a boxcar detector to retrieve information from a repetitive signal. This instrument, like the lock-in amplifier, employs a reference signal derived from the same source as the wave form of interest, but unlike the lock-in, the boxcar is capable of reducing stray light contributions to background. The boxcar uses the reference signal to determine the sampling time of a sample-and-hold circuit. The boxcar arranges the timing of the opening and closing of the sampling "gate window." The gate window is optimized by setting it to coincide with the signal pulse while excluding as much of the noise-producing background as possible. Thus only the signal pulse and the background occurring during the sampling contribute to the output. For fluorescence measurements the excitation source is modulated by either chopping or employing a pulsed radiation source. The boxcar detector has the capability to gate open at a variable time delay after the excitation source has been turned off. This allows the signal due to fluorescence to be measured after specular, Rayleigh, and Raman scatter have disappeared. Therefore the stray light level is reduced and the primary contribution to background comes from impurity fluorescence and dark current, allowing an enhancement of S/N. In addition to improved detectability, gated detection allows time resolution of the fluorescence measurement and thus fluorescence decay times can be determined and differences in decay times can be exploited to improve selectivity.

Furuta and Otsuki (27) employed a boxcar detector to make sensitive determinations of polycyclic aromatic hydrocarbons separated by LC. The gated detection capability allowed them to lower the background and also allowed them to distinguish between analytes with short fluorescence lifetimes such as anthracene and long lifetimes such as benzo[a]pyrene. With a 25-μL flow cell the detection limit for benzo[a]pyrene was as low as 0.9 pg injected onto a 8-mm × 10-cm RadialpakA column. The linearity of response for this analyte extended over two orders of magnitude.

3.2. Commercial Instrumentation

Fluorescence detectors for HPLC are summarized in Table 1. Reasonable effort was made to contact all known manufacturers of fluorescence detectors in order to obtain product information. The manufacturers listed here did provide product specifications, which are listed as provided by the manufacturers. We have avoided stating the price of any detector so that cost will not be an issue when Table 1 is used in evaluating detectors. When the true needs of the analyst have been determined and the number of instruments fulfilling those needs has been reduced to a realistic number, the price then becomes an issue to be considered by the purchaser.

The instruments are divided into two groups. The first and largest group is made up of detectors that are dedicated LC detectors; that is to say, the instrument is designed with LC detection as the principal function of the instrument. The second group is spectrofluorometers with flow cells. These instruments often offer many options that enhances their performance relative to the typical dedicated LC detector.

Before the working chromatographer can effectively use Table 1, the detector requirements dictated by the types of analyses to be performed must first be

Table 1. HPLC Fluorescence Detectors

Manufacturer	Model	Light Source	Wavelength Separation	Wavelength Ranges Excitation	Wavelength Ranges Emission	Flow-Cell Volume (μL)	Optical System	Comments
Dedicated Detectors								
American Research Products Corp.	Fluoro-Tec	4-W Hg lamp	Filters	254–546 nm	254–850	9, 18	Ratio	Xenon and halogen lamps available, portable unit, S-20 PMT needed for detection out to 850 nm
Beckman Instruments Inc.	157	Quartz halogen lamp	Filters	300–800	200–650	9, 0.6	Direct photometry, no ratio	Hg, Zn, Cd lamps available Postcolumn reactor available
EM Science/ Hitachi	F1000	150-W xenon lamp	Monochromators	220–650	220–730	2, 12, 40	Ratio	Monochromators have 15-hm band pass. Lamp has ozone self-dissociation mechanism
Farrand Optical Co.	System 3	150-W xenon lamp	Monochromators	200–900	200–900	1, 10	Ratio	Scanning spectrofluorometer will provide corrected spectra and can be computer interfaced
	R2	85-W Hg lamp	Filters	Filter dependent	Filter dependent	1, 10, 100	Ratio	Digital display, dual PMT ratio
	A4	85-W Hg lamp	Filters	Filter dependent	Filter dependent	1, 10, 100	Direct photometry	Same as R2 without ratio capability
Gilson International	121	50-W Quartz halogen	Filters	320–800	200–600	9	Direct photometry	PMT protected from saturation by electrical circuit

(_Continued_)

Table 1. HPLC Fluorescence Detectors—*(Continued)*

Manufacturer	Model	Light Source	Wavelength Separation	Wavelength Ranges		Flow-Cell Volume (µL)	Optical System	Comments
				Excitation	Emission			
IBM	LC/9524	360-nm lamp standard	Filters	Filter dependent	Filter dependent	30	Ratio	Full range of filters available
ISCO	UAS	Phosphor-coated Hg lamp	Filters	310–390	430–620	As small as 0.25	Ratio	
JASCO	FP 115	Phosphore-coated Hg lamp	Holographic grating	310–390	440–650	19	Ratio	
	FP 210	75-W xenon lamp	Holographic grating	240–650	240–650	15, 6.5	Ratio	Manual wavelength selection
Kratos Analytical	FA 950	Many available	Filters	214–545	300–650	20	Direct photometry	28-µl total volume flow cell can hold standard cuvettes for static volume fluorimetry
	SF 970	D_2 for UV	Excitation monochromators, emission filters	190–700	300–850	1, 5	Direct photometry	Novel 2π steradian emission collection
McPherson	FL 748	Many available	Filter	214–440	300–650	20, 5, 1.5	Direct photometry	
	FL 749	150 W Xe	Holographic grating	0–800	0–800	24, 6	Direct photometry	D_2, Xe–Hg lamps available, A/D converter available. Can accept small cuvettes for static measurements

Manufacturer	Model	Source	Optics	Excitation	Emission		Detector	Features
Perkin Elmer	LC-10	360-nm lamp standard	Filters	340–380	418–700	30	Ratio	Additonal lamps and filters available
	Tri-Det	Hg lamp	Filters	254	>280	2.4	Direct photometry	3-in-1 detector includes absorption and conductivity
	LS-1	Pulsed xenon lamp	Filters	260–650	390–700	7	Ratio	Low-cost detector for repetitive assays
	LS-4	Pulsed xenon lamp	Monochromators	200–720	200–800	3	Ratio	Stop-flow scanning corrected excitation spectra
Shimadzu	RF 530	5-W Hg	Monochromator	240–650	240–650	12		Digital readout
Waters	420 AC	4-W fluorescent	Filters	325–425	254 > 530	8		
JASCO	FP 550	15-W xenon lamp	Monochromators	220–850	220–850	15	Ratio	Scanning spectrofluorometer. Many accessories available
Kontron Analysis	SFM 25	High-pressure xenon	Holographic grating	200–800	200–800	15	Ratio	Keyboard control, full scanning, RS232C interface available
McPherson	FL 750/B	150-W xenon lamp	Monochromators	0–800	0–800	24, 6, 2.5	Ratio	Photon counting, scanning diode array, laser adaptable, TLC scanning, A/D options
Tracor Northern	TN 6053	150-W xenon lamp	Excitation monochromators, emission grating	200–800	Array Dependent	50	Ratio	Self-scanning photodiode array rapid-scan capability

determined. The needs of a quality-control analyst performing routine analysis of a three-component mixture are different from the demands of the environmental analyst doing trace-level analyses of complex samples, perhaps with a microbore LC system, who requires qualitative as well as quantitative data.

Careful consideration of the fluorescence properties of the analytes being detected, the limits of detection required, and sample complexity will assist the analyst in selecting the appropriate detector components. Table 1 attempts to provide the reader with an idea of the general makeup of commercially available detectors. The wavelengths quoted are in nanometers and the flow-cell volumes are usually the illuminated volume of the flow cell. For all the filter detectors offered, a full set of excitation and emission filters is normally available, as well as various options for excitation sources. The PMT tube is employed as the detector for nearly every detector listed. The analyst must keep in mind the spectral efficiency of PMT and select the one that offers the best response for the analyte of his interest. Different PMTs are available for nearly every detector.

Other specifications not listed here that may be of interest include dynamic range, sensitivity (often expressed as the minimum number of moles of quinine sulfate which can be detected), sensitivity range, amplifier time constant, output voltage (for attaching recorder or computer), display (digital or meter), dimensions, and weight. Full sets of specifications must be obtained from the manufacturers; the addresses of manufacturers of fluorescence detectors for liquid chromatography are provided in Refs. 28–30.

4. ANALYTICAL CHARACTERISTICS AND CONCERNS

Among the advantages of fluorometric detection in LC are it is generally nondestructive, it can be employed with solvent gradients without excessive baseline shifts, and it can be adapted fairly easily to microscale LC. The maintenance, cost, and ease of operation of commercial fluorometric detectors is comparable to that of the more commonly used spectrophotometric detectors.

The analytical characteristics of fluorometry that are of primary concern when it is applied to LC detection are (1) sensitivity, (2) selectivity and sample identification capability, and (3) linear dynamic range. In this section we discuss certain analytical methodologies and instrumentation that influence these characteristics.

4.1. Sensitivity

As previously mentioned, fluorescence signal levels increase with increasing excitation radiant power. Obviously it is desirable to employ as powerful an excitation radiation source as possible. Wide excitation monochromator slit widths can be used to pass greater radiant power than with narrow slits, or powerful radiation sources such as lasers (see Section 6.1) can be employed to produce large fluorescence signals. However, at some point problems can arise owing to excited state saturation and photodegradation of the analyte. More importantly, one should recognize that it is S/N, and not strictly signal levels, that determine detectability.

In fluorescence detection, electrical noise (i.e., Johnson noise) originating in the signal recovery equipment is generally less significant than noise resulting from fluctuations in the intensity of the optical background. The base-line noise in a chromatogram is generally some function of this optical background level. In the best case the base-line noise is the result of shot noise (31) and its magnitude is proportional to the square root of the optical background level. The intensity of the optical background exhibits the same dependence on excitation radiant power as fluorescence signals. Hence increasing excitation power, using measures such as described in the preceding paragraph, does not directly lead to a proportionate improvement in detectability. Instead, limits of detection are often determined by the ability of the detector to discriminate spectrally, spatially, or temporally between the fluorescence of the analyte and background radiation.

The major sources of optical background are (1) specular scatter, which results from reflections and refractions of the excitation radiation at the various optical interfaces of the detector, (2) Rayleigh and Raman scattering occurring in the flow-cell material and the mobile phase, and (3) the luminescence of impurities present in the flow-cell material and the mobile phase. Specular and Rayleigh scattering occur coincident with the excitation and are elastic scattering processes (i.e., the scattering occurs without a shift in wavelength). Raman scattering also occurs coincident with the excitation, but is an inelastic scattering process. The difference in energy between the incident and Raman scattered photons corresponds to the energy needed to promote the molecule in question (a mobile phase molecule in this case) from the ground vibrational state of its ground electronic state to an excited vibrational state. This shift to longer wavelengths for the scattered radiation is referred to as a Stokes shift. For example, the common carbon–hydrogen stretching vibration of the organic solvents often used as LC mobile phases produces a Raman band at about 450 nm when excitation occurs at 400 nm. The inherent width of Raman bands is roughly 1 nm; however, the Raman bandwidth is necessarily at least as broad as the excitation bandwidth.

The fluorescence of impurities in the flow-cell material or, more significantly, mobile phase impurities is not a fundamental detector-related contributor to the optical background. High-quality quartz flow cells and purified solvents can be used to minimize the concentrations of these impurities. Unfortunately, the fluorescence of the impurities and that of the analyte are both Stokes shifted and delayed in time (typically 1–100 ns) relative to the excitation. This makes it difficult or impossible to distinguish instrumentally between these two sources of radiation (see below).

Most commercial fluorometric detectors provide only for simple spectra rejection of background radiation. Measures to reject background radiation spatially or temporally can be instrumentally complicated or costly and consequently are not ordinarily employed. Nevertheless, the inherent sensitivity of fluorometric LC detection provides for minimum detectable injection quantities in the picogram range, even when low-cost commercial detectors are employed. When ultimate sensitivity is required, more sophisticated experimental procedures involving spatial or temporal resolution can be employed to reduce the optical background and thereby improve limits of detection.

Table 2 summarizes the characteristics of the various sources of background radiation encountered in fluorometric LC detection and the types of procedures that can be used to minimize their contribution to the optical background. Temporal rejection of "coincidence photons" is accomplished by pulsing the radiation source and employing gated detection (see Section 3.1.6). Because fluorescence lifetimes are typically very short, instrumental requirements for time-resolved detection are very strict. The radiation source must have pulse durations in the low- to sub-nanosecond range and the synchronization signal must have very

Table 2. Sources and Characteristics of Background Radiation in Fluorometric LC Detection

Source	Approximate Intensity (relative to I_o)	Possibilities for Rejecting
Specular scatter	10^{-1}	Spectral, spatial, temporal
Rayleigh scatter	10^{-4}	
Flow cell		Spectral, spatial, temporal
Mobile phase		Spectral, temporal
Raman scatter	10^{-8}	
Flow cell		Spectral,[a] spatial, temporal
Mobile phase		Spectral,[a] temporal
Impurity luminescence	Varies[b]	
Flow cell		Spectral,[c] spatial, temporal[c]
Mobile phase		Spectral,[c] temporal[c]

[a]Depends on the relative spectral positions of the analyte's fluorescence emission and the mobile phase Raman bands. Narrow-band excitation provides larger Raman-free observation windows and analytes with fluorescence Stokes shifts greater than 4000 cm^{-1} can be detected without a Raman background.
[b]Depends largely on the impurity concentration.
[c]Depends on differences in the analyte and impurity excitation and fluorescence spectra or differences in fluorescence lifetimes.

low "jitter" (i.e., there must be very little pulse-to-pulse variation in the time delay between the synchronization pulse and the radiation pulse). Spatial rejection of unwanted radiation is accomplished by carefully imaging the path of the excitation beam as it traverses the flow cell onto the emission monochromator such that the fluorescence of the analyte, which emanates from within the flow cell, passes the entrance slit of the monochromator, while other forms of radiation which originate within the flow cell material or at the flow cell walls (principally specular scatter) are blocked by an optical mask or the monochromator slit plate.

A well-designed flow cell can aid in minimizing the optical background. Several flow cells have been specially designed for use with laser excitation sources. These flow cells produce less scattered radiation or provide for better rejection of the scattered radiation, which is very necessary when the intense radiation of a laser is used for excitation.

4.2. Selectivity and Sample Identification Capabilities

Fluorometry is one of the more selective modes of detection commonly used in LC. This selectivity is the result of two factors. First, relatively few molecules have fluorescence quantum efficiencies that are large enough to make their fluorometric detection analytically useful. Hence the fluorescence signals of molecules that have reasonably large quantum efficiencies (e.g., >0.1) can often be measured without interference. This is actually a negative element of selectivity in that it limits the applications of fluorometry. Secondly, fluorometry possesses a higher degree of "tunable" selectivity, relative to spectrophotometry, because both excitation and fluorescence emission spectra are available for spectral resolution, whereas in spectrophotometry only absorption spectra are available for resolving different sample molecules.

Depending on the sample being analyzed, and the spectral characteristics of its components, reducing the spectral band passes of the fluorometer can enhance the selectivity of the measurement. However, liquid phase excitation and fluorescence emission spectra are rather broad, and reducing spectral band passes below a few nanometers generally has little or no effect on selectivity but can have an adverse effect on sensitivity. It should also be stated that excitation and fluorescence emission spectra are often mirror images of each other and therefore do not always represent totally independent elements for spectral selectivity.

Time-resolved fluorometry can also be employed to distinguish between fluorophores that have significantly different fluorescence lifetimes. The lifetime over which the emission of a fluorescent molecule is observed is equal to the inverse of the sum of the rate constants for the various processes that can depopulate the lowest excited singlet state (S_1) of the molecule (see Eq. 1). At room temperature the range of S_1 lifetimes of fluorescent molecules is fairly small (1–100 ns). Consequently, time-resolved fluorometric detection does not provide a highly selective means of distinguishing between different molecules. Nevertheless, a pulsed excitation source and a gated detector can be used both to reject coincidence photons, which contribute to the optical background, and to provide some aid in distinguishing between different fluorophores in the sample. Time-resolved fluorometry has been employed by Furuta and Otsuki (27) in the fluorescence determination of polycyclic aromatic hydrocarbons (PAH) separated by LC. Employment of a nitrogen laser pumped dye laser and time-resolved detection led to an increase in detection capability of 1–2 orders of magnitude and permitted the discrimination between short and long-lived fluorescent PAHs.

The ability of a spectroscopic detection technique to provide information concerning the chemical structure or identity of a compound is very important in chemical analysis. Spectral identification is usually complicated when the compound under investigation is not in a pure form. This represents the motivation behind combining chromatography with spectroscopic identification techniques. The analytical method of choice is clearly GC–MS. However, MS is not easily combined with LC, whereas fluorometry is very compatible with LC. Because of the broad, featureless nature of condensed-phase fluorescence spectra, the structural information provided by the spectra is rather subtle and limited. The fluorescence spectrum of a molecule does not provide characteristic group frequency information as in IR spectrometry or group fragmentation patterns as in MS. Nevertheless, an important attribute of fluorometric spectral identification is the excellent sensitivity of the technique. This allows "on-the-fly" acquisition of spectra, even for trace-level components of a chromatographically separated sample.

Lyons et al. (32) have used computer file searching and pattern recognition techniques for the structural interpretation of fluorescence spectra obtained in LC detection. Because of instrumental limitations, they were forced to use a stopped-flow procedure to obtain the spectra. The flowchart for their analysis is shown in Fig. 6. An absorbance detector was used to signal the elution of a solute band. At that point the flow was stopped for a few minutes to obtain excitation and fluorescence emission spectra. A synchronous fluorescence (33) scan helped to identify peaks that contained more than a single compound. The eluted compounds were then identified based on comparison of their spectra with computer-stored library spectra (i.e., fluorescence spectra of standard compounds obtained previously, digitized, and stored in the computer).

In order for these comparisons to be valid the spectra of all the compounds must be obtained on the same instrument and under identical experimental conditions. If different instruments or instrumental conditions are employed then the spectra must be standardized or "corrected" for instrumental factors such as source intensity and detector response variations.

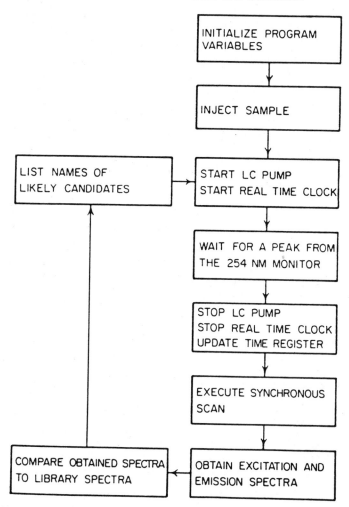

Figure 6. Flowchart for fluorescence spectral acquisition and interpretation in LC. Reprinted with permission of Plenum Publishing.

Lyons et al. corrected their fluorescence spectra for wavelength-related source intensity variations by splitting off a small portion of the excitation beam and using this split-off radiation to excite a "quantum counter." The quantum counter absorbed all the radiation and produced a fluorescence signal, which was monitored by a reference detector, that is proportional to the intensity of the source. The response of this reference detector was used to normalize the fluorescence signal of the eluted compound to the wavelength-dependent intensity of the source. The identification success rate of Lyons' system varied considerably from compound to compound.

4.3. Linearity of Response

The normal linear change in fluorescence signals with concentration can fail if (1) the absorbance of the sample is so large that higher-order terms in the exponential expansion of Eq. 1 (which when ignored reduce Eq. 1 to the linear relationship shown in Eq. 2) can not be ignored or (2) inner-filter effects are observed. The latter effect is generally more significant, although much less so when fluorometry is employed in LC detection. In conventional fluorometry, preabsorption inner filter effects are often observed when greater than a few percent of the radiation incident on the sample cell is absorbed by the sample. In effect this caused the I_0 appearing Eq. 2 to decrease with the depth of penetration of the incident beam into the sample cell. Normally the fluorescence emission collection optics are arranged such that the emission is most efficiently collected from the center of the sample cell, and so this variation in I_0 results in a nonlinear response.

Preabsorption effects can be minimized by imaging the front part of the sample cell onto the detector or by employing a very small sample cell, as is done with the small flow cells used in LC. Unfortunately these measures to minimize preabsorption can reduce signal levels and increase optical background levels.

In LC the common practice is to perform quantitation based on peak height measurements. If the amount of sample injected is increased to the point of column overload, peak asymmetry will occur and the measured peak height will be less than for a symmetric peak. Hence it is often the column and not the detector that produces nonlinearity in LC. In our work we have routinely obtained calibration plots based on fluorescence peak heights which were linear over three to four decades of concentration.

5. APPLICATIONS OF FLUOROMETRIC DETECTION IN LIQUID CHROMATOGRAPHY

5.1. Detection of Compounds that Exhibit Native Fluorescence

Separation and detection of compounds that exhibit significant native fluorescence are straightforward. Of course the compounds of interest must have high quantum yields (generally greater than about 0.1) under the constraints placed by the separation process. A sample containing the analyte of interest is typically "cleaned up" in order to isolate the analyte from compounds which may interfere with the separation and/or the detection. The cleanup step typically involves a solvent extraction or a preliminary chromatographic procedure to prepare the sample adequately for the primary chromatographic separation.

The separation process, regardless of what specific type of column chromatography is being employed, must utilize a mobile phase that provides for adequate chromatographic resolution and fluorescence detectability. As discussed earlier, the mobile phase composition, pH, and dissolved gas content of the mobile phase have a significant effect on fluorescence detection. In certain cases, fluorescent species are added to the mobile phase which experienced an increase or decrease in fluorescence intensities when certain analytes were eluted. This approach has been employed in an assay of lipids (34) and derivatized thiols (35).

Detection of native fluorescent analytes usually employs a commercial detector at the end of the column as described earlier. Once a separation is optimized the selection of the most appropriate wavelength for excitation and emission can be found by obtaining fluorescence spectra of the analytes dissolved in the mobile phase in which they elute. Table 3 provides a listing of some representative examples of the detection of native fluorescent analytes separated by LC.

Table 3. Examples of Native Fluorescent Compounds Separated by LC and Detected by Fluorescence Detection

Class	Analyte	Reference
Clinical	Adrenaline	36
	β-Carboline	37
	Estriol	38
Enzymes	Tryptophan hydroxylase	39
Food testing	Aflatoxins	26,40
	Ethopabate	41
Fungicide	Thiabendazole	42,43
Illicit drugs	LSD	44
	Opiates	45
Neurotoxin	Lolitrem-B	46
Pharmaceuticals	Propranolol	47
	Nabumetone	48
	Sulmazole	49
	Tiodazosin	50
Pollutants	Polynuclear aromatic hydrocarbons	27,51
Steroids	Estriol	52
	Estrogen	53

5.2. Detection of Nonfluorescent Compounds

A drawback in the application of fluorescence detection in LC is the limited number of compounds with large fluorescence quantum efficiencies. Because fluorescence detection does offer distinct advantages over other detection methods, various derivatization schemes have been adapted so that fluorescence detection can be employed for the analyses of compounds that do not exhibit appreciable native fluorescence.

There are two approaches to derivatization for fluorescence detection: precolumn and postcolumn derivatization. There have been reviews of derivatization in LC (54,55), and a recent monograph (56) provides an extensive listing of current derivatizing agents and their application. In this section a short description of both pre- and postcolumn techniques is presented and the merits of both discussed.

5.2.1. Precolumn Derivatization

Precolumn derivatization is the procedure in which the analyte to be detected is derivatized before it is injected onto the column. A typical derivatization procedure involves the covalent bonding of a highly fluorescent molecule to the analyte, usually via reaction between functional groups on the two species. Although this is the usual precolumn approach, other methods exist that can modify the analyte to a strongly fluorescent form. As examples of these alternative approaches, consider the work of Fischer et al. (57), who acetylated salicyclazosulfapyridine with acetic anhydride prior to separation to form a highly fluorescent acetylated analog. Bannister and co-workers (58) employed mercuric acetate to precolumn-oxidize the alkaloid emetine, making it more fluorescent and realizing a 50-fold improvement in detectability. Sternson and co-workers (59) used precolumn irradiation of a sample for 20 min with a 15-W Hg vapor lamp. This transformed the analyte tamoxifen into the more strongly fluorescent phenylphenanthrene. Procedures such as this, which cause the analyte to react to a structurally more favorable form for fluorescence emission (see Section 2.3), are sometimes employed.

For the precolumn procedure the derivatization can be readily optimized by modifying the reaction conditions to provide the optimum yield of derivatized analytes. No special equipment is required and, unlike postcolumn derivatization, long derivatization times are acceptable. If in the process of derivatization there are by-products formed that may affect the separation or detection of the analytes, they can be eliminated via cleanup procedures prior to injection or separated during the chromatographic process. The solvent selection for the derivatization does not depend on the mobile phase of the separation, which can offer advantages over the postcolumn technique.

It may appear that precolumn derivatization provides the answer whenever the analyte is undetectable, but can be derivatized to a fluorescent form. This is not always the case. Separation of structurally similar compounds is often difficult in itself. However, when the compounds are derivatized they become even more similar structurally, complicating the separation. This complication can extend method development time and may ultimately yield separation conditions that are not conducive to good detection. In addition to separation problems it is possible that analytes with more than one functional group will form more than one derivative. This can result in additional peaks in the chromatogram that are very difficult to identify and quantification of analytes becomes a formidable task.

As an aid in understanding the process of precolumn derivatization consider the work of Murray and Sepaniak (60), in which aliphatic amines in beer were derivatized with 7-chloro-4-nitrobenzo-2-oxa-1,3-diazole (NBD-Cl) for the purpose of fluorescence detection. The reaction for the derivatization of amines with NBD-Cl is illustrated in Fig. 7. The optimal derivatization conditions were determined to be pH 9.0 at 60°C for a period of 1 h. The resulting derivatized mixture of amines was diluted and injected onto an analytical scale C_{18} column and the separation effected by an acetonitrile–water gradient elution scheme. A chromatogram of a derivatized beer sample is shown in Fig. 8.

An argon ion laser operated at 488 nm was employed to excite the derivatized amines. This laser-based detection scheme provided a linear response over four decades of concentration with limits of detection of approximately 5 pg injected. This clearly shows the utility of the derivatization of a nonfluorescent species so that sensitive fluorescence detection can be employed.

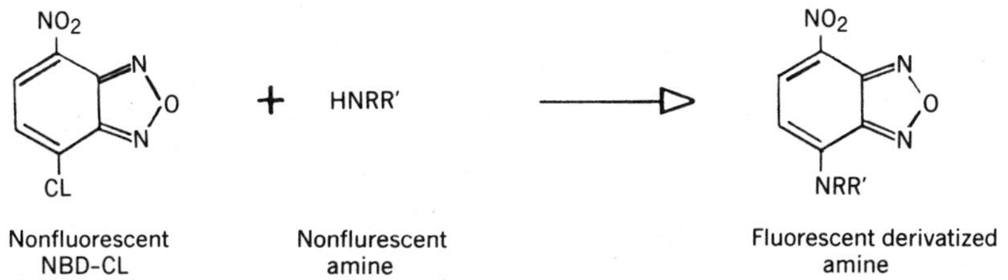

Figure 7. Reaction of NBD-Cl with a secondary amine to form the strongly fluorescent derivatized amine.

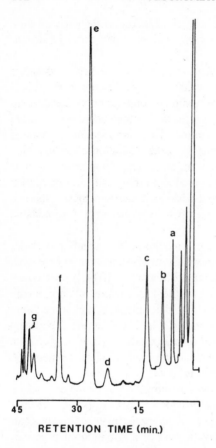

RETENTION TIME (min.)

Figure 8. Chromatogram of a derivatized beer sample: (*a*) ammonia, (*b*) methylamine, (*c*) dimethylamine, (*d*) pyrrolidine, (*e*) proline, (*f*) isobutyl amine, (*g*) isoamyl amine. Reprinted with permission of Marcel Dekker Publishing from Ref. 60.

5.2.2. Postcolumn Derivatization

As the term denotes, postcolumn derivatization involes formation of the derivatized analyte after separation, but before detection. The postcolumn derivatization takes place in a reactor where the appropriate derivatizing reagents are added to the eluent to form the desired derivatives or the proper conditions are provided to yield the desired fluorescent form. Postcolumn derivatization is just like precolumn derivatization in that it is not limited to reactions in which a fluorescent tag is bonded to the analyte. Many postcolumn reactors have been designed to provide a myriad of conditions to modify analytes to more strongly fluorescent forms. Postcolumn oxidation by the use of cerium(IV) has been employed to detect carboxylic acids (61) and phenols (62) by observing the fluorescence of the cerium(III), and sodium nitrite was used to postcolumn-oxidize biopterins to more strongly fluorescent forms (63). The adjustment of pH can yield more strongly fluorescent analytes and has been used in the detection of purines and pyrimidines (64), warfarin (65), and B_6 vitamers (6). Postcolumn irradiation with UV lamps has been employed in the determination of demoxepam (66), cannabinols (67), and clobazam (68). The fluorescence detection of organosulfur compounds was accomplished by a ligand exchange process between palladium(II)–calcein and the sulfur compounds (69). The release of the fluorescent calcein provided indirect measurement of the sulfur compounds. A postcolumn metal–chelate formation provided the means of detection for four 8-hydroxyquinolines, which were separated and detected as magnesium(II) metal-chelates (70). Once again it must be remembered that postcolumn conditions that enhance fluorescence are desirable, provided the effect of the postcolumn reactor on chromatographic resolution is acceptable. Theoretical and design considerations in the use of postcolumn reactors have been discussed by Scholten and co-workers (71).

Postcolumn derivatization offers the advantage of being able to employ previously determined or literature-accessed separation conditions without having to change these conditions to accommodate the derivatized forms of the analyte. A postcolumn derivatization reaction does not have to be complete in order for quantitative analyses to be performed. However, it is generally observed that reproducibility suffers when the derivatization is incomplete. A postcolumn reactor can be placed in series with other types of detectors in order to increase the information available to the analyst.

Postcolumn derivatization has some important disadvantages. There is the problem of solute band dispersion within the reactor due to the mixing of reagents to form the derivative. This dispersion can destroy resolution, and the dilution of peaks may render the detection advantage of derivative formation insignificant. The mobile phase that is the optimum for separation is often not the best medium in which to perform the derivatization reaction. This can limit the ability to perform gradient elution schemes. Because the derivatization process often requires the addition of reagents to the eluent before detection there is always the possibility of excess reagents interfering with the signal produced by the analyte via absorption of fluorescence emission or contributing to the background by fluorescing at the monitored wavelength.

An example of the utility of postcolumn derivatization for fluorescence detection is provided by Seki and Yamaguchi (72). The compounds of interest were the 17-hydroxycorticosteroids in a urine matrix. Detection of these analytes was typically accomplished via UV absorbance monitoring or by ineffective derivatization procedures. A new derivatization procedure was desired. The authors employed the reaction of benzamidine and a ketolic group to form a strongly fluorescent derivative. The postcolumn derivatization was accomplished by employing a mixing chamber to add the analyte, benzamidine, and sodium

hydroxide together to form the derivative. The reaction requires base, and because
the most strongly fluorescent form of the derivative requires a pH of 13–14, the
use of basic conditions was optimal for both reaction of reagents and detection.
The reaction mixture was passed through a 0.5 mm ID Teflon tube 30 m long
for 5 min at 95°C in a water bath. This provided the necessary temperature and
time conditions in order to form the desired derivatives completely. The solution

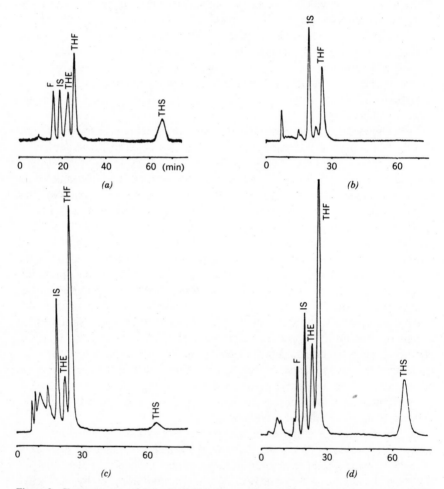

Figure 9. Chromatograms of derivatized 17-hydroxycorticosteroids: (*a*) Chromatogram of standard
solution. (*b,c*) Chromatograms of urine sample from a normal subject. (*d*) Chromatogram of urine
sample of a patient with Cushing's disease. F = cortisol, THE = tetrahydrocortisone,
THF = tetrahydrosortisol, THS = tetrahydro-11-deoxycortisol, IS = β-methasone (internal standard). Reprinted with permission of Elsevier Science Publishers from Ref. 72.

was cooled to room temperature before fluorescence detection. The chromatogram shown in Fig. 9 shows that good chromatographic efficiency was maintained despite the mixing of reagents. The added sensitivity provided by derivatization with fluorescence detection allows the use of this technique to assist in the diagnosis of Cushing's disease.

In summary, when selecting a derivatizing agent the working chromatographer must consider the following criteria in order to achieve a successful fluorescence derivatization procedure (56):

1. Fluorophore must possess intense absorption bands.
2. Fluorophore must have large quantum yield.
3. Derivatizing agents must be stable.
4. Derivatizing agent and by-products formed during derivatization should not be fluorescent or must be separable from the analyte.
5. Analyte must be reactive with derivatizing reagent under reasonable conditions.
6. Derivatives formed should be able to be isolated from the derivatization reagents if desired.
7. Reagents should be nontoxic if possible.

Table 4 provides some examples of both precolumn and postcolumn derivatization of analytes for fluorescence detection in LC. These examples are provided to illustrate the broad range of analytes that can be effectively derivatized for detection in LC. This table merely skims the surface of the vast number of techniques employed in derivatizing analytes for detection in LC.

6. MODERN RESEARCH APPLICATIONS

Preceding sections of this chapter have concentrated on describing routine or established methodologies and instrumentation used in fluorometric LC detection. It is generally observed that lower and lower limits of detection are being sought in chemical analysis. Determinations of sample components at the part-per-billion (ppb) or even part-per-trillion (ppt) level are found to be important to many areas of science. Even an apparently clean sample, such as country air or water from a mountain stream, can contain many components (possible interferences) at concentrations equal or greater than these. This can make ultra-trace analyses extremely formidable. In order to meet this analytical challenge many researchers have extended the conventional bounds of LC and fluorometry, often by employing new methodologies or state-of-the-art instrumentation. In this section

Table 4. Examples of Compounds that have been Derivatized or Modified for Purposes of Fluorescence Detection

Class	Analyte	Pre- or Postcolumn	Derivatizing Agent or Procedure	Ref.
Antibiotics	Ampicillin	Pre	Degradation of analyte	73
	Penicillin V	Post	o-Phthalaldehyde (OPA)	74
Beverage testing	Amines	Pre	7-Chloro-4-nitrobenzo-2-oxa-1,3-diazole (NBD-Cl)	54
Clinical	Amino acids	Post	Pentane-2,4-dione formaldehyde system	75
		Pre	OPA	76
		Post	OPA	77
	Histamine	Pre	Fluorescamine	78
	Guanidino compounds	Pre	Benzoin	79
Illicit drugs	Cannabinoids	Pre	5-Dimethylaminonaphthalene-1-sulfonyl chloride (DNS-Cl)	80
	Barbiturates	Pre	4-Bromomethyl-7-methoxycoumarin	81
Lipids	Phosphatidic acids	Pre	Dansylethanolamine	82
Pesticides	Carbaryl	Post	Catalytic hydrolysis to a fluorescent residue	83
	Methoxuron	Pre	DNS-Cl	84
Pharmaceuticals	Chloroxamine	Post	DNS-Cl	85
	Melphalan	Pre	OPA	86
	Morphine	Post	Oxidation	87
	Phenylpropanolamine	Post	OPA	88
Pollutants	Phenols	Pre	2-Fluorenesulfonyl chloride	89
		Pre	4-Aminobenzonitrile	90
Steroids	Cortisol	Pre	9-Anthroylnitrile	91
	17-Hydroxycorticosteroids	Post	Benzamidine	72

we endeavor to discuss several modern areas of analytical research involving fluorometric LC detection.

6.1. Laser Excitation

The application of laser fluorometry to LC detection has been reviewed by Yeung and Sepaniak (92). A discussion of the instrumental principles of lasers is inappropriate for this chapter. However, we briefly discuss the unique properties of laser radiation that, under the proper experimental conditions, can provide for improvements in the analytical characteristics of fluorometric LC detection.

The most notable property of lasers is high output power (e.g., the peak powers for many pulsed lasers are in the megawatt range). Even with large band passes it is generally not possible to isolate similar powers from conventional radiation sources. Two benefits that accrue from the high powers of lasers are larger fluorescence signal levels and the possibility of utilizing nonlinear excitation (see Section 6.3). One possible adverse effect of high excitation power is a large optical background. Fortunately, certain other properties of laser radiation can be exploited to minimize optical background.

The output beams of most lasers are highly collimated. Beam divergences are often in the low milliradian range. Provided a properly designed flow cell is employed, this laser property can facilitate spatial rejection of stray radiation. Highly collimated beams can also be efficiently focused into the very small flow cells used in microscale LC (see Section 6.2). The pulse durations of most pulsed lasers are in the nano- to picosecond range. This makes pulsed lasers convenient excitation sources when time-resolved fluorometry is used to reject coincidence photons.

Another important property of laser radiation is monochromaticity. The output bandwidths of most lasers are in the sub-angstrom range. This high degree of monochromaticity is more than is necessary in condensed-phase fluorometry. However, it can provide for better spectral rejection of Rayleigh and Raman scattering, because the bandwidths for these scattering processes are very narrow with laser excitation. It should be mentioned that although laser outputs have very narrow bandwidths, they are seldom pure. For example, a gas laser such as an argon ion laser can be aligned to have lasing action at just one wavelength, but many potentially interfering "plasma" lines, corresponding to nonlasing electronic transitions of argon ions within the laser's plasma tube, are also present in the output beam. Thus an excitation wavelength selector is generally needed even when a laser is employed.

Replacing a conventional radiation source by a laser will not improve detection unless the flow cell can accommodate the intense laser radiation without damage and unless the laser properties mentioned in the preceding paragraph can be effectively utilized to minimize optical background. The fiberoptic-based flow cell shown in Fig. 10 was used in conjunction with an argon ion laser by Sepaniak and Yeung to determine the naturally fluorescent antitumor drugs doxorubicin and daunorubicin (93). By positioning the focused laser beam (G in the figure) close to the fiberoptic terminus, it was possible spatially to reject scatter that occurred within the flow-cell material or at the optical interfaces of the flow cell. This scatter was outside the "acceptance cone" of the fiber and consequently was not transmitted to the photodetector. Despite the high intensity of the laser excitation source, relatively low background levels were observed. The limits of detection for the antitumor drugs were in the low picogram injected range.

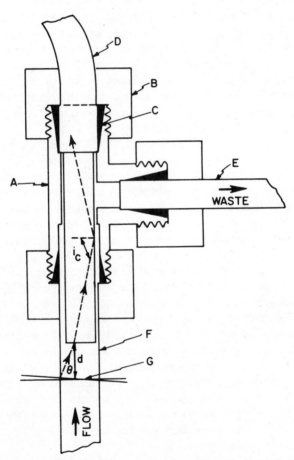

Figure 10. Close-up of fiber-optic flow cell. A, 1/16-in. stainless steel tee; B, 1/16-in. stainless steel nut; C, graphite ferrule; D, optical fiber (1.0-mm core diameter, 1.5-mm OD); F, quartz capillary tube (1.05-mm ID, 2.0-mm OD); G, focused laser beam. Reprinted with permission of Elsevier Science Publishers from Ref. 93.

Diebolt and Zare also developed a flow cell for use with laser excitation (26). Their "flowing droplet" detector was composed of a droplet of column effluent suspended between a stainless steel tube, connected to the column, and a stainless steel rod. This design eliminates the possibility of flow-cell fluorescence. Moreover, because there are no flow-cell "windows," problems with burn spots and cracks in the flow cell, which result from the large power densities of focused laser beams, are also eliminated. Limits of detection for certain aflatoxins were in the sub-picogram injected range when a HeCd laser was used for excitation. Further descriptions of flow cells designed for laser excitation are provided in the next section.

6.2. Fluorometric Detection In Microscale LC

Column technology in LC has improved dramatically over the past decade. One very active area of research is the development of microscale LC columns (94,95). When compared to conventional analytical scale columns, microscale columns have the advantages of very small sample and solvent consumption and the potential for very high efficiency, which is sometimes attained at the expense of long analysis time. A comparison of the physical characteristics of conventional analytical and microscale LC columns is shown in Table 5.

It is the extremely small "peak volumes" of eluted solute bands for microscale columns, particularly open capillary columns, that represent the largest experimental obstacle to the routine use of these columns. The average flow-cell volume for the commercial fluorometric detectors listed in Table 1 is roughly 10 μL. Obviously these detectors would seriously disperse the solute bands eluted from microscale columns, particularly the open capillary columns. Two approaches to circumventing this detector dispersion problem are (1) employing a solvent

Table 5. Typical Physical Characteristics of Conventional Analytical and Microscale LC Columns

| | Dimensions | | | Peak Volumes for |
Column Types	Id	Length	Flow Rates	Eluted Solute Bands
Conventional analytical	4.6 mm	25 cm	1.0 mL/min	0.5 μL
Microbore	1.0 mm	25 cm	0.1 mL/min	30 μL
Packed fused silica capillary	0.3 mm	1.0 m	10 μL/min	2 μL
Prepacked drawn capillary	60 μm	5.0 m	1.0 μL/min	200 nL
True open capillary	20 μm	5.0 m	<1.0 μL/min	50 nL

"makeup" flow, which mixes with the column effluent and produces a much larger peak volume, and (2) miniaturization of the fluorometric flow cell. The former approach is not generally useful in chemical analysis because it severely dilutes the eluted band and thereby dramatically increases minimum detectable injection quantities.

Concentration sensitivity suffers when the flow-cell miniaturization approach is employed because the miniaturized flow cells generally have shorter optical path lengths than conventional-scale flow cells. However, if the microscale column is highly efficient, there will be very little chromatographic dilution of the solute band. This can compensate for shorter pathlength and the resulting minimum detectable injection quantities can be very low. For example, Guthrie and Jorgenson were able to detect 10 pg of perylene when injected into a 64-μm ID by 3-m-long open tubular column (96). Detection was performed "on-column" using a mercury arc lamp as the excitation source. The advantage of "on-column" detection is that there is essentially no flow cell and hence no possibility of detector-related dispersion of solute bands.

Because of the extremely small dimensions of open capillary columns, on-column detection is best performed using a laser excitation source. A comparison of excitation with a HeCd laser and a mercury arc lamp for on-column detection in open tubular LC has been reported (97). Although the detection limits for 9-methylanthracene were slightly better for the conventional excitation source, no effort was made to utilize the collimation of the laser beam to reject specular scatter spatially. We have found it relatively simple to focus laser beams "cleanly" through small-bore open capillary columns and then image the path of the beam through the column onto the emission monochromator, such that the large amount of specular scatter induced by the intense laser radiation does not pass the monochromator entrance slit (98). This is not possible with nonlaser light sources because they can not be focused to sufficiently small spot sizes. Detection limits obtained in our laboratory for compounds such as the antitumor drugs doxoribicin and daunorubicin (see the chromatogram in Fig. 11) are typically in the low picogram injected range (99).

On-column detection is difficult or impossible when the spectral regions involved are strongly absorbed by the capillary column or when packed capillary columns are used. In those situations an ultra-low volume flow cell is needed. Folestad et al. (100) employed a "free falling jet" flow cell arrangement for laser fluorometric detection and, impressively, they were able to detect injected quantities of fluoranthrene as small as 20 fg. Unfortunately, their free-flowing jet required volumetric flow rates greater than or equal to 1 mL/min and therefore was not useful for microscale LC. Instead, they employed high-quality quartz tube flow cells with microbore columns.

When on-column detection is not possible, simple capillary tube flow cells can also be used with open capillary columns. However, flow-cell dimensions

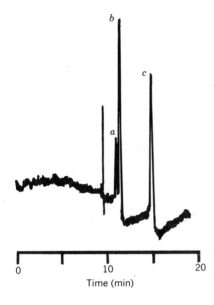

Time (min)

Figure 11. Open capillary LC separation of anthracyclic drugs. *a* and *b* are doxorubicin and *c* is an impurity daunorubicin. The 15 μm × 5.0-m column used here was etched, then dynamically modified by incorporating 1 g of trimethylhexyldecylammonium bromide per liter of mobile phase (pH 6.8 buffer/acetonitrile/methanol, 65: 26: 10 by vol). For fluorometric on-column detection the 488-nm output line of an argon ion laser (power approximately 400 mW) was used for excitation and a small monochromator set to 585 nm was used to isolate the fluorescence emission. Reprinted with permission of Clinical Chemistry, Winston-Salem, NC

and shape, and connections to the column, influence band dispersion much more critically. We studied the contribution of certain capillary tube flow cells to solute band dispersion in open tubular LC (98). The observed band dispersions were much greater than predicted based on Eq. 5 (see Section 3.1.4). This illustrates the importance of designing flow cells, and making column–flow cell connections, that provide for efficient washout.

A fluorometric flow cell, which has been used with laser excitation and should be generally useful for microscale LC, is the sheath flow cell shown in Fig. 12 (101). This type of flow cell restricts the column effluent to a narrow channel, by confining it with a sheathing solvent. Under conditions of laminar flow, the column effluent and the sheathing solvent do not mix. By adjusting the flow rates involved it is possible to control, to a limited extent, the dimensions of the flow channel. Using an argon ion laser excitation source, illumination volume for the sheath flow cell shown in the figure was adjustable between 150 and 6 nL, making it compatible with the eluted peak volumes given in Table 5.

In addition to its small volume, the sheath flow cell has the advantage of having a constantly regenerated sheathing fluid-column effluent interface. This virtually eliminates the previously mentioned problem of laser "burn spots" on the flow cell. Furthermore, if the sheathing solvent is chosen to have the same RI as the effluent, there is very little scatter of the laser beam. The obvious disadvantages of this flow cell is that it is much more complicated to build and use than a simple quartz capillary tube flow cell.

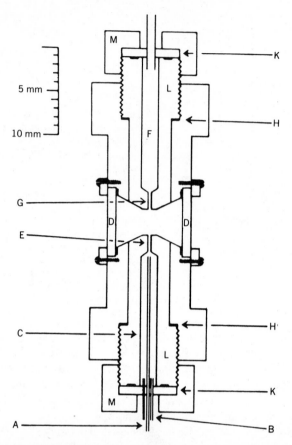

Figure 12. Schematic diagram of the sub-microliter fluorescence flow-through cuvette. (A) effluent entry tube; (B) sheath entry tubes; (C) sheath inlet channel; (D) 8-mm diameter quartz windows; (E) 500-μm diameter inlet alignment bore; (F) exit channel; (G) 500-μm diameter exit alignment bore; (H) Teflon O-ring; (K) inlet or outlet tube holder; (L) inlet or outlet probe; and (M) stainless steel nut. Reprinted with permission from L. W. Hershberger, J. B. Callis, and G. D. Christiansen, *Anal. Chem.*, **51**, 1444 (1979). Copyright 1979 American Chemical Society.

6.3. Nonlinear Excited Fluorometric Detection

The possibility of attaining extremely large photon densities with focused laser beams facilitates the observation of certain nonlinear spectroscopic phenomena under conditions that are realistic to LC detection. Two nonlinear excited fluorescence techniques, two-photon excited fluorescence (TPEF) and sequentially excited fluorescence (SEF), have been used for detection in LC (102,103). The two-photon excitation process involves the simultaneous absorption of two photons to produce a resonant transition in a molecule. This is a very inefficient excitation process and requires large photon densities to be observed. Because of the quadratic power dependence of signals in TPEF (104) the technique is best performed using high peak power pulsed lasers. Emission in TPEF is monitored from the lowest excited singlet state of the molecule. Excitation in the SEF technique involves the resonant absorption of two photons. The molecule is promoted to a highly excited electronic state and emission is also monitored from a highly excited state. This emission process contradicts Kasha's rule (105) and hence is also very inefficient. Again large photon densities are required to observe reasonably large SEF signals.

The inefficiency of the optical processes in these nonlinear techniques results in relatively poor sensitivity. However, the large blue shift between excitation and emission wavelengths (anti-Stokes shift) makes efficient spectral rejection of stray radiation possible. This compensates somewhat for the inefficiency of the nonlinear optical processes. Limits of detection in LC for these techniques are considerably poorer than for conventional fluorescence, but are generally comparable to those obtained with commercial absorbance detectors. The principal analytical advantage of these nonlinear excitation processes is that they involve unique selection rules (104). The nonlinear excitation processes tend to populate symmetric (i.e., gerade) excited electronic states whereas the conventional (one-photon) excitation process tends to populate asymmetric (ungerade) excited electronic states. Although these selection rules are strictly valid only for centrosymmetric molecules, the aforementioned tendencies do provide for an added measure of spectral selectivity in fluorometric detection.

The unique selectivity of TPEF is illustrated in Fig. 13 for the separation of a mixture of oxadiazole molecules and polynuclear aromatic molecules. The compounds in the chromatogram have excited states that can be populated via one-photon absorption of UV radiation from a mercury lamp (254 nm). However, only the oxadiazole molecules absorb two photons from the 515-nm line of a focused argon ion laser beam and produce the TPEF signals in chromatogram *b*. The energy of two 515-nm photons is about the same as a single UV photon from the mercury lamp and the polynuclear aromatic molecules, which appear as interfering peaks in Fig. 13*a,* have reasonably large fluorescence quantum efficiencies. This illustrates the unique selectivity characteristics that are possible with TPEF.

6.4 Multidimensional LC Detection that Includes Fluorometry

Despite the excellent resolving power of modern LC columns, the complexity of many samples can make the complete separation of all the components of interest in a sample difficult or impossible. Under these conditions, multidimensional methods of detection can be employed to increase the differentiating

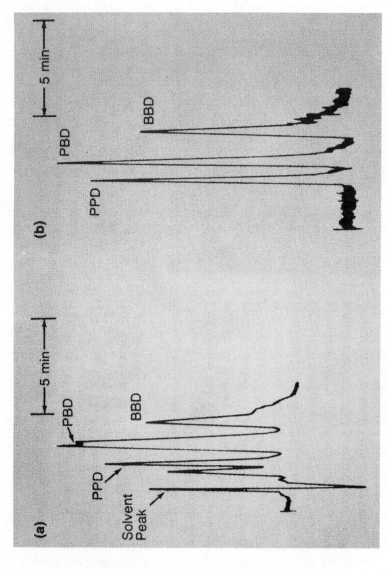

Figure 13. Simplification of chromatograms by two-photon excited fluorimetry. All seven compounds in a mixture show up and interfere with one another in a standard UV absorption detector (*a*), whereas only the oxadiazoles appear in the two-photon detector (*b*). Reprinted with permission from M. J. Sepaniak and E. S. Yeung, *Anal. Chem.*, **49**, 1554 (1977). Copyright 1977 American Chemical Society.

capability of the analysis. The large number of "hyphenated"methods that have been described in the literature attest to the success of this approach to solving complex analytical problems (106). When more than one LC detector is employed it is often desirable that they provide as much exclusive information as possible. It is also important to be able to arrange the detectors in series without causing excessive solute band dispersion. Alternatively, it should be possible to use a single, low-volume flow cell for more than one mode of detection (e.g., the Perkin Elmer Tri-Det detector listed in Table 1 combines three modes of detection in a single flow cell). Of course, if the detectors are to be arranged in series all but the last detector must be nondestructive.

The LC detector most commonly used in conjunction with fluorometric detection is the absorbance detector. Because excitation via the absorption of radiation must precede fluorescence emission, all the fluorescent components in a sample will also give a response on the absorbance detector. However, the sensitivity of the absorbance detector is typically 2–3 orders of magnitude less than that of the fluorometric detector. Thus this combination can provide the chromatographer with a highly sensitive, specific detector and a moderately sensitive, general detector. Recall that Lyons et al. (32) used a fixed-wavelength (254 nm) absorbance detector to indicate the elution of a solute, which was not necessarily fluorescent, and signal the initiation of an attempt to obtain fluorescence spectra of the solute (see the flowchart in Fig. 6).

The DuPont Model 836 absorbance-fluorescence detector provides for simultaneous absorbance and fluorescence detection in a single 16-μL flow cell. One of the demonstrated advantages of this detector is its extended linear dynamic range (107). The fluorescence mode of detection can be employed to obtain a calibration plot at very low solute levels. At higher solute levels, where inner filter effects can lead to nonlinearity, the absorbance detector is usually sufficiently sensitive to be used.

The combination of absorbance and laser fluorometric detection using both conventional and nonlinear excitation was found to be useful in the LC classification of the asphaltene fractions of coal liquefaction products (108). As mentioned in Section 6.3 the conventional and nonlinear excitation processes tend to probe different excited states of a molecule. This complementary property of these two fluorometric modes of detection proved useful when computer pattern recognition techniques were used with the chromatograms to classify the asphaltene samples (109).

An interesting multidimensional detector flow cell was developed by Voightman et al. (110). They modified the flowing droplet flow cell discussed in Section 6.1 to provide for simultaneous fluorescence, photoacoustic, and photoionization detection. Using a nitrogen laser excitation source, a comparison was made of the limits of detection for a number of polynuclear aromatic hydrocarbons. Sensitive detection for all these modes of detection requires strong absorption by the analyte at the laser wavelength. However, the different processes responsible for generating signals for these detection techniques provided for compound to compound differences in response (i.e., the detectors provided complementary information). For example, the fluorescence process involves the radiative deactivation of an excited molecule whereas photoacoustic detection involves the detection of an acoustic wave generated by the nonradiative (heat dissipating) deactivation of an excited molecule.

6.5. Other Luminescence Modes of Detection

Certain nonfluorometric modes of luminescence detection have been successfully applied to LC and are described very briefly here. As stated in Section 2.2, phosphorescence involves the spin-forbidden emission from an excited triplet state and consequently does not represent an efficient excited state deactivation process. Phosphorescence measurements are generally obtained in frozen matrices where the rate constants for nonradiative modes of deactivation are greatly reduced (111). Naturally such sample preparation techniques are not easily performed in conjunction with LC. However, room-temperature phosphorescence in liquid solution is observed for a limited number of molecules, particularly if a heavy atom is included in the solvent to enhance intersystem crossing or micelles are incorporated into the solvent to organize the environment around the phosphor microscopically (112).

The principal advantage of phosphorescence detection is that the long radiative lifetimes that are observed facilitate efficient temporal rejection of background radiation. High sensitivity is possible. For example, Scypinski and Love obtained detection limits for certain polynuclear aromatic hydrocarbons in the 10^{-11}–10^{-13} M range using an aqueous measurement solvent that contained both cyclodextrins and heavy-atom-containing species (113). It should be possible to use a solvent such as this as a mobile phase for reverse-phase LC.

Sensitized and quenched room-temperature phosphorescence of biacetyl has been employed by Donkerbroek et al. in the detection of mixtures of polychloronapththalenes and polychlorobiphenyls following separation by LC (114). With this detection technique the phosphorescent compound biacetyl is incorporated into the mobile phase. An eluted compound can increase (sensitize) the phosphorescence background of the biacetyl by absorbing incident radiation, intersystem crossing to the triplet state, then exciting a biacetyl molecule via an intermolecular triplet–triplet energy transfer. Alternatively, an eluted compound can decrease (quench) the phosphorescence background of the biacetyl by depleting the excited triplet state population of the biacetyl again via intermolecular triplet–triplet energy transfer. Limits of detection for the compounds tested were in the low to sub-nanogram injected range. The relatively high limits of detection are a result of the signal acquisition procedure. A change in a relatively large, and possibly fluctuating, phosphorescence background was measured rather than a positive phosphorescence signal over a near-zero background, as would be the case in a direct phosphorescence measurement.

The products of chemical reactions are often formed with excess energy. This excess energy is usually disposed of thermally via energetic collisions with other molecules or the walls of the container in which the reaction takes place. Occasionally reaction products are left in an excited state from which radiative emission, termed chemiluminescence, occurs. Alternatively, the excited product can transfer its energy to a fluorosphor, promoting it to an excited state from which radiative emission occurs. Seitz has recently reviewed the principles, instrumentation, and modern applications involving chemiluminescence detection (115).

Chemiluminescence measurements are generally very sensitive and, owing to the specificity of the chemical reaction involved, are also very selective. The high sensitivity is primarily the result of the low optical background levels that are observed with the technique. Because no excitation radiation source is used stray radiation does not contribute to the background. Instead, the limiting background is determined by the level of reactive impurities in the solvents and reagents employed. The low background levels permits the use of sensitive photon counting (see Section 3.1.6) signal recovery equipment.

As with fluorescence derivatization performed postcolumn, chemiluminescence detection in LC requires postcolumn mixing chambers and reactors. Unlike derivatization, the reaction that produces the chemiluminescent product must take place within, not prior to, the flow cell. Because of these complications detector-related solute band dispersion can be a significant concern. Nevertheless, several researchers have developed and characterized chemiluminescence detectors for LC (116–119).

Much of the work involving chemiluminescence detection in LC has utilized the peroxyoxalate system (118); with this system the peroxyoxalate [bis(2,4,6-trichlorophenyl)oxalate] molecule reacts postcolumn with hydrogen peroxide to form the reactive intermediate 1,2-dioxetanedione, which further reacts with fluorophores in the column effluent to form an excited fluorophore. The advantage of this system is that it permits the detection of a fairly wide range of fluorophores. However, it is not applicable to all fluorophores and, in fact, the observed chemiluminescence signal levels have been shown not to correlate well with fluorescence quantum efficiencies (120). Sigvardson et al. (119) used the peroxyoxalate system to detect amino polycyclic aromatic compounds separated by LC (see the chromatogram in Fig. 14). A Kratos URS 051 postcolumn mixing device and a Kratos FS 970 fluorescence detector were used in that work. For chemiluminescence detection the detector was operated without the lamp on and

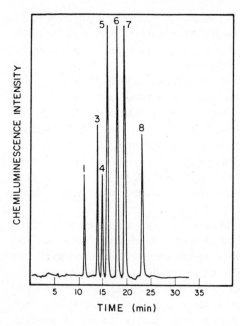

Figure 14. HPLC chromatogram of seven amino-PAH standards with chemiluminescence detection, 75% acetonitrile at 0.7 mL/min with two Zorbax-ODS columns in series: (1) 1-aminonaphthalene, 250 pg; (3) 2-aminofluorene, 3.0 ng; (4) 9-aminophenanthrene, 150 pg; (5) 1-aminoanthracene, 110 pg; (6) 3-aminofluoranthene, 50 pg; (7) 1-aminopyrene, 80 pg; (8) 6-aminochrysene, 140 pg. Reprinted with permission from K. W. Sigvardson, J. M. Kennish, and J. W. Birks, *Anal. Chem.*, **56,** 1096 (1984). Copyright 1984 American Chemical Society.

without any emission wavelength selection. Detection limits were in the sub-picogram injected range and 1–2 orders of magnitude better than those obtained for fluorometric detection using the same instrumentation.

This work was supported by the Division of Chemical Sciences, Office of Basic Energy Sciences, U.S. Dept. of Energy, under Contract no. DEASO5-83ER13127 with the University of Tennessee (Knoxville).

REFERENCES

1. D. M. Hercules, Ed., *Fluorescence and Phosphorescence Analysis,* Wiley-Interscience, New York, 1966.

2. G. H. Schenk, *Absorption of Light and Ultraviolet Radiation: Fluorescence and Phosophorescence Emission,* Allyn and Bacon, Boston, 1973.

3. G. G. Guilbault, *Practical Fluorescence: Theory, Methods, and Techniques,* Dekker, New York, 1973.

4. J. D. Winefordner, S. G. Schulman, and T. C. O'Haver, *Luminescence Spectrometry in Analytical Chemistry,* Wiley-Interscience, New York, 1972.

5. D. M. Hercules, in D. M. Hercules, Ed., *Fluorescence and Phosoporescence Analysis,* Wiley-Interscience, New York, 1966, p. 12 and references cited therein.

6. D. E. Burton and M. J. Sepaniak, "Determination of B_6 Vitamers in Blood by Ion-Pair Reverse Phase HPLC With Laser Excited Fluorescence Detection," paper presented at The Pittsburgh Conference and Exposition on Analytical Chemistry and Applied Spectroscopy, Atlantic City, NJ, 1983.

7. T. R. Hopkins and R. Lumry, *Photochem. Photobiol.*, **15**, 555 (1972).

8. P. M. Froehlich and M. Yeats, *Anal. Chim. Acta*, **87**, 185 (1976).

9. A. T. Balchunas, M. J. Capacci, M. J. Sepaniak, and M. P. Maskarinec, *J. Chrom. Sci.*, **23**, 381 (1985).

10. W. Bragg, *The Universe of Light*, Dover, New York, 1959.

11. H. H. Willard, L. L. Meritt, J. A. Dean, and F. A. Settle, *Instrumental Methods of Analysis*, 6th ed., Van Nostrand, New York, 1981, p. 45.

12. J. W. Lyons and L. R. Faulkner, *Anal. Chem.*, **54**, 160 (1982).

13. J. H. Knox, *J. Chromatogr. Sci.*, **18**, 453 (1980).

14. J. H. Knox and M. T. Gilbert, *J. Chromatogr.*, **186**, 405 (1979).

15. Robert Weinberger, Kratos Analytical Instruments, personal communication, June 1985.

16. R. W. Engstrom, *Photomultiplier Handbook*, R. C. A. Corporation, Lancaster, PA, 1980.

17. Ref. 11, p. 120.

18. H. V. Malmstadt, M. L. Franklin, and G. Horlick, *Anal. Chem.*, **44**, 63a (1972).

19. Y. Talmi, D. C. Baker, J. R. Jadmec, and W. A. Saner, *Anal. Chem.*, **50**, 936a (1978).

20. G. D. Christian, J. B. Callis, and E. R. Davidson, "Array Detectors and Excitation-Emission Matrices in Multicomponent Analysis," in E. L. Wehry, Ed., *Modern Fluorescence Spectroscopy*, Vol. 4, Plenum, New York, 1981.

21. E. F. Hilinski, S. V. Milton, and P. M. Rentzepis, "Multichannel Image Detectors Volume 2," in Y. Talmi, Ed., *ACS Symposium Series Vol. 236*, American Chemical Society, Washington, DC, 1979.

22. J. R. Jadmec, W. A. Saner, and Y. Talmi, *Anal. Chem.*, **49**, 1316 (1977).

23. L. W. Hershberger, J. B. Callis, and G. D. Christian, *Anal. Chem.*, **53**, 971 (1981).

24. G. M. Hieftje, *Anal. Chem.*, **46**, 69a and 81a (1972).

25. A. J. Difenderfer, *Principles of Electronic Instrumentation*, Saunders, Philadelphia, 1972.

26. G. J. Diebold and R. N. Zare, *Science*, **196**, 1439 (1977).

27. N. Furuta and A. Otsuki, *Anal. Chem.*, **55**, 2407 (1983).

28. 1984–1985 LabGuide, *Anal. Chem.*, **56**, (1984).

29. 1985 Buyers Guide Edition, *American Laboratory*, **16** (1985).

30. 1984–1985 Guide to Scientific Instruments, American Association for the Advancement of Science, Washington, DC 1985.

31. H. V. Malmstadt, C. G. Enke, S. R. Crouch, and G. Horlich, *Optimizatin of Electronic Measurements*, W. A. Benjamin, Menlo Park, CA, 1974, p. 14.

32. J. W. Lyons, P. T. Hardesty, C. S. Baer, and L. R. Faulkner, "Structural Interpretation of Fluorescence Spectra by Automated File Searching. Implementation and Applications in Liquid Chromatography," in E. G. Wehry, Ed., *Modern Fluorescence Spectroscopy*, Vol. 4, Plenum, New York, 1981.

33. T. VoDinh, "Synchronous Excitation Spectroscopy," in E. L. Wehry, Ed., *Modern Fluorescence Spectroscopy*, Vol. 4, Plenum, New York, 1981.

34. P. A. Asmus, J. W. Jorgenson, and M. Novotny, *J. Chromatogr.*, **126**, 317 (1976).

35. H. Nakamura and Z. Tamura, *Anal. Chem.*, **53**, 2190 (1981).

36. G. Kamperman and J. C. Kraak, *J. Chrom.*, **337**, 384 (1985).

37. J. DeJong, J. P. Schouten, R. G. Muusze, and U. R. Tjaden, *J. Chromatogr.*, **319**, 23 (1985).

38. F. Andreolini, C. Borra, A. DiCorcia, A. Lagana, R. Samperi, and G. Rapini, *Clin. Chem.*, **30**, 742 (1984).

39. F. Geeraerts, L. Schimpfessel, and R. Crokaert, *Chromatographia*, **15**, 449 (1982).

40. D. M. Takahashi, *J. Chromatogr.*, **131**, 147 (1977).

41. T. Nagata, M. Saeki, H. Nakazawa, M. Fujita, and E. Takabatake, *J. Chromatogr.*, **281**, 367 (1983).

42. D. M. Victor, R. E. Hall, J. D. Shamis, and S. A. Whitlock, *J. Chromatogr.*, **283**, 383 (1984).

43. Y. Kitada, M. Sasaki, and K. Tanigawa, *J. Assoc. Off. Analy. Chem.*, **65**, 1302 (1982).

44. K. J. Harzer, *J. Chromatogr.*, **249**, 205 (1982).

45. J. A. Glasel and R. F. Venn, *J. Chromatogr.*, **213**, 337 (1981).

46. R. T. Gallagher, A. D. Hawkes, and J. M. Stewart, *J. Chromatogr.*, **321**, 217 (1985).

47. N. Terao and D. D. Shen, *Chromatographia*, **15**, 685 (1982).

48. J. E. Ray and R. O. Day, *J. Chromatogr.*, **336**, 234 (1984).

49. W. Roth, *J. Chromatogr.*, **278**, 347 (1983).

50. B. A. Mico, R. A. Baughman, Jr., and L. Z. Benet, *J. Chromatogr.*, **230**, 203 (1982).

51. M. Novotny, A. Hirose, and D. Wiesler, *Anal. Chem.*, **56**, 1243 (1984).

52. F. Andreolini, C. Borra, A. DiCorcia, A. Lagana, R. Samperi, and G. Raponi, *Clin. Chem.*, **30**, 7442 (1984).

53. G. J. Krol, C. A. Mannan, R. E. Pickering, D. V. Amato, B. T. Kho, and A. Sonnenschein, *Anal. Chem.*, **49**, 1836 (1977).

54. J. F. Lawrence, *J. Chromatogr. Sci.*, **17**, 1447 (1979).

55. J. F. Lawrence and R. W. Frei, *Chemical Derivatization in Liquid Chromatography*, Elsevier, Amsterdam, 1976.

56. H. Lingeman, W. J. M. Underberg, A. Takadata, and A. Hulshoff, *J. Liq. Chromatogr.*, **8**, 789 (1985).

57. C. Fischer, K. Maier, and U. Klotz, *J. Chromatogr.*, **225**, 498 (1981).

58. J. S. Bannister, J. Stevens, D. Musson, and L. A. Sternson, *J. Chromatogr.*, **176**, 381 (1979).

59. L. A. Sternson, N. Meltzer, and F. Shih, *Anal. Lett.*, **14**, 583 (1981).

60. G. M. Murray and M. J. Sepaniak, *J. Liq. Chromatogr.*, **6**, 931 (1983).

61. S. Katz, W. F. Pitt, Jr., and G. Jones, Jr., *Clin. Chem.*, **19**, 817 (1974).

62. A. W. Wolkoff and R. H. Larose, *J. Chromatogr.*, **99**, 731 (1974).

63. K. Yawazwa and Z. Tamura, *J. Chromatogr.*, **254**, 327 (1983).

64. S. P. Assenza and P. R. Brown, *J. Chromatogr.*, **289**, 355 (1984).

65. S. H. Lee, L. R. Field, W. N. Howald, and W. F. Trager, *Anal. Chem.*, **53**, 467 (1981).

66. U. A. Th. Brinkman, P. L. M. Welling, G. deVries, A. H. M. T. deScholten, and R. W. Frei, *J. Chromatogr.*, **217**, 463 (1981).

67. P. J. Twitchett, P. L. Williams, and A. C. Moffat, *J. Chromatogr.*, **149**, 683 (1978).

68. M. Uihlein and E. Schwab, *Chromatographia*, **15**, 141 (1982).

69. C. E. Werkhoven-Goewie, W. M. A. Niessen, U. A. Th. Brinkman, and R. W. Frei, *J. Chromatogr.*, **203**, 165 (1981).

70. K. Miura, H. Nakamura, H. Tanaka, and Z. Tamura, *J. Chromatogr.*, **210**, 536 (1981).

71. A. H. M. T. Scholten, U. A. Th. Brinkman, and R. W. Frei, *Anal. Chem.*, **54**, 1932 (1982).

72. T. Seki and Y. Yamaguchi, *J. Chromatogr.*, **305**, 188 (1984).

73. K. Miyazaki, K. Ohtani, K. Sunada, and T. Arita, *J. Chromatogr.*, **276**, 478 (1983).

74. W. Buchberger, K. Winsauer, and F. Nachtmann, *Fres. Z. Anal. Chem.*, **315**, 525 (1983).

75. K. Kakehi, T. Konishi, I. Sugimoto, and S. Honda, *J. Chromatogr.*, **318**, 367 (1985).

76. S. J. Prince, T. Palmer, and M. Griffin, *Chromatographia*, **18**, 62 (1984).

77. T. D. Schlabach and T. C. Wehr, *Anal. Biochem.*, **127**, 222 (1982).

78. A. Bettero, M. R. Angi, F. Moro, and C. A. Benassi, *J. Chromatogr.*, **310**, 390 (1984).

79. M. Kai, T. Miyazaki, and Y. Ohkura, *J. Chromatogr.*, **311**, 257 (1984).

80. S. R. Abbott, A. Abu-Shumays, K. O. Loeffer, and I. S. Forrest, *Res. Commun. Chem. Pathol. Pharmacol.*, **10**, 9 (1975).

81. W. Duenges and N. Seiler, *J. Chromatogr.*, **145**, 483 (1978).

82. P. J. Ryan, K. McGoldrick, D. Stickney, and T. W. Honeyman, *J. Chromatogr.*, **320**, 421 (1985).

83. L. K. She, U. A. T. Brinkman, and R. W. Frei, *Anal. Lett.*, **17**, 915 (1984).

84. J. Lantos, U. A. T. Brinkman, and R. W. Frei, *J. Chromatogr.*, **292**, 117 (1984).

85. C. E. Werkhoeven-Goewie, U. A. Th. Brinkman, and R. W. Frei, *Anal. Chim. Acta*, **114**, 147 (1980).

86. D. J. Sweeney, N. H. Greig, and S. I. Rapoport, *J. Chromatogr.*, **339**, 434 (1985).

87. P. E. Nelson, S. L. Nolan, and K. R. Bedford, *J. Chromatogr.*, **234**, 407 (1982).

88. D. Dye, T. East, and W. F. Bayne, *J. Chromatogr.*, **284**, 457 (1984).

89. R. M. Carlson, T. A. Swanson, A. R. Oyler, M. T. Lukasewycz, R. J. Liukkonen, and K. S. Voelkner, *J. Chromatogr. Sci.*, **22**, 272 (1984).

90. C. Baiocchi, E. Campi, M. Gennaro, E. Mentasti, and P. Mirti, *Chromatographia*, **15**, 661 (1982).

91. J. Goto, F. Shamsa, N. Goto, and T. Nambra, *J. Pharm. Biomed. Anal.*, **1**, 83 (1983).

92. E. S. Yeung and M. J. Sepaniak, *Anal. Chem.*, **52**, 1465A (1980).

93. M. J. Sepaniak and E. S. Yeung, *J. Chromatogr.*, **180**, 337 (1980).

94. M. Novotony, *Anal. Chem.*, **53**, 1294A (1981).

95. D. Ishii, K. Asai, K. Hibi, T. Jonokuchi, and M. Nagaya, *J. Chromatogr.*, **144**, 157 (1977).

96. E. J. Guthrie and J. W. Jorgenson, *Anal. Chem.*, **56**, 483 (1984).

97. E. J. Guthrie, J. W. Jorgenson, and P. R. Dluzneski, *J. Chromatogr. Sci.*, **22**, 171 (1984).

98. M. J. Sepaniak, J. D. Vargo, C. N. Kettler, and M. P. Maskarinec, *Anal. Chem.*, **56**, 1252 (1984).

99. M. J. Sepaniak, *Clin. Chem.*, **31**, 671 (1985).

100. S. Folestad, L. Johnson, B. Josefsson, and B. Galle, *Anal. Chem.*, **54**, 925 (1982).

101. L. W. Hershberger, J. B. Callis, and G. D. Christian, *Anal. Chem.*, **51**, 1444 (1979).

102. M. J. Sepaniak and E. S. Yeung, *Anal. Chem.*, **49**, 1554 (1977).

103. P. B. Huff, B. J. Tromberg, and M. J. Sepaniak, *Anal. Chem.*, **54**, 946 (1982).

104. W. M. McClain, *Acc. Chem. Res.*, **7**, 199 (1974).

105. M. Kasha, *Farady Soc. Discuss.*, **9**, 14 (1950).

106. T. Hirschfeld, *Anal. Chem.*, **52**, 297A (1980).

107. D. R. Baker, R. C. Williams, and J. R. Steichen, *J. Chromatogr. Sci.*, **12**, 499 (1974).

108. M. J. Sepaniak and E. S. Yeung, *J. Chromatogr.*, **211**, 95 (1981).

109. M. J. Sepaniak and E. S. Yeung, *Fuel Preprs*, **26**(2), 1 (1981).

110. E. Voigtman, A. Jurgensen, and J. D. Winefordner, *Anal. Chem.*, **53**, 1921 (1981).

111. M. Zander, *Phosphorimetry*, Academic, New York, 1968, p. 118.

112. R. J. Hurtubise, *Anal. Chem.*, **55**, 669A (1983).

113. S. Scypinski and L. J. Cline Love, *Anal. Chem.,* **56,** 322 (1984).

114. J. J. Donkerbroek, N. J. R. Van Eikeme Hommes, C. Gooijer, N. H. Velthorst and R. W. Frei, *J. Chromatogr.,* **255,** 581 (1983).

115. W. R. Seitz, *CRC Crit. Rev. Anal. Chem.,* **13,** 1 (1981).

116. G. Melbin, *J. Liq. Chromatogr.,* **6,** 1603 (1983).

117. R. Weinberger, C. A. Mannan, M. Cerchio, and M. L. Grayeski, *J. Chromatogr.,* **288,** 445 (1984).

118. M. S. Gandelman and J. W. Birks, *J. Chromatogr.,* **242,** 21 (1982).

119. K. W. Sigvardson, J. M. Kennish, and J. W. Birks, *Anal. Chem.,* **56,** 1096 (1984).

120. M. M. Rauhut, *Acc. Chem. Res.,* **2,** 80 (1969).

POLARIMETRIC DETECTORS

EDWARD S. YEUNG

Ames Laboratory-USDOE and Department of Chemistry
Iowa State University, Ames, Iowa

1. INTRODUCTION

The measurement of rotation in the polarization plane of light provides unique information about the analytes separated by LC. The most obvious application is the measurement of the optical activity. This molecular characteristic is common to chiral species, that is, those that can be distinctly classified as either left- or right-handed. In many situations, nature makes use of the special conformation to provide specificity to biochemical interactions. The presence of optical activity is thus usually a good indication of biological activity. An optical activity detector for LC can then provide a special kind of selectivity for studying complex samples. With the growing interest in biotechnology and genetic engineering, this type of selectivity can be an important consideration for chemical analysis.

There is also a place for an optical activity detector in routine applications of LC. Even though the absorption detector is suitable for a large number of situations in LC, there are cases where the analyte does not absorb strongly in a convenient region of the spectrum. The common RI detector has limited utility because of the poor overall detectability. One must then try to monitor a totally different molecular property, such as its optical activity. The additional advantage of an optical activity detector in LC is that most common chromatographic solvents are optically inactive and will not contribute to the detected signal. This makes solvent purity a less important consideration and allows gradient elution to be used.

The rotation in the polarization plane of light is not restricted to optically active molecules (see below). With the proper design, it is possible to obtain a signal from optically inactive species. The function of the detector is then substantially broadened in scope.

It is obvious that the usefulness of an optical activity detector depends on its detectability. A theoretical consideration based on common commercial polarimeters gives a detectability of 3×10^{-4} g/cm^3 in the optical path (1), corresponding to a rotation of 1.5×10^{-3} deg. Because dilution occurs during separation, this level is of little practical value to HPLC. With a larger ion-exchange column, a modified microcell of 110-μL volume has been used with a Bendix-NPL automatic polarimeter to study the chromatographic eluates (2). For a rotation of $0.01°$, the S/N level is only 4. A Perkin-Elmer 241 Spectropolarimeter has also been used for LC (3). Even with the use of a short wavelength (302 nm) to take advantage of a higher specific rotation, and with the use of a special flow cell of 10-cm path and 33-μL volume, the detectability was 60 μg of material (4). It is clear that the instrumentation has to be improved drastically for routine LC work.

2. THEORETICAL CONSIDERATIONS

The basic components of a polarimeter are a light source, a pair of plane polarizers, and a photoelectric detector. With the polarizers oriented at 90 degrees relative to each other, very little light reaches the detector. If an optically active material is placed between the polarizers, light is transmitted because the polarizers are effectively no longer "crossed." Qualitatively, it can be seen that the intensity of the light source must be stable, and should be large enough to provide good photon statistics, so that the intensity measurement is precise. It is also apparent that the transmission at the crossed orientation should be as low as possible, so that small increases in light can be readily observed.

In the existing literature, there is some confusion as how best to optimize the S/N ratio for measurements of polarization rotation. One report indicates that low-quality polarizers are sufficient to produce good results (5). Another report (6) finds that the quality of the polarizers is the key to high sensitivity. Some experimental arrangements call for a small offset from maximum extinction (7), but others depend on working at the point of maximum extinction (6). It is apparent that a detailed analysis of the S/N ratio for each type of experimental arrangement is needed to allow proper optimization of the measurements.

The signal obtained from a polarimeter can be derived using the Jones matrix formalism (5). Essentially, one measures the intensity of light transmitted through a pair of "crossed" polarizers aligned along the x axis and the y axis, respectively. Neglecting any attenuation in the polarizing element owing to absorption or natural reflection, there is still a nonideal behavior, characterized by a parameter k (5), which causes transmission even if the polarizers are perfectly crossed. The extinction ratio, defined as the fraction of light transmitted at the crossed alignment, is

$$\varepsilon \simeq |k_x|^2 + |k_y|^2 \tag{1}$$

For the purpose of understanding optimization, one can set $k_x = k_y = k$. If I_0 is the intensity of light of the correct polarization that enters the first polarizer, then the intensity of light transmitted, I_T, is given by

$$I_T = I_0[2|k|^2\cos^2(\theta + \chi) + (1 + |k|^4)\sin^2(\theta + \chi)] \qquad (2)$$

Here θ denotes the offset of the second polarizer from the crossed position, and χ denotes the amount of rotation caused by the sample in between the polarizers. In Eq. 2 we have neglected any depolarization due to circular dichroism in the sample. A more convenient form is obtained by combining Eqs. 1 and 2:

$$I_T = I_0\left\{\varepsilon \cos^2(\theta + \chi) + \left[1 + \left(\frac{\varepsilon}{2}\right)^2\right]\sin^2(\theta + \chi)\right\} \qquad (3)$$

The signal due to the presence of the sample is readily identified as

$$S = I_0[f(\theta, \chi) - f(\theta, 0)] \qquad (4)$$

where f is the functional relationship inside the curly brackets in Eq. 3. If modulation is used in conjunction with a lock-in amplifier (6), the output is proportional to

$$S' = I_0\{[f(\theta + \alpha, \chi) - f(\theta - \alpha, \chi)] - [f(\theta + \alpha, 0) - f(\theta - \alpha, 0)]\} \qquad (5)$$

where α is the modulation extent. For $\theta = 0$ and small rotations χ in the sample, one can see that Eq. 4 gives a signal proportional to the square of the rotation, whereas Eq. 5 gives a signal directly proportional to the rotation.

A complete account of the noise in a given experiment is usually very difficult. One can, however, sort out different types of noise according to their functional behavior. Shot noise is simply $\sqrt{I}$, if I is in units of number of photons observed during the time of data acquisition. Flicker noise including certain amplifier noise, laser intensity fluctuations, and beam misalignment variations is given by AI, where A is a fractional constant that depends on the experiment. Other types of noise from the electronics may show no intensity dependence, and can be characterized by a parameter B. Thus the total noise is

$$N = \sqrt{I + A^2I^2 + B^2} \qquad (6)$$

Explicitly, I is the signal obtained without the sample rotation χ, in number of photons. For Eqs. 4 or 5,

$$I = I_0 f(\theta, 0) \qquad (7)$$

We note that by using a lock-in amplifier in Eq. 5, one can optimize A and B by providing measurements at a frequency that can be selected. For small modulation (α) and small signals (χ), Eq. 5 reduces to $S' = 4I_0\alpha\chi \cos \theta$. One should then always work at null ($\theta = 0$) where S' is maximized and I, and thus N, is minimized. When modulation is used, instabilities in the modulation depth will also contribute to the total flicker noise.

The implication of Eqs. 4, 6, and 7 is that we can calculate the S/N for a given set of experimental conditions. We note that the S/N ratio normalized to χ is inversely proportional to the detectability of the polarimeter. For simplicity, we can neglect the contributions from B. Even for the extreme case of a 1 μW of visible light and $\varepsilon = 10^{-10}$, the typical photocurrent available from a photomultiplier tube is about 1 nA. A good picoammeter will allow readings down to 10^{-13} A, so that B will be important only if A is smaller than 10^{-4}. It has been shown (5) that in the shot-noise limit, that is $A = B = 0$, the best S/N ratio that can be achieved is independent of ε for $\varepsilon < 10^{-3}$. The optimum angle, however, depends on ε, such that

$$\theta_{opt} = \arctan \varepsilon^{1/4} \qquad (8)$$

In such cases, low-quality polarizers will suffice. With $A \neq 0$, the situation is quite different. This is because for 500 nm light, the shot noise for a 1-μW light beam is only 1 part in 10^6 for 1 s. This is substantially smaller than the fractional flicker noise, A, in most light sources. To maintain good detection in polarization rotation, one needs to maintain as low a light flux reaching the phototube as possible, so that the total flicker noise can be reduced. For the more stable laser sources, a fractional flicker noise $A = 10^{-3}$ is typical. This implies one should try to approach 0.4 pW of power at the phototube while maintaining as high a power level as possible at the sample. An excellent set of polarizers at the crossed alignment is thus needed.

Figure 1 shows the dependence of θ_{opt}, the offset angle at which the S/N ratio is maximized, on the flicker noise A. The data points are generated by using Eqs. 4, 6, and 7 for a range of θ values at a given A, and then comparing the S/N ratios. At large values of A, the contributions of shot noise can be neglected. It is then desirable to operate near the crossed alignment position to minimize the light intensity reaching the detector. Because ε is not zero, one can operate slightly away from the crossed alignment, that is, at $\theta \neq 0$, without substantially increasing the residual transmitted intensity. The signal level can thus be increased without increasing the noise. Figure 1 shows that when A is small, the optimum offset angle will eventually increase to the value predicted by Eq. 8. For higher laser powers, the fractional flicker noise A must be even smaller before θ_{opt} can increase to larger values. Figure 1 shows that even for low-power light sources with typical values of $A = 10^{-2}$–10^{-4}, the shot-noise limit (5) is

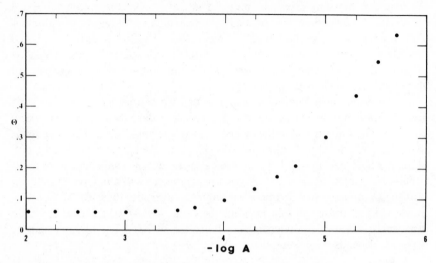

Figure 1. Dependence of the optimum offset angle (degrees) on the flicker noise: $\varepsilon = 10^{-6}, I_0 = 5$ μW.

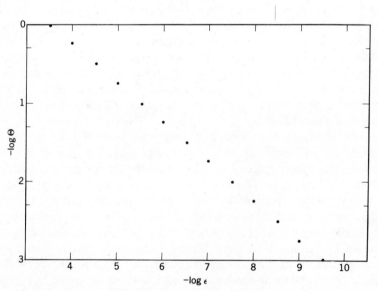

Figure 2. Dependence of the optimum offset angle (degrees) on the extinction ratio of polarizers: $A = 10^{-2}, I_0 = 5$ mW.

not important. A final point about Figure 1 is that even though θ is not affected much by typical values of A, the actual S/N ratio depends strongly on A. To a first approximation, the S/N ratio is inversely proportional to A.

Figure 2 shows the dependence of θ_{opt} for various extinction ratios, ε, when A is fixed. Within the range displayed, a linear dependence is found, with a slope of -0.510. Eq. 8 predicts a slope of -0.250. The difference of a factor of 2 is accidental. The slope gradually increases from -0.250 to larger negative values as A increases. Also, in contrast to the shot-noise limit (5), the S/N ratios for the data points in Figure 2 depend on ε. Some representative values are tabulated in Table 1. For a light source with significant flicker noise ($A = 10^{-2}$), substantial improvements can be made by using high-quality polarizers. For a stable light source ($A = 10^{-4}$), however, the advantage of high-quality polarizers will become significant only if a high light intensity is available. Conversely, one does not require a high-intensity light source unless the polarizers are of high quality and the source intensity is stable. It is interesting to note that the optimal S/N ratio is achieved for $\varepsilon = 10^{-10}$, $I_0 = 5$ mW, and $A = 10^{-4}$, when $\theta = 1.6 \times 10^{-3}$ deg, which is an insignificant offset from the crossed alignment, given the mechanical rigidity of typical optical mounts. At the crossed alignment (i.e., $\theta = 0$), however, the S/N ratio drops from 3610 to 340. This implies that finely adjustable rotational stages are desirable in polarimeters. Table 1 also shows that further gains in S/N ratio are possible by stabilizing the intensity of the laser.

Thus the above S/N analysis shows that it is desirable to have good polarizers, moderate light intensities, and good intensity stabilization in the construction of polarimeters.

Table 1. Optimized S/N Values per Millidegree of Sample Rotation in Polarimeters

	I_0		
	5 μW	5 mW	50 mW
$\varepsilon = 10^{-6}$			
$A = 10^{-2}$	1.75	1.75	1.75
$A = 10^{-4}$	87.5	174	176
$\varepsilon = 10^{-10}$			
$A = 10^{-2}$	92	190	190
$A = 10^{-4}$	124	3610	9200

3. INSTRUMENTATION

A schematic of the laser-based polarimetric detector is shown in Fig. 3. About 0.5 W of 514.5-nm radiation was used from an argon ion laser. A 1-m fl crown glass lens was used to collimate the light through the flow cell. A pair of selected 8-mm aperture Glan prisms served as the polarizer and analyzer, respectively. The prisms were mounted in rotational stages with a resolution of 10^{-3} deg. The rotational stages were in turn mounted rigidly on a vibration-isolated optical table. The separation between the prisms was kept at about 2 m to reduce stray light. Appropriate apertures were also used at various points in the optical path for the same reason. The flow cell was machined from an aluminum cylinder 10 cm long and 5 cm in diameter. A 1.58-mm-diameter hole was drilled along its axis as the active region, which was intercepted at both ends at 60° by 0.89-mm bores to interface with the chromatographic plumbing. The cell was held rigidly with spring-loaded positioners at each end for optical alignment. Cell windows were selected microscope cover slips, 25-mm square, held onto the cell by a rubber cement. A modulating and a compensating Faraday rotator were placed in the optical path. These were based on air as the active medium and were constructed by winding 8000 turns of #30 gauge magnet wire along 10 cm of 4.8-mm OD, thin-wall, nonferromagnetic stainless steel tubing. Each Faraday rotator was mounted at both ends with positioners for alignment purposes. Light is detected after suitable apertures through a narrow-band interfer-

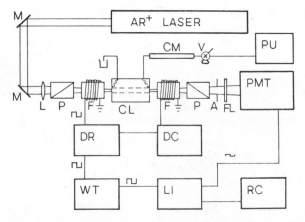

Figure 3. Experimental arrangement for the optical activity detector. M, Mirrors; P, Glan prisms; F, Faraday rotator; CL, flow cell; A, aperture; FL filter; PMT, photomultiplier tube; DR, driver; DC, power supply; WT, wave generator; LI, lock-in amplifier; RC, recorder; PU, pump; V, injection valve; CM, column; L, collimation lens. Reproduced with permission from E. S. Yeung, L. E. Steenhoek, S. D. Woodruff, and J. C. Kuo, *Anal. Chem.*, **52**, 1399 (1981). Copyright 1981 American Chemical Society.

ence filter centered around 514.5 nm, by a photomultiplier tube. The output of the photomultiplier tube was terminated in a 100-kΩ resistor, and the voltage was monitored by a lock-in amplifier. The output of the lock-in amplifier was displayed on a chart recorder. Modulation was derived from a switching amplifier driven by a wave generator. Typically a 50–100-V, 210-Hz square wave was applied to the modulating Faraday cell. The compensating Faraday cell was driven by a variable, 0–25-V power supply for fine alignment and to provide a calibration.

A cursory inspection of Fig. 3 leads to the conclusion that the laser-based polarimeter consists of components similar to those of commercial spectropolarimeters. There are, however, very important differences that allow improved detectability.

3.1. Polarizers

Commercial polarizers of the highest quality are generally of the Glan–Thompson design and provide an extinction ratio, ε, of 10^{-6}. We have confirmed Moeller and Grieser's report (8) that using a second tandem polarizer and a second tandem analyzer does not improve the extinction ratio, because the major contributions come from imperfections (scattering centers and strain) within the calcite crystals. The half of the polarizer away from the source and the half of the analyzer toward the source are the contributing parts. Improvement was observed (8) by limiting the aperture so that "good" localized regions were used, and an extinction ratio of 3×10^{-8} was reported. We found that by selecting specific combinations of crystals as analyzers and polarizers, and specific combinations of entrance and exit faces for each, an extinction ratio of 10^{-10} could be obtained in one of the many combinations for four Glan prisms. The worst case among these combinations, however, still met the manufacturer's specifications of 10^{-6}. An important observation is that the 10^{-10} extinction ratio can be routinely achieved after the initial trial-and-error process, even after a full year of use.

It is interesting to note that an extinction ratio of 10^{-10} is equivalent to having "noise" from a 5.7×10^{-4} deg misalignment using perfect polarizers and analyzers. For this reason, the prisms must be rigidly mounted and must have fine angular adjustments. Our original mounting plates of aluminum show enough deformation when torque is exerted manually to one end that one can actually detect a larger transmitted intensity.

3.2. Faraday Rotators

Normally Faraday rotators in polarimeters are made of quartz or a liquid such as water. In order not to introduce any additional scattering centers or strain-induced birefringence, which result in large values for A, the Faraday rotators

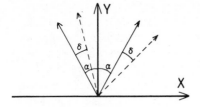

Figure 4. Vector diagram for polarization modulation. Reproduced with permission from E. S. Yeung, L. E. Steenhoek, S. D. Woodruff, and J. C. Kuo, *Anal. Chem.*, **52**, 1399 (1981). Copyright 1981 American Chemical Society.

are based on air as the medium. A rough calculation shows that the solenoids produce 1000 G/A of current, or about 1×10^{-3} deg of rotation per ampere. This is an unusually small amount of modulation, but can be shown to be adequate for LC applications. The reason that modulation is required is that A can be improved by choosing an observation frequency free from environmental noise in the laboratory. Modulating the polarization is preferred (as opposed to amplitude modulation) because one can then average out all other background contributions.

We can calculate the signal strength that can be obtained. In Fig. 4, the axes are chosen as the laboratory frame of the analyzer; that is, light polarized in the x direction is unobstructed, but light in the y direction is rejected, to the extent of the extinction ratio. For a modulation angle of α, the maximum amount of light transmitted can be found as the projection of that vector on the x axis, or $E \sin \alpha$, where E is the field amplitude. The intensity, being the square of the amplitude, is thus $I_0 \sin^2 \alpha$, where I_0 is the incoming light intensity. The optimum modulation is from an angle α to an angle $-\alpha$ from the y axis, so that the lock-in signal is zero. When an optically active sample is present, the modulation is then from $\alpha + \chi$ to $-\alpha + \chi$, with χ being the extent of rotation of the sample. It is easy to see that the lock-in signal is proportional to

$$I = I_0[\sin^2(\alpha + \chi) - \sin^2(\alpha - \chi)] \tag{9}$$

where I is the transmitted intensity. For small angles, which is always the case here, we have

$$I = 4I_0\alpha\chi \tag{10}$$

It is therefore desirable to use as large a modulation α as possible, provided that the stability of the modulation is better than χ. The modulation frequency is arbitrary, and was chosen in this work to coincide with a relatively noise-free region in our laboratory environment.

3.3. Flow Cell

The dimensions of the flow cell are a compromise between having a long light path and a small volume. For focusing a Gaussian beam such as a laser, we have

$$z = \frac{\pi w^2}{\lambda} \tag{11}$$

where z is the range of collimation (the Rayleigh range), w is the beam radius at the $1/e^2$ intensity points, λ is the wavelength, and all three have the same units. For a path length of 10-cm and 500-nm light, we find that the beam radius at the waist is 126 μm. Because $2w$ gives the 1/99 intensity point and the beam size at the extremes of the Rayleigh range is twice that at the waist, the radius of the cell bore should be 500 μm, corresponding to a cell volume of 80 μL. Naturally, depending on whether one is interested in trace analysis or microanalysis, the optimum dimension will change. It can be shown (9) that the cell volume is proportional to the square of the path length for Gaussian beams.

The windows on the flow cell introduce additional scattering centers and birefringence, which in turn increases A, and these must be reduced to a minimum. We have tested various window materials including the highest-quality laser optical compenents using the selected polarizers. One solution is to use thin window material so that the optical path is minimized. Standard commercial microscope cover slips were used, and we found that about 1 in 10 showed good enough optical quality that the 10^{-10} extinction ratio was maintained with these in place. In fact, in certain cases, the resulting extinction ratio was improved, probably due to a cancellation of the birefringence in the polarizing prisms. Mounting these cover slips presented another problem, because they must remain strain free; a silicone cement that remains slightly flexible even when cured can be used. Once the windows were in place little deterioration occurred in a month.

3.4. Other Developments

After the original laser-based polarimeter was developed (6), several improvements were made in the instrumental design. Laser intensities as low as 5 mW (from a HeNe laser) have been used without degrading the S/N ratio (10). This is consistent with the results in the last row of Table 1. The longer wavelength used does in general lead to smaller specific rotations for molecules, but the higher stability in intensity in HeNe lasers is advantageous. The simplicity of a HeNe laser makes it an attractive light source as a routine instrument in the analytical laboratory.

The original Faraday coils are quite bulky and require large currents to drive. Heat from the coils causes additional instabilities in the system. We later realized that there is always chromatographic solvent in the optical regions and that can become the medium for Faraday rotation without increasing stray light. The larger number density of a liquid compared to a gas is about 10^3, so that lower currents (0.1 A) and fewer turns on the solenoid (200 turns) are sufficient (11). In fact, the solenoid can then be driven directly by the wave generator, and better modulation stability is achieved.

Various flow cells have been used successfully. For analytical-scale LC (4.6-mm ID columns), a 5-cm-long cell with a 80;μL volume seems to be optimum. For microbore LC (1-mm ID columns), a 1-cm-long cell with a 1-μL volume can be used (12). The cell has been tested with a 25-cm-long commercial microbore column, and the full separatory power of the column (100,000 plates/m for $k' = 2.1$) was preserved. The shorter cell implies a poorer concentration detectability, but the smaller dilution factor for microbore LC partially offsets the problem. The best detectability is found to be 4 ng (at S/N = 3) of injected material with a specific rotation of 100°. This corresponds to a rotation of 4×10^{-6} deg, which is 3 orders of magnitude better than standard polarimeters. It is interesting to compare the experimental noise level of 1.3×10^{-6} deg with the predicted noise level of 3.4×10^{-7} deg (last row of Table 1). Considering that there is some degradation in ε owing to the extra liquid cell and a flowing sample, the agreement is good. One can also see that further improvements should be possible if the laser intensity can be stabilized to better than 1 part in 10^4.

4. APPLICATIONS

4.1. Detection of Chiral Molecules

The higher degree of selectivity of polarimetry allows determinations in complex mixtures without extensive separation. In addition, the signal is related to the enantiomeric purity of the analyte. Such information is extremely important to process control, and can be obtained from polarimetry without chiral separation.

4.1.1. Carbohydrates

The carbohydrates that are present in body fluids are an interesting class of compounds, and are related to metabolic processes in general. Much attention has been given to the routine clinical screening for glucose (13), because of the disease diabetes mellitus. The other simple sugars are often neglected, primarily because of the lack of reliable quantitative methods at the low concentration levels typical of body fluids. Automated high-resolution analyzers with sensitivities in the microgram range have been used to study carbohydrates in body fluids (14,15), but hours are required per analysis. The problem is the lack of convenient UV absorption bands for the carbohydrates. Several uses of carbohydrates other than glucose for physiological profiling are known. Excess fructose in the urine can be a sign of an inherited metabolic defect (16). Lactose is present in urine in late pregnancy and during lactation, but excess can indicate a rare metabolic disease (17). The inherited disease galactosemia causes the presence of galactose in urine, but galactose can also be an indication of severe hepatitis or biliary atresia in neonatal infants (17). The last-named condition can lead to liver damage, mental retardation, and cataracts. The presence of xylulose is related to yet another familial disorder (18). The D-xylose absorption test can be used to diagnose either enterogenous steatorrhea (19) or kidney malfunction. Because the carbohydrates are directly involved in the metabolic cycles of the body, one would expect dietary and metabolic deficiencies to affect carbohydrate profiles in serum and urine. It is therefore important to have reliable methods for analysis, so that correlations can be studied. Separation of the monosaccharides can be accomplished on a commercial heavy metal cation-exchange column. A complication is that the column operates at 85°C. Thus a 50-cm length of stainless steel tubing (0.001-in. ID) is placed between the column and the detector. This prevents turbulence in the optical region which can distort the laser beam. To protect the column from the high ionic concentration in urine, deionization is necessary either before injection or by using a guard column. A chromatogram of human urine (100-μL sample) is shown in Fig. 5. The chromatographic efficiency can be further increased by using a second column in series, but the simplicity of the chromatogram makes such a compromise with analysis time not necessary. Table 2 shows quantitative results for the six urinary sugars, confirming the applicability of this method for screening carbohydrates. We note that a UV absorption detector placed in series with the system did not show any peaks corresponding to these sugars (10), as expected. There are, however, broad UV absorption features at those retention times, implying that a RI detector, even if the sensitivity is high enough, will suffer from interferences in this application.

4.1.2. Cholesterol and its Esters

Cholesterol is a major component of all mammalian plasma membranes. Though it is vital to cell growth and survival (20), there is a statistically significant correlation between elevated serum cholesterol levels and cardiovascular diseases in general and atherosclerosis (21) in particular. Cholesterol is found in two forms in the serum: either as free cholesterol, or esterified with long-chain fatty acids such as palmitic acid. Typically about 75% of the serum cholesterol is

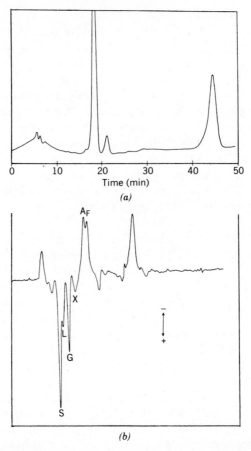

Figure 5. Separation of components in human urine (100 μL injected) by HPX-87 heavy metal column (85°C) with absorbance detection (*a*) and optical activity detection (*b*). Peaks: S, sucrose; L, lactose; G, glucose; X, xylose; A, arabinose; F, fructose. Mobile phase is water at 0.64 mL/min. Reproduced with permission from Ref. 10.

esterified (22). This ratio has been shown to be related to hormones (23), diet (24), toxic conditions such as ethanol poisoning (25), and disorders such as familial lecithin:cholesterol acyltransferase deficiency (26). The most commonly used methods for the measurement of serum cholesterol are photometric and are based on the color reaction of cholesterol. These are accurate but are mainly for total serum cholesterol determination. Chromatographic methods are capable of separating free and esterified cholesterol. GLC (27), TLC (28), and HPLC using a variable-wavelength UV detector (29) are available. Although cholesterol can be detected at 200 nm, the choice of solvents becomes severely limited owing

Table 2. Concentrations of Sugars in Urine

Sugar	Concentration (μg/mL)[a]		
	This Work	Ref. 14	Normal Range
Sucrose	22	12	0–50
Lactose	10	49	0–100
Glucose	16	24	10–120
Xylose	10	10	0–30
Arabinose	7	26	0–30
Fructose	7	2.7	0–50

[a]All concentrations are ± 10%.

to the transparency requirement in this region. A particularly serious problem is the interference from triglycerides (29).

For demonstrating polarimetric detection (30), fresh serum from a healthy individual was obtained. For each 2.4 mL of serum sample, 7.6 mL of tetrahydrofuran was added drop by drop with vigorous stirring. The mixture was then centrifuged for 20 min at 11,000 g in a clinical centrifuge. The clear and yellowish liquid phase was thus separated from the residue and was used as the sample without further treatment. A typical chromatogram is shown in Fig. 6. The selected experimental conditions did not allow clean separation of every individual component. In each pair of unresolved components, the presence of an additional double bond is the main structural difference. It should be possible to develop a set of chromatographic conditions (e.g., by introducing silver ions) to resolve each species fully. The main feature of Fig. 6 is that triglycerides are not expected to interfere. Only when different fatty acid substituents are present in the triglyceride can there be an asymmetric center. Even then, the differences in the substituents are most likely several carbons away from the asymmetric center, resulting in only negligible amounts of optical activity. Another feature of Fig. 6 is that the specific rotations of the species are expected to be quite similar. This is because they share the same chiral center but different chain lengths in the substituents. This makes quantitation straightforward.

4.1.3. Fossil Fuels

Of the many physical and chemical properties suitable for characterization of fossil fuels, perhaps the most interesting is the associated optical activity. Unlike the more reactive functional groups in the molecule, the chiral centers may be retained despite the hostile conditions that led to the formation of fossil fuels. Optically active materials have been found in the montan wax of brown coal

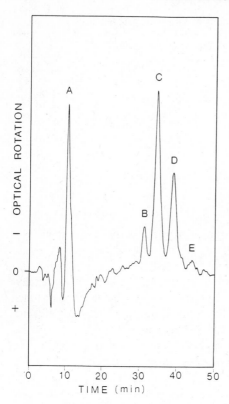

Figure 6. Separation of cholesterol and cholesterol esters in human serum. Peaks: A, cholesterol and cholestanol; B, cholesteryl linolenate and arachindonate; C, cholesteryl palmitoleate and linoleate; D, cholesteryl palmitate and oleate; E, cholesteryl stearate. Mobile phase, tetrahydrofuran–water (76:24, v/v); flow rate 0.5 mL/min, column, 10 μm C_{18}. Reproduced with permission from Ref. 30.

(31), oil distillates from coal (32), petroleum distillates (33), lubricating oils (34), and shale oil (35). The observed bulk optical rotation has been correlated with retorting conditions (35), geologic source (34), aromatic/aliphatic ratio (33), geologic age (33), distillate fraction (36), and thermal history (35). Because of the relatively small amounts of optically active materials that are present, typically very small rotations are observed. And, because of the highly colored nature of most of these materials, measurements are sometimes impossible to obtain. The use of bulk optical rotation for correlation is questionable. The reason for this is that optical rotation can be dextrorotatory or levorotatory, and both types have been found in fossil fuels (31). The presence of materials of both classes causes a cancellation in the observed quantity. The bulk optical rotation is then only an indication of the minimum amount of chiral materials in the sample. It is therefore highly desirable to perform some separation on the sample before the measurement of optical activity. Separation also allows one to determine the individual contributions of the various components to refine any correlations further.

The chromatograms that are obtained naturally depend heavily on the mode of extraction for the material injected. The particular procedure in the examples below is chosen to favor the saturated hydrocarbons, which are known contributors to observed optical activity in fossil fuels. Acetonitrile was chosen as the solvent and the eluent to eliminate the solvent front associated with the injection process. This way, events very early on in the chromatogram can be recorded faithfully. There is, however, no guarantee that all interesting components will be eluted during a run of 40 min, or that this choice of eluent/stationary phase is the best. Still, within these guidelines, the studies here provided some interesting results.

Figure 7 shows chromatograms of three different shale oils (11) while Fig. 8 shows chromatograms of extracts of these different coals and one solvent-refined coal (37). The first observation is that these chromatograms are extremely rich in structure. Potentially, these chromatograms can be used for characterizing fossil fuels. The second observation is that there are dextrorotatory as well as levorotatory components in all the samples. This confirms that bulk optical activity is not a useful parameter in the characterization of the samples because of cancellation of the effects. In fact, the better the separation, the more reliable is the interpretation. The third observation is that in general more features are present in the optical activity chromatogram than in the RI chromatogram, showing the high degree of selectivity of the former. The fourth observation is that some qualitative trends can be seen in the chromatograms. For example, the Dietz coal contains more chiral components than the others (per unit weight). This is consistent with its being a younger coal, which has gone through less structural degradation. The amount of chiral components is also found to decrease as the coal is processed (37), which again is consistent with structural degradation.

4.2. Universal Detection

The standard universal detector in LC is the RI detector, which is limited by its poor detectability and the relatively large volumes in commercial units. Recently, the concept of indirect polarimetry was demonstrated (12,38) as a universal LC detector. If an optically active solvent is used for LC, there will be a large constant background rotation in the absence of any analyte. This background can be compensated for by physically rotating the second polarizer (analyzer) so that once again a low light intensity reaches the phototube. The polarimeter still functions as before, but now a zero (base-line) signal is observed when the optically active solvent is present. When an optically inactive analyte elutes from the column, it displaces an equal amount of the solvent in the optical region. Therefore fewer solvent molecules are present in the optical region, and a decrease in optical rotation results. This then registers as a negative signal in the polarimeter, that is, opposite in sign to the rotation of the solvent. It can be seen

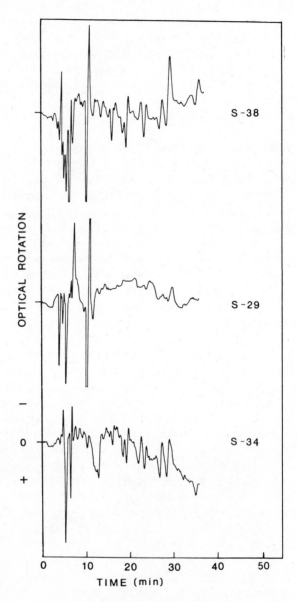

Figure 7. Chromatograms of shale oil extracts for 0–2.5-cm particle sizes and different origins. S-34, Utah Shale (12 gal/ton); S-29, Anvil Points Shale (25 gal/ton); S-38, Colony Shale (35 gal/ton). Mobile phase, acetonitrile at 0.8 mL/min, column, 10 μm C₁₈. Reproduced with permission from Ref. 11.

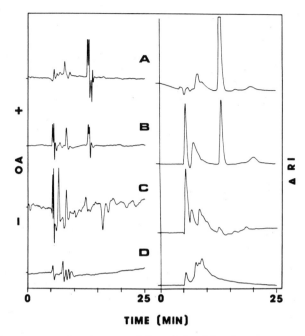

Figure 8. Chromatograms of coal extracts: left, optical activity chromatograms; right, RI chromatograms. Top to bottom: Illinois No. 6 coal; Kittanning coal: Dietz coal; Pamco SRC. Mobile phase, acetonitrile at 0.6 mL/min, column, 10 μm C_{18} (37). Reproduced by permission of the publishers. Butterworth Co. (Publishers) Ltd. 1985.

that unless the analyte has exactly the same specific rotation as the solvent, a signal is observed. This is analogous to the RI detector, where a difference in RI between the analyte and the solvent is required for detection. In general, the chance of having identical specific rotations is much lower than that of having identical RI values, because most substances are optically inactive. Thus the indirect polarimetric method is "more universal" than RI methods.

An important difference between indirect polarimetry and indirect photometry (39) is the dynamic reserve of the system. This is the ability to detect a small signal on top of a large background. In photometry, a background absorbance of larger than unity will seriously degrade the ability to detect small changes in absorbance, because the amount of light reaching the phototube is reduced. The dynamic reserve of an absorbance detector is about 2×10^3 (40), which implies that a change of 5×10^{-3} a.u. is detectable on top of a unit absorbance background. In contrast, the polarimeter can retain its low noise level in the presence of a large background because of mechanical compensation from the analyzer.

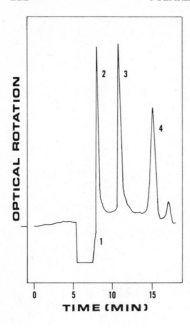

Figure 9. Detection by indirect polarimetry; mobile phase, 95:5 (R)-(+)-limonene/isooctane; flow rate, 24 μL/min; column, 25-cm × 1-mm ID 10-μm silica; (1) injection peak; (2) dioctyl phthalate; (3) dibutyl phthalate; (4) diethyl phthalate. Reproduced with permission from D. R. Bobbitt and E. S. Yeung, *Anal. Chem.*, **57**, 271 (1985). Copyright 1985 American Chemical Society.

The dynamic reserve is about 10^7 (38). This is also better than the RI detector, where a 10^{-7} RI change can be detected for a solvent–analyte difference of 0.1 RI, that is, a dynamic reserve of 10^6. The only reason that indirect photometry (39) gives reasonable detectability is that ion chromatography employs very dilute eluting ions (10^{-3}–10^{-4} M). Thus even a moderate dynamic reserve will provide nanogram level detectabilities.

Figure 9 shows a chromatogram obtained by indirect polarimetry, for analytes that are not optically active. The detectability of this system is 4 ng of injected material, which is identical to that for a chiral analyte in conventional polarimetry. The solvent essentially provides the response factor, corresponding to a specific rotation of 150 degrees. Below this level of concentration, there is a problem with thermal noise, analogous to the RI detector. The indirect polarimetric method is a number-density detector, and temperature effects eventually limit its usefulness. Pump pulsations lead to similar base-line fluctuations. Such effects, naturally, do not affect the polarimetric detector when optically inactive solvents are used.

The success of indirect polarimetry is a direct result of the low solvent consumption in microbore LC. Natural products such as limonene are available in a variety of elution strengths. In fact, the cost of running a chromatogram such as in Fig. 9 is about the same as using UV-grade acetonitrile in analytical-scale

(4.6-mm ID) LC. Purity of the solvent is not critical in indirect polarimetry, because one simply loses a bit of sensitivity. The small volume and the good detectability may eventually allow indirect polarimetry to replace RI detection for many LC applications.

4.3. Absorption Detection

The concept of indirect polarimetry can be taken one step further. If an analyte absorbs light in the optical region, heat is generated. If an optically active solvent is used, the heating effect produces an *expansion* in the liquid and a *decrease* in optical rotation is detected due to the decrease in number density in the optical path.

To determine the relationship between the amount of light absorbed and the observed rotation, two assumptions are necessary. First, the amount of light absorbed is equal to $2.303AI$ where A is the absorbance of the sample and I is the intensity of the source in joules. This is true only for small absorptions, but this covers most of the interesting cases in LC. Second, it is assumed that all the light absorbed eventually becomes heat. With these two conditions met, the temperature increase, δT (K), in the interaction region of cross-sectional area a (cm^2) and unit length is

$$\Delta T = \frac{2.303AI}{C_p \rho a} \tag{12}$$

where C_p is the specific heat of the solvent (J/g · K) and ρ is the density of the medium (g/cm^3).

The rotation α observed for a mixture is related to the individual specific rotations $[\alpha]_i$, such that (12)

$$\alpha = \sum [\alpha]_i l \rho_i V_i \tag{13}$$

where l is the path length (dm) and V is the volume fraction of the material at the detector. For low solute concentrations (which is the more interesting case), one can assume that the absorption-induced rotation is entirely due to the eluent, so that $\alpha = [\alpha] l \rho$. If B is the coefficient of expansion of the eluent (K^{-1}), assuming small changes, there will be a change in the measured optical rotation:

$$\Delta\alpha = [\alpha] l \rho B \Delta T \tag{14}$$

Finally, substituting into Eq. 12, we obtain an explicit relation between the absorbance of the solution and the rotation observed:

$$A = \frac{\Delta \alpha a C_p}{2.303[\alpha] l I B} \tag{15}$$

To assess the potential of measuring absorptions via indirect polarimetry, Eq. 15 can be used to estimate the minimum detectable absorbance. The micropolarimeter has a detectability of 4×10^{-6} °(S/N = 3). Using typical values for the other parameters, $a = 9 \times 10^{-4}\,cm^2$, $C_p = 1.8\,J/g \cdot K$, $[\alpha] = 192.0\,cm^3/g$ dm, $l = 0.1$ dm, $I = 8.0 \times 10^{-2}$ J, and $B = 1.24 \times 10^{-3}\,K^{-1}$, one finds that the minimum detectable absorbance is 1.2×10^{-6}. This is superior to conventional absorption detectors in LC.

Figure 10 demonstrates the use of the indirect polarimetric scheme as an absorption detector. The three chromatograms represent data taken at three different laser powers; each has been normalized against the standard rotation produced by a DC solenoid (6) to maintain constant sensitivity. Peak 3 is the indirect polarimetric signal (displacement of solvent) resulting from the elution of dimethyl phthalate, and the peak height is approximately equal in the three chromatograms. Peak 2 is the signal for N-methyl-o-nitroaniline by the method

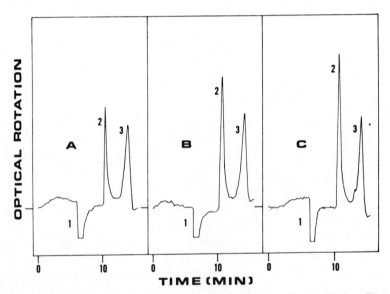

Figure 10. Normal-phase separation of N-methyl-o-nitroaniline and dimethyl phthalate. Chromatographic conditions were identical with those in Fig. 9, except that 100% limonene was used: λ = 488.0 nm, (1) injection peak, (2) absorption detection of 890 ng of N-methyl-o-nitroaniline, (3) indirect polarimetric signal from 2 μg of dimethyl phthalate, (a) 31 mW, (b) 48 mW, (c) 55 mW. Reproduced with permission from D. R. Bobbitt and E. S. Yeung, *Anal. Chem.*, **57**, 271 (1985). Copyright 1985 American Chemical Society.

of absorption via indirect polarimetry. The peak height increases linearly with increasing laser power as predicted by Eq. 15. Figure 10 shows that one can use the same experimental setup for universal detection and for absorption detection by simply changing the laser power. Peaks that do not change with laser power correspond to indirect polarimetry, and peaks that vary with laser power are due to absorption processes. At 488 nm, N-methyl-o-nitroaniline has a molar absorptivity of 112 L/mol · cm. The detectability in this scheme with 32 mW of radiation is 7 ng. At 458 nm, the molar absorptivity is 1040 L/mol · cm, and $[\alpha]$ becomes larger. With 73 mW of radiation, the detectability is 36 pg of injected material. This corresponds to 1.8×10^{-6} a.u. (peak volume 28 μL, S/N = 3). Compared to other photothermal techniques, the absorption detectability is about the same. The limit seems to be that of solvent absorption. The advantage of polarimetry is that the bulk heating effect is probed, so that beam quality and beam alignment create fewer problems.

4.4. Quantitation without Standards

The general scheme for quantitation without standards is presented in Chapter 9. For the polarimeter, Eq. 13 gives the relationship between the observed rotation of a mixture and the properties of the individual components. Thus, if the same sample is first eluted with an optically active solvent and then its racemic mixture,

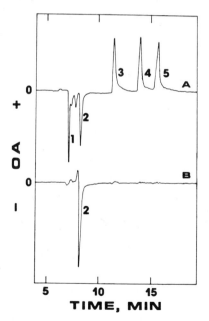

Figure 11. Chromatograms of a mixture in the two mobile phases: (a) ($-$)-2-methyl-1-butanol, (b) ($\pm$)-2-methyl-1-butanol; (1) injection peak, (2) l-2-octanol, (3) decane, (4) tetradecane, and (5) hexadecane. Column, 5-μm C_{18}, 25-cm $\times$ 1.0-mm ID, flow rate, 20 μL/min. Reproduced with permission from D. R. Bobbitt and E. S. Yeung, *Anal. Chem.*, **56**, 1577 (1984). Copyright 1984 American Chemical Society.

Table 3. Quantitation of Sample Components in Two Different Mixtures

		l-2-Octanol	$C_{10}H_{22}$	$C_{14}H_{30}$	$C_{16}H_{34}$
True	(1)	4.00	8.00	8.00	8.00
$C_x(\times 10^{-2})$	(2)	2.00	4.00	4.00	4.00
True $[\alpha_x]_{590}^{20}$	(1)	-9	0	0	0
	(2)	-9	0	0	0
Calcd	(1)	3.93 ± 0.38	7.74 ± 0.34	7.83 ± 0.36	8.36 ± 0.27
$C_x(\times 10^{-2})$	(2)	1.98 ± 0.24	3.88 ± 0.23	3.87 ± 0.19	4.06 ± 0.22
Calcd $[\alpha_x]_{488}^{27}$	(1)	$-8.0° \pm 2.8$	0.07 ± 0.2	0.01 ± 0.2	0.01 ± 0.1
	(2)	$-7.9° \pm 3.0$	0.02 ± 0.3	0.02 ± 0.2	0.01 ± 0.2

two expressions of the form of Eq. 13 are available to determine the concentration of the analyte, without standards and without identification.

Figure 11 shows the pair of chromatograms obtained from the $(-)$ and then the $(\pm)$ form of 2-methyl-1-butanol. Quantitation of the four components, 2–5, is shown in Table 3 for two mixtures. Chromatogram A in effect is identical to indirect polarimetry for peaks 3 through 5. The zero areas in chromatogram B for these peaks are perfectly good numbers for using Eq. 13 to solve for the concentrations. The agreement in Table 3 is good, with errors that are due to uncertainties in measuring peak areas. Of all the variations on the scheme for quantitation without standards in LC, polarimetry appears to be the best. The dynamic reserve is the largest for polarimetry, leading to the best detectability. The elution characteristics of the chiral solvent and its racemic counterpart are expected to be closely matched, so that peak correlation is not a problem. Such is always the case if the solvent molecules are small and no reaction occurs with the analytes.

SUMMARY

It is interesting that the same instrumentation can be used in four different modes depending on the operation, as a chiral detector, as a universal detector, as an absorption detector, and as a means for quantitation without standards. The system still needs engineering to become a "black box" for the routine laboratory. However, at least the cost of assembly is within the acceptable price range of other LC detectors. Future improvements can be expected if the laser intensity can be stabilized further. Furthermore, it may even be possible to extend the application of polarimetry to include other physical properties of analytes that eventually lead to optical rotation.

ACKNOWLEDGMENTS

I thank the many co-workers in my group who have contributed to various parts of this work, particularly S. D. Woodruff, J. C. Kuo, D. R. Bobbitt, and B. H. Reitsma. The Ames Laboratory is operated by the U.S. Department of Energy by Iowa State University under Contract No. W-7405-eng-82. This work was supported by the Office of Basic Energy Sciences.

REFERENCES

1. A. Z. Baumann, *Anal. Chem., **284,** 31 (1977).

2. P. Rossi, *Analyst (London),* **100,** 25 (1975).

3. W. Boehme, G. Wagner, U. Oehme, and U. Priesnitz, *Anal. Chem.,* **54,** 709 (1982).

4. W. Boehme, *Chromatogr. Newsl.,* **38,** 8 (1980).

5. J. J. Kankare and R. Stephens, *Talanta,* **31,** 689 (1984).

6. E. S. Yeung, L. E. Steenhoek, S. D. Woodruff, and J. C. Kuo, *Anal. Chem.,* **52,** 1399 (1981).

7. H. W. Schrotter, in W. Kiefer and D. A. Long, Eds., *Nonlinear Raman Spectroscopy and Its Chemical Applications,* Reidel, New York, 1982, pp. 603–611.

8. C. E. Moeller and D. R. Grieser, *Appl. Opt.,* **8,** 206 (1969).

9. E. S. Yeung, in M. V. Novotny and D. Ishii, Eds., *Microcolumn Separations,* Elsevier, Amsterdam, 1985, pp. 135.

10. J. C. Kuo and E. S. Yeung, *J. Chromatogr.,* **223,** 321 (1981).

11. J. C. Kuo and E. S. Yeung, *J. Chromatogr.,* **253,** 199 (1982).

12. D. R. Bobbitt and E. S. Yeung, *Anal. Chem.,* **56,** 1577 (1984).

13. I. Davidsohn and J. B. Henry, *Clinical Diagnosis by Laboratory Methods,* 14th ed., Saunders, Philadelphia, 1969, pp. 57.

14. R. L. Jolley and M. L. Freeman, *Clin. Chem.,* **14,** 538 (1968).

15. S. Katz, S. R. Dinsmore, and W. W. Pitt, Jr., *Clin. Chem.,* **17,** 731 (1971).

16. E. R. Froesch, H. P. Wolf, and H. Baitsch, *Am. J. Med.,* **34,** 151 (1963).

17. C. U. Lowe and V. H. Auerback, in W. E. Nelson, Ed., *Textbook of Pediatrics,* 8th ed., Saunders, Philadelphia, 1964, 298.

18. I. M. Freedberg, D. S. Feingold, and H. H. Hiatt, *Biochem. Biophys. Res. Commun.,* **1,** 328 (1959).

19. J. H. Roe and E. W. Rice, *J. Biol. Chem.,* **173,** 507 (1948).

20. M. S. Brown and J. L. Goldstein, *Science,* **191,** 150 (1976).

21. J. R. Sabine, *Cholesterol,* Dekker, New York, 1977, pp. 245–276.

22. R. L. Searcy, *Diagnostic Biochemistry,* McGraw-Hill, New York, 1969.

23. R. A. Jungmann and J. S. Schweppe, *Steroids,* **17,** 541 (1971).

24. N. Takeuchi and Y. Yamamura, *Atherosclerosis,* **17,** 211 (1973).

25. N. Takeuchi, M. Ito, and Y. Yamamura, *Lipids,* **9,** 353 (1974).

26. K. R. Norum and E. Gjone, *Scand. J. Lab. Clin. Med.,* **20,** 231 (1967).

27. R. Watts, T. Carter, and S. Taylor, *Clin. Chem.,* **22,** 1692 (1976).

28. J. C. Touchstone, T. Murawec, M. Kasparow, and W. Worthmann, *J. Chromatogr. Sci.,* **10,** 490 (1972).

29. I. W. Duncan, P. H. Culbreth, and C. A. Burtis, *J. Chromatogr.,* **162,** 281 (1979).

30. J. C. Kuo and E. S. Yeung, *J. Chromatogr.,* **229,** 293 (1982).

31. V. Jarolim, M. Streibel, M. Horak, and F. Sorm, *Chem. Ind. (New York),* 1142 (1958).

32. C. Zahn, S. H. Langer, B. D. Blaustein, and I. Wender, *Nature (London),* **200,** 53 (1963).

33. W. D. Rosenfeld, *J. Am. Oil Chem. Soc.,* **44,** 603 (1967).

34. R. E. Hersch, M. R. Fenske, H. J. Matson, E. F. Koch, E. R. Booser, and W. G. Braun, *Anal. Chem.,* **20,** 434 (1948).

35. D. R. Lawlor in O. P. Strausz, Ed., *Oil Sands and Oil Shale,* Verlag Chemie, New York, 1978, p. 267.

36. M. R. Fenske, F. L. Carnahan, J. N. Breston, A. H. Caser, and A. R. Rescorla, *Ind. Eng. Chem.,* **34,** 638 (1942).

37. D. R. Bobbitt et al., *Fuel,* **64,** 114 (1985).

38. D. R. Bobbitt and E. S. Yeung, *Anal. Chem.,* **57,** 271 (1985).

39. H. Small and T. E. Miller, *Anal. Chem.,* **54,** 462 (1982).

40. S. A. Wilson and E. S. Yeung, *Anal. Chim. Acta,* **157,** 53 (1984).

CHAPTER

7

DETECTION BASED ON ELECTRICAL AND ELECTROCHEMICAL MEASUREMENTS

STEPHEN G. WEBER

Department of Chemistry, University of Pittsburgh, Pittsburgh, Pennsylvania

1. INTRODUCTION

My experience in teaching instrumental analysis to undergraduate students has taught me that electrochemistry is a subject that is not well understood. There are probably several reasons for this. One can see things related to spectroscopy (colors, luminescence), and one can see separations (thin-layer chromatography), but electrochemistry is rarely experienced first hand. Electrochemistry deals in a variety of quantities with complicated units taken from chemistry, electromagnetism, and of particular importance to this readership, hydrodynamics. Finally, there is the dynamic aspect of electrochemistry. Certainly spectroscopy and separations involve dynamics, but at the undergraduate level it is easy to explain a large part of the disciplines in terms of steady states or equilibria. In electrochemistry the dynamics occur on the experimental time scale, so very often they are visible.

It is for these reasons that this chapter consists of two sections. The first section covers some basic electrochemistry and related electrical concepts. It is intended that this section be enough to provide the newcomer to electrochemistry with sufficient background to appreciate the principles of the various detectors. The second section deals with the detectors. Current research in the various areas is reviewed critically. Promising new techniques are discussed as well.

2. AN INTRODUCTION TO DETECTION BASED ON THE ELECTRICAL PROPERTIES OF SOLUTIONS

2.1. Overview

Permittivity, conductance, potentiometric, amperometric, and coulometric detectors are covered in this chapter. The electrical basis of each is demonstrated. There is value in this because the detector may then be viewed simply as an

229

equivalent circuit. Some readers who are not conversant with basic electricity will not find this to be of value. To be sure, electrochemistry is at the interface between chemistry and electricity. Thus some basic electrical concepts are given to educate the reader who has been insulated from electrical ideas. In each section the physical basis for the measurement is covered. Also, the relationship between the measured quantity and the concentration of the analyte is discussed. Of all the sections, the most space is devoted to amperometry because it is both widely used and based on complex phenomena.

2.2. Electrical and Chemical Concepts

2.2.1. Capacitance and Permittivity

A capacitor is formed from two conducting plates or surfaces separated by an insulator or dielectric. This is shown in Fig. 1. Charges residing on the plates create a potential gradient across the dielectric. Experimentally one can measure a potential difference across the plates, V, due to a charge, Q, in a capacitor of capacitance C, that is given by

$$V = \frac{Q}{C} \tag{1}$$

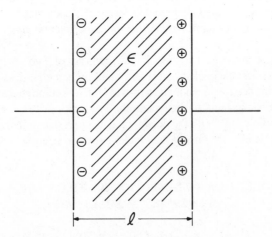

Figure 1. A schematic diagram of a capacitor. The charge resides on the plates. By convention the charge on only one plate is used in computation of the voltage. The plates, of area A, are separated by a distance l. The space between them is filled with a dielectric with permittivity ε. For capacitors that are not planar, equations other than Eq. 2 apply.

Potentials are usually measured in volts (V). If two regions of space differ by 1 V, then it will require (or yield, depending on the sign of the charge) 1 joule (J) to move 1 coulomb (C) of charge from one region to the other. Thus the units of a volt are joules per coulomb. The capacitance is given by Eq. 2 for parallel planar plates. Note that the capacitance has a geometric

$$C = \frac{A\varepsilon}{l} \tag{2}$$

part, area A divided by separation distance l, and a material-dependent part, the permittivity ε. The units of permittivity are farads per meter (F/m). The farad is a unit of capacitance equal to coulombs per volt or coulombs squared per joule. The dielectric constant of a material or a mixture is the ratio of the permittivity of the material to the permittivity of free space (a vacuum or, practically speaking, air). In the Gaussian system of units the permittivity of free space is exactly 1 (units esu^2/erg · cm), so in this system of units permittivity and dielectric constant are rather freely interchanged. In the MKSA system of units (1) the permittivity of free space is 8.854×10^{-12} F/m, so in this set of units the dielectric constant (unitless) of a material is multiplied by this factor to arrive at the permittivity of the material.

Measurements of permittivity are made by many methods (2), but in all cases it comes down to estimating the capacitance of a capacitor in which the dielectric material is the sample being studied. Capacitors have the property of passing alternating signals but not steady signals. Thus most capacitance measurements, and therefore permittivity, employ an alternating voltage. In detection work one is interested in the presence of small quantities of a solute in a much larger quantity of solvent. Under these conditions the measured permittivity, or the measured capacitance of a capacitor in which the sample is the dielectric, changes linearly with the concentration of the solute, as shown by

$$\varepsilon = \varepsilon_s + \delta c \tag{3}$$

In this equation c is the solute's molar concentration, ε_s is the permittivity of the solvent, and δ is the dielectric increment (F/m · M), the change in permittivity per molar change in analyte concentration (3).

The permittivity of a material depends on the frequency of the measurement. At low frequencies (in and below the MHz range) when one impresses an alternating voltage across the capacitor, the dielectric responds in two distinctly different ways to the field. There is relative displacement of the nuclei and electrons in the material. The molecular property of importance in the material's response in this way is its polarizability. Also, if the molecules making up the fluid have a permanent dipole, then they will orient themselves according to the applied field. The molecule's dipole moment is important in this case. At much higher frequencies (GHz or microwave and above) only the polarizability contributes to the permittivity because of the damping of the reorientation of permanent dipoles owing to the viscosity of the fluid containing the dipoles. At these frequencies the dielectric constant is the square of the index of refraction. The situation is far more complicated if the system resonates with the applied field, that is, if it absorbs the radiation.

2.2.2. Conductance and Conductivity

An excellent textbook chapter has recently appeared covering conductance (4). A conductor is a material that contains mobile charge carriers. Of course the electrons in metals make metals conductors, but there are ionic conductors too. Ionizable species dissolved in polar liquids conduct electricity because the ions are mobile. One can measure the conductance of a sample by measuring current across the conductor at some applied voltage difference (see Eq. 4).

$$G = \frac{I}{E} \tag{4}$$

In this equation the conductance, G, is given in units of ohm^{-1} or mho or Siemen's units (S), all of which are coulombs squared per joule per second. The current, I, is in amperes, and the potential difference across the resistor, E, is given in volts. Recall the coulomb is $1 \, A \cdot s$. The conductance is in turn related to geometric and physical or chemical properties of the sample. The conductance is given by the product of the cell constant, A/l, and the specific conductance of the medium in the conductor, K. The cell constant (A/l) in centimeters

$$G = K \left(\frac{A}{l} \right) \tag{5}$$

is the ratio of the cross-sectional area of the conductor to the length of the conductor. In a measurement apparatus it is easier and more precise to measure this quantity by measuring the conductance of a solution of a material of known specific conductance and calculating the cell constant from Eq. 5 than it is to measure the actual distances involved. The specific conductance, K, is related to the composition of the material. K is related to the number of charge carriers per unit volume of solvent (concentration of salt), the charge on the ions, the mobility of the ions in the fluid, and the presence of any ion pair or complex equilibria that alter the total number of charged species in the solution. For a solution of a simple salt K is given by Eq. 6 in which c is the molar concentration of ion, z is the number of charges per ion, F is the Faraday, 96484.7 C per mole of charges, and u is the mobility of the ion [units, (cm/s)/(V/cm)].

$$K = \frac{(c_+|z_+|F|u_+ + c_-|z_-|F|u_-)\alpha}{1000} \tag{6}$$

The term α is the fraction of the salt that is dissociated. The number 1000 enters in as the number of milliliters per liter. Note that, like the permittivity detector, a detector based on the measurement of conductance responds linearly to the concentration of ionic species in the solution, assuming that α is not affected by concentration. For species such as weak acids this assumption is not necessarily true.

In some cases the conductance of a mixture of ions is measured. For example, in the case of so-called unsuppressed ion chromatography, in which one may be measuring the conductance of a solution containing an analyte ion, a coion (the eluent ion), and a counterion (5–9). In this case the detector responds to the change in conductance due to the replacement of the coions by the analyte ions. For ions of the same charge the quantity measured is therefore related to the concentration of the analyte species and the difference in the mobilities of the analyte ion and the coion (5). Ionic mobilities are usually tabulated as equivalent ionic conductance, λ, where λ is Fu.

At low (0 to kHz) frequencies the conductance of a solution should be independent of frequency. Because of the experimental apparatus this may not be the case. For reasons that will become clear below, the measurement is usually made in the kilohertz region. Square waves, sine waves, and other waveforms are used; these are discussed in the next section.

2.2.3. Potentials and Potentiometry

It is an unavoidable fact of nature that charged particles respond to potential *differences*. Thus any mention of an electrical potential either explicitly or implicitly states an initial state and a final state. The change in potential energy of a mole of z charged particles on moving between potential E_1 and E_2 is given by

$$\Delta\mu = zF(E_2 - E_1) \tag{7}$$

Here $\Delta\mu$ is the change in (electrochemical) potential in joules per mole. Measurements of the potential energy released in transferring electrons from one chemical species to another have led to the establishment of a scale of reduction/oxidation potentials. Under a set of well-defined conditions, usually very dilute solutions, the potential of the electrons in a chemical system is given by Eq. 8, due to Nernst, for the reaction shown in Eq. 9.

$$E = E° - \frac{RT}{nF} \ln \frac{(A_{red})^{\gamma_r} (P)^{\pi}(R)^{\rho} \cdots}{(A_{ox})^{\gamma_o} (B)^{\beta}(D)^{\delta} \cdots} \tag{8}$$

$$\gamma_o A_{ox} + \beta B + \delta D + \cdots \overset{+ne^-}{\longrightarrow} \gamma_r A_{red} + \pi P + \rho R + \cdots \tag{9}$$

The measured difference in potential between the redox system under study and a reference redox system is E. $E°$ is the same quantity under conditions representing the standard state, extrapolated infinite dilution. In this standard system the species are at unit activity, yet there is no interaction between any of the species except that between solvent and the dissolved species. This is an imaginary construct, because at concentrations of about 1 M, collisions between analyte species, with the chemical consequences that may result, are not that unlikely. R is the gas constant and T is temperature, in Kelvin. For a simple system involving no equilibria other than the electrochemical one, Eqs. 8 and 9 become Eqs. 10 and 11. The measured energy of the system is

$$E = E° - \frac{RT}{nF} \ln \frac{(A_{red})^{\gamma_r}}{(A_{ox})^{\gamma_o}} \tag{10}$$

$$\gamma_r A_{red} \overset{+ne^-}{\longrightarrow} \gamma_o A_{ox} \tag{11}$$

the average of the highest occupied orbital energy of the reduced species and the lowest unoccupied orbital of the oxidized species. The implication is that in order to make this measurement one must probe *both* energies; that is, one must put electrons into the chemical system and remove them. In solution self-exchange reactions are occurring, as shown in Eq. 12, in which the primed species are chemically the same as the unprimed species; the prime serves to identify a particular nucleus.

$$A_{red} + A'_{ox} \rightleftarrows A_{ox} + A'_{red} \tag{12}$$

In the presence of a chemically inert metal electrode (that is not connected to any external circuit that would add or remove electrons) the exchange reaction can occur using the metal as a mediator. This is shown in Scheme 1. In this scheme e_m is an electron in the metal electrode. Note that the electrode both donates and accepts electrons

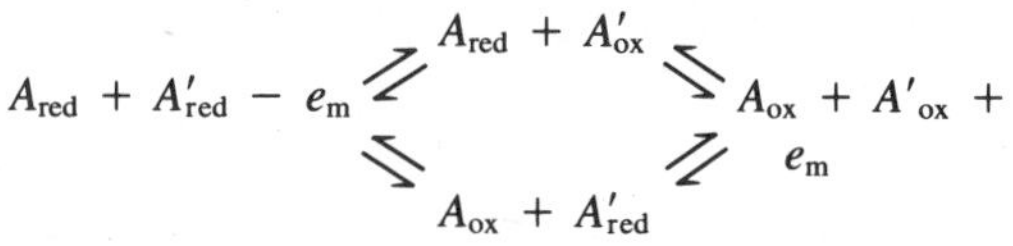

$$A_{red} + A'_{red} - e_m \underset{\searrow}{\overset{\nearrow}{}} \begin{array}{c} A_{red} + A'_{ox} \\[6pt] A_{ox} + A'_{red} \end{array} \underset{\searrow}{\overset{\nearrow}{}} \begin{array}{c} A_{ox} + A'_{ox} + \\[6pt] e_m \end{array}$$

SCHEME 1

so that there is an indeed an exchange occurring. In this way, at equilibrium the energy of the electrons in the metal is equal to the energy of the electrons in the redox system, and the difference in potential can be measured between this system and a reference system.

It is appropriate to consider two common laboratory reference electrodes. They work on the same principle as just described. The silver/silver chloride electrode has the electron energy defined by the equilibrium between Ag metal (the electrode proper) and Ag^+ ion. The activity of the Ag^+ ion is controlled by the activity of Cl^- through the appropriate solubility product equilibrium. Thus if the activity of Cl^- is kept constant, the electrode potential is constant. Likewise, electrodes based on calomel, Hg_2Cl_2, rely on the redox equilibrium between the mercury electrode and Hg_2^{2+} in solution and, in turn, on the activity of Cl^- through a solubility product equilibrium. Note that this is perfectly general. Thus a bismuth electrode responds to the presence of species that react with Bi^{3+}, a copper electrode responds to species that react with Cu^{2+}, and so on. The requirement for good analytical data is that the equilibria be established rapidly.

Because the Nernst equation is the governing equation, it can be seen that the signal is proportional to the logarithm of the activity of the various species. Because mass-balance relationships come into play when one has to determine the potential, given a limited quantity of an analyte, the final analytical equations can be more complex than Eq. 8. These details are covered below.

There is a subset of potentiometric measurements in which the potential of a membrane is measured. The membrane has a potential across it that depends on the activities of ions that are both soluble and mobile in the membrane. It also responds to species that alter the activity of any ion to which the membrane is sensitive. These electrodes have not been used extensively in chromatographic detection because of their high selectivity. For more details on the technique consult the review by Buck (10).

2.2.4. Charge Transfer and Amperometry

The essential feature of the interface between an electrode and a solution containing a redox species is that it represents the point at which current changes from being caused by the motion of electrons to being caused by the motion of ions. This event is of primary importance when current flows at relatively low frequencies through electrodes into electrolytes.

The useful feature of electrodes is that the potential energy of the electrons in the electrode can be controlled by an external circuit. This potential depends on the excess positive or negative charge density on the electrode. Because excess charge on a conductor resides on the surface of that conductor (11) and the electrode surface is at the solution–electrode interface, it is fair to ask about the effect of the charges in the electrode on the structure of the nearby solution. A detailed treatment of this is beyond the scope of a chapter on detectors. Further details can be found in the treatise by Bockris and Reddy (12). As a whole, the interface is electroneutral. At the very least, then, one expects ions of a sign opposite to the electrode charge to predominate in the solution at the interface. In fact, this is the case. If the ions in solution do not interact chemically with the electrode surface, then there will be a concentration of ions with sign opposite to the electrode charge that is high at the electrode surface and decreases away from the electrode surface until it becomes equal to the bulk concentration of the ions. For ions with the same charge sign as the electrode, the opposite is true—that is, the concentration is low at the surface, rising to the bulk concentration. The distance over which most of this change occurs is called the double-layer thickness, where "double layer" refers to the charges that have been discussed. This number can be calculated with Eq. 13, which is for a symmetrical electrolyte at 25°C.

$$\text{thickness} = 3.58 \times 10^{-11} \left(\frac{D}{z^2 C} \right)^{1/2} \tag{13}$$

In this equation D is the dielectric constant or relative permittivity of the solvent, which is about 78.5 for water, C is the molar concentration of the electrolyte, the ions of which have a charge $\pm z$. For a univalent–univalent electrolyte at 0.1 M the thickness is about 10^{-9} m.

Owing to the charge segregation there is an electrostatic potential change in the solution near the interface. For small differences in potential between the electrode and the bulk solution the potential changes exponentially from its value at the electrode surface to the value in the bulk of solution. The distance over which this change occurs is the same as that over which the concentration of ions occurs.

If charged species interact chemically with the surface, or if the surface has covalently bound charges that may be changed by the solution conditions (i.e., the effect of pH on the dissociation equilibria of surface-bound hydroxyl groups in metal oxide electrodes), then the situation is more complicated. The only reason that charge distributed itself as it did in the preceding example was to satisfy electrostatic requirements. Chemical forces may, for example, cause the adsorption of a charged species, such as Br⁻. This then contributes to the surface charge density so the surface potential of the electrode and the distribution of charges in the double layer is correspondingly altered.

A simple physical picture of the interface has charges on the electrode surface with a corresponding excess of the opposite charge in the solution. The local excess charge in solution is smeared out, being highest at the electrode surface. Finally, the local excess of charge in the solution is due to an increase in the concentration of counterions to the electrode charge, and a decrease in the concentration of coions. This picture is simplified but contains the essential features of the interface. Refer to Fig. 2 as this discussion develops.

Any time there is a charged "plate" separated by an insulator or a dielectric from an oppositely charged "plate" the result is a capacitor. It will be recalled from above that a capacitor stores charge with a resulting potential difference between the two plates. One can then see that the double layer at the interface creates a capacitor. From the equation for a capacitor (Eq. 2), it can be seen that a small interplate distance results in a large capacitance. Indeed, the typical

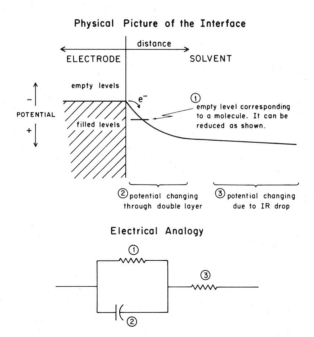

Figure 2. The interface. The top, physical, view of the figure shows the potential versus distance relationship, including effects due to the resistance of the solution and a Faradaic reaction. The analogous electric circuit is also shown. 1, Faradaic resistance; 2, double-layer capacitance; 3, solution resistance.

value for the capacitance for a 1 cm^2 electrode with a smooth surface in an electrolyte of 0.1 or 1.0 M concentration or so is about 10–20 μF.

Let us assume that one establishes a potential between an electrode and a reference electrode. The former electrode is called the working or indicator electrode. Recall that the reference electrode is a metal, the electrons of which are in equilibrium with well-controlled activities of a redox couple. Thus, in effect, when one makes use of a reference electrode, one is taking electrons from a reproducible chemical system. Once the potential difference between the working electrode and the reference electrode (this phrase is usually shortened to "the working electrode potential," in which the existence and function of the reference electrode are implicit) has been established, one would like to see how it progresses. One can measure it with a high-input impedance voltmeter. The phrase "high-input impedance" means that the voltmeter requires very small current for its operation. Current is the passage of charge with time (I [amperes] = dQ [coulombs]/dt [seconds]). Thus if the voltmeter required current, it would pass the charge on the capacitor through itself, removing that charge. Because the voltage on the capacitor is proportional to the charge on it (see Eq. 1), the measurement process itself would cause a change in the measured quantity. This is a situation to be avoided at all costs. To avoid it in this case, make sure that the product of the current required for the measurement multiplied by the time of the measurement is much less than the charge stored on the capacitor. Very good meters have input impedances in the 10^{15}-Ω range. Ordinary meters have input impedances in the 10^7-Ω range. These numbers are for low frequencies; for higher frequencies the capacitive component of the input impedance becomes important. The ohm is the MKSA unit of resistance and it comes from Ohm's law,

$$E(V) = I(A)R(\Omega) \tag{14}$$

$$E(J/C) = I(C/s)R(J \cdot s/C^2) \tag{15}$$

which is given in Eqs. 14 and 15, showing both the common laboratory units and the more basic units of the quantities. It is easily appreciated that the voltage measurement on a capacitor with a 1-V potential difference between its plates, if made with a meter having a 10^{12}-Ω input impedance at low frequencies, would take 1 pC of charge per second from the plates. Assume that this will not measurably alter the voltage on the capacitor that we are measuring. Then the voltage on the capacitor will remain a constant.

So the interface is like a capacitor, and if the voltage on the capacitor is measured carefully, one finds that it does not change. How can it be made to change? Another electrical digression is in order. The analog to Ohm's law for electrical potential and current at any frequency is given by Eq. 16. The new term, z, is the impedance, still with the units of ohms.

$$e = iz \qquad (16)$$

The impedance of a resistor is its resistance, R, at all frequencies. The impedance of a capacitor is given by Eqs. 17 and 18.

$$z_{cap} = \frac{-j}{2\pi f C} \qquad (17)$$

$$|z_{cap}| = \frac{1}{2\pi f C} \qquad (18)$$

Equation 17 includes the imaginary j which is $\sqrt{-1}$. Using this equation one can derive phase information about alternating signals. The use of Eq. 18 will only provide amplitude information. The term f is the signal frequency in hertz (units, s^{-1}). Equation 18 embodies the information gained in the preceding paragraph. When a constant ($f = 0$) voltage is impressed on a capacitor, its impedance is effectively infinite, so even though there is a voltage difference between the plates, Eqs. 18 and 16 say that the current (at 0 Hz) will be zero. Thus one cannot cause charge to pass in this simple system. Chemical intuition says that the reason charge will not pass is that there are not low-lying, unfilled orbitals on the solution side of the interface that can accept electrons from the electrode, nor are there filled levels at energies appropriate for donation of electrons to the electrode. Clearly what is needed is to inoculate the solution with a certain concentration of a species with either or both of those properties. Then electron exchange between solution and the electrode will occur, and the voltage across the interface will change as the charge on the capacitor "leaks" across the interface.

To summarize, the potential on an electrode penetrates into the solution as a decaying exponential function with a characteristic distance. This distance is of the order of tens of angstroms. The interfacial charge balance requires charge of one sign in the electrode and charge of an opposite sign in the solution, resulting in a capacitor, the double-layer capacitor, at the interface. The potential across this capacitor, once established, and if measured correctly, is constant. The presence of chemical species in solution that can accept or donate electrons to the electrode causes the charge maintaining the potential on the capacitor to be withdrawn and the potential difference between the electrode and the solution to change. Electrically the leakage of charge across the interface is represented as a resistor. This resistance is called the Faradaic resistance and is shown in

Fig. 2. Further discussion below of the Faradaic resistance will show its connection to the chemistry at the electrode surface.

The experiment just described in which the double-layer capacitor is charged and then the leakage of the charge across the interface due to the presence of a redox active species is followed by observing the shift in the potential difference across the double-layer capacitance is called a coulostatic experiment. It is infrequently (but very effectively, see below) used in electrochemical detection. However, it serves as a good point of departure to understand the functioning of the more common amperometric and coulometric detectors.

Consider the amperometric detector shown in Fig. 3. A mass balance can be performed on the electroactive species.

$$C_{in}U = C_{out}U + \frac{i}{nF} \tag{19}$$

Equation 19 states that the number of moles of analyte per second flowing into the cell equals the number of moles of analyte per second flowing out of the cell plus the number of moles per second being electrolyzed at the electrode. If the concentrations are expressed in millimolar (μmol/cm^3), then the volume flow rate of the fluid, U, should be in cubic centimeters per second. The last term for electrolysis is from Faraday's law expressed in Eqs. 20 and 21.

$$Q = nF \text{ moles} \tag{20}$$

$$\frac{dQ}{dt} = i = nF \frac{(d \text{ moles})}{dt} \tag{21}$$

Using concentration in millimolar units allows the use of microamperes for the current (μC/s). The number n is the number of electrons transferred per molecule

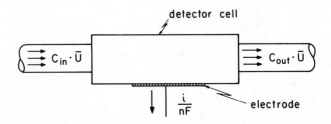

Figure 3. Mass balance in a flow-through cell. C is analyte concentration, U is the mean volume flow rate, and i is the current at steady state.

of analyte electrolyzed. Note that this is a steady state; as long as the concentration going in and the flow rate remain constant, then the concentration emanating from the detector will be constant, if the electrolysis rate remains constant. The current, i, is the analytical signal, and because it is part of a steady-state system, the way to increase the current is to increase the rate of the current-producing reaction while holding the rate of analyte efflux constant. Equation 21 can be rearranged to Eq. 22, showing that the sum of the fraction of the species electrolyzed plus the fraction leaving the cell is 1.

$$1 = \frac{C_{out}}{C_{in}} + \frac{i}{nFC_{in}U} \tag{22}$$

The electrolytic efficiency of the cell is the second term on the right-hand side of Eq. 22. As this term approaches 1, the cell is called "coulometric" even though it is the current that is measured. The term coulometric has been used because of the analogy with a coulometric titration in which an aliquot of sample is titrated with an electrogenerated reagent. The charge passed during the titration is related, through Faraday's law, to the mass of the analyte in the sample. In using the coulometric detector, it is possible to integrate the current over the time corresponding to the elution of a peak. If the electrolytic efficiency is 1, and if the stoichiometry of the reaction is known, then the mass of analyte in the peak is calculable without standards (13).

The possible steps that limit the current and therefore the sensitivity are transport, electrochemical, and chemical. These are discussed in turn.

Because analyte can be electrolyzed only if it is very close to the electrode surface, the transport of analyte molecules to the interface is important. Transport may occur in three ways, migration (the motion of charged particles in an electric field), diffusion (the random motion of molecules driven by thermal energy), and convection (the motion of solution).

It is customary to minimize the effects of migration by adding an excess of an inert (supporting) electrolyte to the solution. To understand the effect of this consider the generalization of Eq. 6 for specific conductance in a mixture. If the product czu for the analyte in its oxidized or reduced form

$$K = \left(\frac{F}{1000}\right) \sum_i c_i z_i u_i \tag{23}$$

is much less than the sum of all other $(czu)_i$, then it will contribute little to the conductance. This experimental environment allows us to neglect migration.

Diffusion occurs in all directions. It is a random process, and so it is understandable in statistical terms. The probability that a molecule takes a jump in any given direction in some specified time is the same for all molecules of that species in a homogeneous system. If there is a concentration gradient or inhomogeneity in the solution then, because the probability of an individual molecule's moving is constant, an area of low concentration beside one of high concentration will suffer an increase in concentration because of diffusion. The region of higher concentration will suffer a decrease in concentration. This is because the total number of species that move is proportional to the probability that a species moves (a constant) multiplied by the number of species present. Thus the number moving out of the region of high concentration into the region of low concentration is greater than the number moving in the opposite direction.

Convection is the process of bulk fluid motion. The laws governing fluid flow are complex, but for most detector work the results are simply stated. Virtually all detectors that employ single open spaces through which chromatographic effluent passes as it goes by the electrode experience laminar flow. The flow in packed beds may be laminar or turbulent. In laminar flow the fluid near the walls of the container does not move, and the stationary wall exerts its influence out into the solution far from the wall; see Fig. 4. The depth of penetration of the influence of the wall depends on the fluid and the time that the penetration has had to occur. The region affected by the walls is called the boundary layer. It is useful to view this process as the diffusion of momentum from the wall into the solution (1,14). The diffusion coefficient of the momentum is the kinematic viscosity, often given the symbol ν, with dimensions of centimeters squared per second. It is the ratio of the fluid's viscosity to its density. This is a relatively fast process in liquids, faster than the diffusion of molecules in liquids. If the fluid is passing between two walls, then the effects of the two walls diffuse into the fluid and meet in the center. If the walls are parallel (a

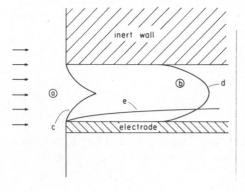

Figure 4. Hydrodynamics and diffusion in an open conduit. Fluid in region ⓐ is at a uniform velocity unaffected by the presence of the conduit. Fluid in region ⓑ is affected by the conduit. A boundary between the two, c, exists. The fully developed laminar flow velocity profile is parabolic, d. If electrolysis is occurring at an electrode, then there is depletion of the analyte near the electrode, creating a diffusion layer bounded by curve e.

rectangular conduit or channel, or a tube) this leads to the familiar bullet-shaped velocity profile, the fluid velocity being zero at the walls and a maximum at the center. This steady condition is called fully developed flow. If an electrode is in one of these walls, there is no direct convective motion of analyte to the interface because the fully developed flow is parallel to the surface. However, there is an effect of the flow because fresh solution can flow into a region that has been emptied by electrolysis. More complex geometries may be useful for detectors. These are discussed in the second part of the chapter.

It should be noted that only migration can create an inhomogeneous solution from a homogeneous solution and that this transport process is of minor importance. Practically speaking, then, the transport processes act to destroy concentration inhomogeneities that have been created in solution because of other processes, namely, electrolysis. We now consider electrolysis.

When a species reaches the interface, it may be oxidized or reduced. The equilibrium conditions are given by the Nernst equation, where it is understood that the discussion of the equilibrium is confined to the electrode itself and the species in solution in the immediate vicinity of the electrode. Thus, given a potential on the electrode of E, the equilibrium situation for redox species Q (for quinone) and H_2Q (for hydroquinone) is to have an activity ratio given by the Nernst equation shown in Eq. 23.

$$\frac{(Q)}{(H_2Q)} = \left(\frac{1}{H^+}\right)^2 \exp\left\{\frac{(E - E^\circ)nF}{RT}\right\} \tag{23}$$

If the activity ratio that is presented to the interface is different from that given by Eq. 23, then the system will move to equilibrium by altering the ratio. Let us say that the electrode potential is much more positive than the standard potential of the quinone/hydroquinone couple, and the activity of hydrogen ion is 1. Then Eq. 23 demands that $(Q) \gg (H_2Q)$. If one injects a solution of hydroquinone into this system, then the electrode will oxidize (remove electrons from) the hydroquinone, forming quinone plus protons. By convention (among chemists) electrons move to systems of more positive potential. The electrode's being more positive than the E of the Q/H_2Q couple, resulting in the oxidation of H_2Q, is an example of that convention. One could follow this oxidation coulostatically, but usually a potentiostat is used to measure the signal. The potentiostat holds the working electrode/reference electrode potential difference at a constant value by supplying the current necessary to carry out the electrolysis and preventing the electrolysis from taking the charge that is present to maintain the potential across the double layer. The charge that passes is supplied by an external circuit; it travels in a circular path from the working electrode to the external circuit where it is measured, then through another electrode, the auxiliary electrode, into the solution (where it is carried by supporting electrolyte) and through the solution to the working electrode interface. Also note that the electrolysis creates a concentration inhomogeneity by making conditions at the electrode different from those in the bulk of solution. Diffusion and convection work to counteract the effects of the electrolysis.

Electrode reactions in which the Nernst equation is obeyed at the interface are called reversible. There are electrochemically irreversible reactions in which the oxidation or reduction reaction of interest is slow in the kinetic sense (12,16). In this case a molecule must be at the interface a long time in order to react. The implication is that an electrochemically irreversible reaction can be brought to equilibrium given sufficient time, and this is true. For irreversible reactions the concentrations of species at the electrode surface do not obey the Nernst equation, but those concentrations may be different from those of the bulk solution, for the system will attempt to move toward equilibrium. Irreversible reactions may be hastened by the application of a potential to the working electrode that is greater in absolute value than the potential that is required to accomplish the objective according to the Nernst equation. Increasing the potential increases the thermodynamic driving force for the reaction, in effect making the reaction more exergonic and therefore more likely to occur.

There are other sorts of irreversibility (12,16). A common phenomenon is the reaction of the just-produced electrolysis product. Thus the oxidation of a molecule, making it electron poor, is often accompanied by the attack of the electron-rich solvent molecule to yield a product that will not undergo reduction back to the starting material.

Chemical reactions may also limit the rate of electrolysis (12,16). For example, a molecule may not be reduced until it undergoes a homogeneous (i.e., solution phase) reaction, and if the rate of the homogeneous reaction is slow, then the rate of reduction is slow.

Let us summarize the working of the amperometric (current measurement) detector. A steady state exists in a flow cell; what is not electrolyzed is passed out of the cell. Important to improving the signal is increasing the rate of electrolysis. One establishes a potential at the interface that is sufficient to accomplish a desired reaction. After the double layer is charged, nothing further happens until the flow stream is inoculated with the analyte. A potentiostat maintains the electrode potential by replacing charge withdrawn from the electrode in the course of electrolysis. When analyte flows through the detector, molecules at the surface have a thermodynamic tendency to reach electrochemical equilibrium with the electrons in the working electrode by undergoing oxidation or reduction. The current from this electrolysis is the signal. The current may be limited by the rate of chemical or electrochemical steps at the interface. This limitation can be avoided by increasing the driving force for the reaction. Because the electrolysis causes a concentration inhomogeneity to exist, diffusion, assisted by convection, acts to restore the homogeneous condition.

In amperometric detectors the signal is, in principle, a linear function of the concentration of analyte. The governing law is based on the mass-transport behavior of the system. In linear systems (diffusion coefficients independent of concentration, fluid viscosity independent of fluid velocity) the flux of matter between two regions of space, one of concentration C_1 and the other of concentration C_2, is proportional to the difference in the two concentrations, as expressed in Eq. 24.

$$J = k(C_1 - C_2) \tag{24}$$

The flux, J, has units of moles per square centimeter per second. It is related to the current by Faraday's law; see Eqs. 21 and 25.

$$i(A) = nFA \, (cm^2) J (mol/cm^2 \cdot s) \tag{25}$$

Note that this equation is consistent with the discussion of transport processes above. The proportionality constant k is called the mass-transfer coefficient, and it has the units of velocity. In the context of a detector, C_2 may be viewed as the concentration in the bulk of the solution flowing through the cell, whereas C_1 is the concentration at the surface of the electrode. From a consideration of mass balance (oxidized form of analyte + reduced form of analyte = a constant) and Eq. 23 one arrives at the equation for current as a function of $\theta = nF(E - E°)/RT$ shown in Eq. 26.

$$i = nFAkC_2 \left(\frac{1}{1 + e^{-\theta}} \right) \tag{26}$$

The resulting relationship, known as a hydrodynamic voltammogram is shown in Fig. 5. For cells of low electrolytic efficiency (which are common) C_2 is the concentration of the analyte as it enters the flow cell. In most analytical appli-

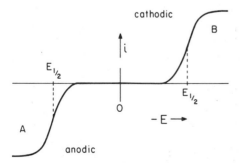

Figure 5. Hydrodynamic voltammograms for a species that is oxidized, A (this corresponds to Eq. 26), and for one that is reduced, B (change the sign in the exponent of Eq. 26). The potential corresponding to the point halfway up the wave is characteristic of a compound, given constant conditions, and is called the "half-wave potential." At the wave plateau the current is limited by transport processes.

cations C_1 is driven practically to 0 by applying the appropriate potential. Thus Eq. 27 is seen to hold.

$$i = nFAkC_{in} \qquad (27)$$

3. DETECTION

3.1. Introduction

A general discussion of figures of merit is now presented. Each of the four types of detectors is discussed with consideration of the following properties shown in Table 1: the property measured, the limiting background process, the detection limit, the linear dynamic range, the sample suitability, control over detector selectivity, and important or potential representative applications. A more detailed discussion of each detector type follows discussion of Table 1.

3.2. Figures of Merit

There have been much discussion and a proliferation of nomenclature involving the best performance one can get from an analytical system. Most of this concentrates on understanding the minimum detectable quantity, because trace analysis has been and will continue to be an important area. An essential figure of merit is the detection limit (dl) (17,18). The dl is based on a statistical concept. The dl is often defined as the concentration or quantity of analyte that allows one to identify the presence of analyte a certain fraction of the time correctly.

All analytical detection instruments are based on transducers. It is natural to ask what is the minimum detectable *signal* as opposed to the minimum detectable *quantity*. This unambiguously directs attention to the functioning of the apparatus itself. To determine the minimum detectable signal, one records a segment of base line. The conditions of the measurement must be the same as those during chromatography. Because much of the noise in chromatography (and especially in amperometric detection) seems to be of a $1/f$ (f = frequency) character rather than white noise, there are differences in how the problem should be treated in comparison to the case in atomic absorption spectrometry, where the noise is indeed white and where one has control over the length of time one measures the signal (17).

The term $1/f$ noise implies that a plot of noise power versus frequency (noise power spectrum) is hyperbolic; that is, noise power is proportional to $1/f$. Examples of white and $1/f$ noise traces are shown in Fig. 6a and b. One can see from the plots that the major difference in the traces is the correlation evident in the $1/f$ trace. A "quick and dirty" way to see this is to note that there are fewer

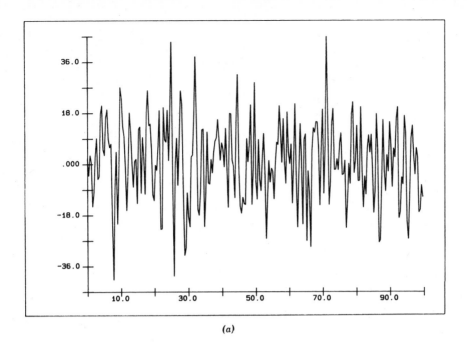

(a)

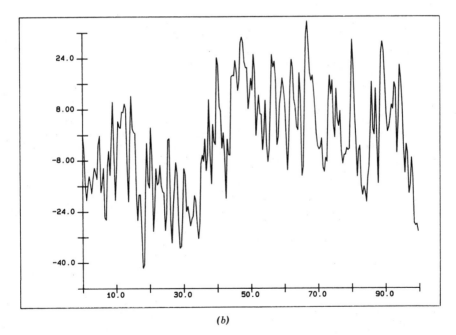

(b)

Figure 6. Noise (*a*). White noise signal is plotted versus time (s). (*b*). 1/*f* noise. (*c*) Autocorrelation of white (solid) and 1/*f* (dashed) noises.

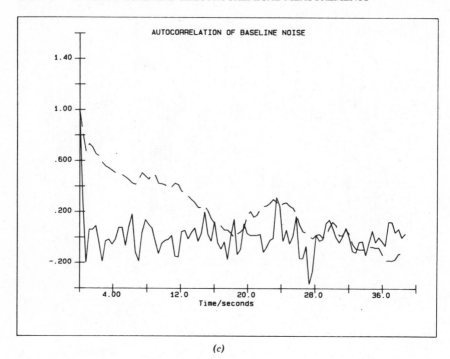

(c)

Figure 6—*Continued*

zero crossings in the $1/f$ trace. This means that at any time $t + \Delta t$ the pen position depends on its former position at t. This is what is meant by correlation. This is always true for some Δt, usually given by the response time of the electronics or the recording device. In the example traces given, a low-pass filter was employed with a cutoff frequency of 1 Hz. Thus for the white noise one expects the pen positions to be correlated for times on the order of 1 s; however, the $1/f$ trace shows correlations over longer periods. This is graphically demonstrated in Fig. 6c, which shows the autocorrelation plots for white and $1/f$ noise. An autocorrelation is a graphic way of seeing influence of the signal at $t - \Delta t$ on the signal at t. In this sense view the plot as "influence" versus "Δt." The plot for the white noise shows the expected positive correlation for short times, after which the correlation randomly bounces around zero. On the other hand, the $1/f$ autocorrelation plot does not drop to 0.5 until about 5 s. In all other respects (magnitude of the [root-mean-square, rms] noise, ratio of peak-to-peak to rms noise [about 5.5], distribution [Gaussian]) the two noise traces are very similar. (One must avoid the word "identical" in statistical systems!)

Current practice is to determine the detection limit based on statistical considerations (17,18). One employs Student's t test. This test is used to compare means given standard deviations of two populations. In the case under consideration the "population" consists of pen position on the chart paper as a function of time. A usual assumption in the S/N ratio problem is that when a signal at the dl is being measured, the noise is the same intensity as it is for the base line alone. One then determines the base-line noise in terms of the standard deviation of the pen position around the average position. This average position of the base line can be assigned a value of y_0. Then one measures a small signal, repeating the calculation of standard deviation and calculating the average pen position. The t test is used to answer the question, "Is the mean calculated from the experiment statistically significantly different from y_0?." Equation 28 shows the mathematical form for the calculation of t.

$$t = \frac{(y - y_0)(n/2)^{1/2}}{s} \tag{28}$$

In this equation y is the signal amplitude, y_0 is the background amplitude, n is the number of pairs of signal/background measurements, and s is the standard deviation of the measurements of the background, equivalent to the rms background noise. The magnitude of t is then compared to the (one-tailed) t statistic in a table for the number of degrees of freedom in the experiment, $2n - 2$ (one degree of freedom for each measurement minus the number of restrictions, i.e., each mean value). If t is less than the statistic in the table then, with a stated confidence, one answers the question "No." However, if the calculation in Eq. 28 leads to a number larger than that in the table, then the answer is "Yes." If one rearranges Eq. 28 using $y - y_0 =$ signal and $s =$ noise, then one has Eq. 29.

$$\text{S/N} = t \, (2/n)^{1/2} \tag{29}$$

It can be seen that the value of the S/N ratio depends on the value of t. In Eq. 29 choose t from the table for the appropriate number of degrees of freedom and confidence limits. For example, for $n = 2$ and 99.9% confidence in the presence of a "signal" S/N must be 31! For 99.5% confidence and $n = 5$, S/N must be 1.53.

A major assumption in the use of Student's t statistic is that the measurements are uncorrelated. This may not be the case for electrochemical detectors, which display a $1/f$ type of noise, implying a correlation between points. At the least one must assume a smaller number of degrees of freedom than $2n - 2$; at the worst it is possible that the use of the t statistic is incorrect, and other methods will be required for the estimation of the dl.

It would seem to be optimal for chromatography to employ a detector the output of which is filtered with a filter having a time constant on the order of 1/10 or 1/20 of a chromatographic peak width. Then one measures the base-line noise for a length of time corresponding to the width of a peak. It is ordinarily done by measuring the peak-to-peak noise and dividing by 5 to get the rms noise. Then one has a measurement of the noise *in the measured quantity*. The filter causes the base-line measurement to correspond to $n = 10$ or 20. Thus $2n - 2$ is 18–38. Making an ad hoc guess about the effects of the correlation, one can estimate the number of degrees of freedom as about 5–10. Then the signal to (rms) noise ratio required for 99.5% confidence in peak detection is about 1.6. This can be rounded to S/N = 2. Note that N/S, the inverse of S/N, is the relative standard deviation, which for the detection limit is 0.5. This means that quantitation at the detection limit is not possible. To obtain a relative standard deviation of 0.05 (5% relative error) one must have a concentration 10 times the detection limit (this assumes that "base-line" noise dominates the measurement). This is sometimes called the determination limit. Because the two are so obviously related, having a term for each seems unnecessary.

The sensitivity of the detector is the slope of the calibration curve. It has units [measured quantity]/[concentration of analyte injected]. Alternatively one may choose to work with units of mass; thus the calibration curve would be [measured quantity]/[amount of analyte injected]. This figure of merit is not necessarily related to the detection limit. The dl and the sensitivity together provide the analytically useful minimum detectable quantity or concentration; see Eq. 30.

$$\text{min. det. concn.} = \text{dl} \times \text{sensitivity} \tag{30}$$

To determine the concentration at which quantitative analysis can be done with about 5% precision multiply the minimum detectable concentration by 10.

The linear dynamic range is the range of concentrations of quantities over which the calibration curve is linear. It is important for simplicity in data analysis, though not essential for quantitative analysis. A derived quantity may be used, as in absorbance detection, to linearize the calibration curve.

Table 1 proves an overview of the important properties of the four types of detector that have been discussed so far. A discussion of Table 1 follows.

3.3. Permittivity

3.3.1. Discussion of Table 1

The permittivity detector measures permittivity which is in turn related to a molecule's polarizability, α, and the presence of permanent dipoles, μ (2). In

Table 1. Properties of Detectors

Property Measured	Detector Type			
	Permitivity $\varepsilon(\alpha + \mu)$	Conductance λ	Potentiometric E	Amperometric I
Control	T, cell const.	T, cell const.	$i(=0)$, electrode material	E, electrode material
Limiting background	$d(\varepsilon_{solvent})$	$d(\lambda_{solvent})$	d (side reactions)	d (solvent electrolysis)
Routine detections limits	10μg	10 ppb in 100 μL	vwe[b]	10 pg in 20 μL
LDR	vw[a]	vw[a]	Depends on ligand concn.	10^3–10^5
Suitable sample and eluent	Nonionized eluent	Ionized analyte	Donor analyte	Redox analyte
Application	GPC	Ion chromatography	Amino acids	Phenols, catechols, thiols, aromatic amines

[a]Varies depending on electronics and medium.
[b]Varies depending on equilibrium used for detection.

its role as a detector of solutes, the important property is the difference between the permittivity of the solvent and the permittivity of the solution. Like the closely related RI detector, the permittivity detector is a bulk property detector, and it is very temperature sensitive. Changes in temperature alter both the geometry of the cell and the permittivity of the solution in the cell (19–21). It has recently been shown that automatic software compensation for temperature drift can be very effective in lowering detection limits (22). As might be expected, the limiting background process is the fluctuating permittivity of the background solvent. Detection limits of about 1 μg are obtained when the temperature drift problem is addressed (20,22). There are insufficient data in the literature to specify a linear dynamic range. For maximum sensitivity the eluent should have a low permittivity. There should not be ionized species in the eluent that would have the effect of altering the purely capacitive nature of the cell. Applications of this detector are few. It would seem ideally suited to gel permeation chromatography of polymers and oligomers where the chromatographic demands on the solvent are few, and trace detection is not a requirement.

3.3.2. Design Considerations

The theory for permittivity detection has been given detailed treatment (19–21). There are many ways to measure changes in permittivity, and these have also been discussed (2,19). The most precise method seems to be the heterodyne method; data on other methods (as applied to detection work) are rare. In the heterodyne method the cell makes up part of an oscillating inductor–capacitor circuit, and the output from another inductor–capacitor circuit is mixed with it. The second circuit has a constant output frequency. The results of mixing the two signals are the sum and difference frequencies. The output of the mixer is filtered to remove the sum frequency, leaving the difference frequency. If the inductor–capacitor circuit containing the cell is initially tuned with a variable inductor or capacitor so that the output of the mixer is 0 when the eluent is passing through the cell, then a change in capacitance caused by a change in the permittivity of the eluent, causes the output of the mixer to be a nonzero frequency. The output frequency change, Δf, for a given concentration of analyte in an eluent of permittivity, ε_1, is given by Eq. 31 (21).

$$\frac{\Delta f}{m_2} = \frac{f_1 \delta Z}{2\varepsilon_1} \text{ (units Hz/}M\text{)} \tag{31}$$

In this equation f_1 is the frequency of the constant oscillator, and Z is a term made up of the various capacitances in the cell, that is, the detector cell, and series and parallel capacitances. δ is the dielectric increment of the solute, and the solute's concentration is m_2 (see Eq. 3); Z increases as the detector cell capacitance decreases. Note that the highest sensitivity is obtained with high f_1 and a low detector cell capacitance. Also, because the permittivity of the eluent may change depending on the particular chromatographic experiment, the sensitivity will change. A cell with adjustable inter-electrode spacing has been used for this reason (21).

There are upper limits on f because at high frequencies the stray capacitances in the system play a larger role. Alder et al. (21) have demonstrated a rms noise of 5 Hz. This is quite low with respect to the operating frequency of 10^7–10^8 Hz.

Many workers have pointed out that thermal fluctuations limit detection (19–23). It will be recalled (Eq. 2) than the capacitance of the detector cell is determined by both the cell geometry and the permittivity of the eluent. Temperature may influence both. It has been shown theoretically that the thermal effects on the permittivity are far more important than the geometric effects (22). By considering the heat-transfer problem in the cell one can design a system for low sensitivity to temperature fluctuations (20). The most successful apparatus (submicrogram detection) is one in which the temperature is measured and the base-line drift due to temperature drift is corrected with software, after analog to digital conversion (22).

There is a close parallel between the design of the permittivity and the conductance cell. Furthermore, the permittivity detector ceases to function when the eluent is conductive, so the two detectors are complementary. This has led some workers to combine the two detection systems into one. A useful circuit is presented by Alder et al. (23). In this design the difference in the "interrogation" frequencies is used advantageously. The permittivity detector, operating in the megahertz region, is connected to the cell through a 33-pF capacitor. The conductance detection circuit (operating in the kilohertz region) is connected to the cell through an inductor. The inductor blocks the higher-frequency signal from getting into the conductance detection system, and the small capacitor prevents the lower-frequency conductance signal from getting into the permittivity measurement circuit. In another clever circuit (19) the cell (acting as a resistance and a capacitance) is part of a circuit with a resistor and an inductor. From Eq. 31 and 4 it can be seen that the permittivity measurement based on the heterodyne principle depends on the exciting signal's frequency but not its magnitude, whereas the conductance signal depends on the exciting signal's magnitude, not its frequency. Thus a self-balancing circuit that has control over both the frequency and voltage of the exciting wave yields both permittivity and conductance signals from one cell.

3.4. Conductance

3.4.1. Discussion of Table 1

The conductance detector measures the conductance of a solution which is in turn related to the presence of charged species in the solution. Although not a detector of a bulk property per se, the conductivity response does depend on the viscosity of the solvent. Hence this detector is also quite temperature sensitive. The temperature dependence is approximately given by Eq. 32. In this equation T' is the operating temperature and λ is

$$\lambda_{T'} = \lambda_{25}\exp\{k(T' - 25)\} \tag{32}$$

the equivalent ionic conductance at the specified temperature. The constant k is about 0.02. Thus there is about a 2% change in conductance per degree change in the system temperature (24). The importance of temperature control has been pointed out (25,26). In particular, it has been shown that virtually the entire system should be thermostatted (26). This minimizes heat flux from one area of the instrument to the other (15,20). Using a dual cell with a reference side that determines the conductance of the eluent alone and then measuring the differential signal may be useful (27). The limiting background process is the fluctuating conductance of the eluent. The conductance detector has its widest application in the field of ion chromatography. Both suppressed (28) and unsuppressed (9) procedures are used. In the latter, the background conductance is higher than in the former, so temperature control is more important. Detection limits are given in many ways. The few that give limits of the apparatus tend to show a mimimum detectable change in conductance near 10 nS/cm (25,29). Many workers demonstrate detection limits in the 10-ppb range for chloride (25,26,30). It can be lower if preconcentration of the sample is employed (31). The linear dynamic range can vary. It can be low with ordinary (few kilohertz AC square or sine-wave excitation) detection due to stray capacitance (24). With appropriate attention to this problem, ranges of 10^4 (32) to 10^6 (33) can be obtained. The bipolar pulsed conductance technique has been suggested to avoid the nonlinearity problem (7,24). (The details of bipolar pulsed conductance are discussed below.)

Of course the suitable sample/eluent combination is an ionic sample in a polar solvent. There is a high degree of interaction between chromatographic and detection considerations in the ion chromatography–conductance detection field. The major difficulty in unsuppressed ion chromatography is the high conductivity in the eluent. One would like to decrease the salt concentration in the eluent but doing so reduces the strength of the eluent. To permit the use of low-conductivity eluents, low-capacity resins must be used for the analytical separation (9,34). The pH (35) and organic modifier concentration (29) are important for chromatographic optimization, but their influence on conductance cannot be overlooked. The pH can alter conductance sensitivity if the eluent ion or analyte is acidic or basic. Decreasing the solvent polarity inhibits ionization. Tetrabutylammonium salicylate has been found useful for dynamic ion chromatography on a PRP-1 resin column (6). The linearity of conductance detection allow the use of Yeung's technique for quantitative analysis of unknowns. For accuracy one must be able to do the separation with eluents of widely differing conductance (5).

Certain reaction detectors employ conductance. Several reports have appeared in which procedures similar to the Coulson and Hall conductivity detectors for GC have been employed in LC (36–38). Good detection limits for a variety of heteroatom-containing species have been demonstrated. Halogen-containing species can be selectively detected after combustion and dissolution of the combustion products. The technique was optimized and applied to the analysis of mixtures of PVC oligomers (36). A sulfur-selective detector was designed with a detection limit of 600 pmol of sulfur. Its selectivity for S was limited, however, because it responds well to Cl, Br, I, P, and N (37). A detector useful for agricultural chemicals was designed and optimized. It had a detection limit of 5–50 ng for lindane and was more effective at quantitating aldrin and dieldrin than UV detection at either 254 or 210 nm. The detector responds to Cl, F, N, and S heteroatoms (38). Photolysis can lead to ionic products from neutral precursors, and this has been made the basis of a detection technique (32). Halogenated species, some sulfur-containing species, and nitro compounds are detectable with this detector.

3.4.2. Design Considerations

There are three methods employed to determine the conductance of the solution in a conductance cell. Traditionally the excitation has been a sine wave or a square wave in the kilohertz range. There are difficulties with this. The effect of capacitance in the circuit is to alter the cell's impedance, making determination of the purely resistive component difficult. The kilohertz region is chosen because the effects of the interfacial capacitance are minimized and the influence of the capacitance of the cell due to the dielectric behavior of the eluent (i.e., as in a permittivity detector) is not large (4). Some detection circuits employ balancing capacitors to offset the effects of stray capacitance (19,41). A side effect of the measurement of conductance in this way is to heat the sample by Joule heating. This can be minimized by using a small excitation signal, but of course this decreases the measured signal and consequently makes the method less sensitive. Thus an optimal voltage exists. The magnitude of the optimal voltage depends on the cell constant and the thermal characteristics of the cell.

An obvious way to measure conductance is to impress a DC voltage across a pair of electrodes and measure the current. Because the measurement is DC there is no effect due to stray capacitance. However, owing to the potential drop that occurs at interfaces, much of the potential measured would not be due to *IR* drop in solution. Four-probe measurements, in which one impresses a current through a solution with a pair of electrodes and then measures the voltage difference between two reference electrodes in the current path, obviate the interfacial potential drop problem because the drop does not occur at the measurement electrodes. This principle has been used in a Japanese detector (25). A very clever detector based on simple DC conductance has been built for detection in microbore LC (29). Advantage was taken of the small diameter of the column. A long section of the narrow-bore column was used for the cell to provide a large resistance to measure. This was necessary for the following reason. In order to obviate the problem of the interfacial potential drop one can use a large potential difference between electrodes. This ensures that the bulk of the applied potential occurs across the solution. However, this leads to excessive Joule heating if the measured resistance is low. By making the cell constant small, excessive Joule heating, even with a 30-V potential difference, was avoided. The cell volume was only 0.1 μL, and the peak height required for detection was 20 nS/cm.

The technique of bipolar pulsed conductance (42) can be used to eliminate the problems due to capacitance and Joule heating. An ancillary advantage is that the technique is naturally wed to a computer; thus signal manipulation is simple. In a measurement of conductance that relies on the determination of the current flowing because of an applied potential, one must be aware that some of the applied potential exists across the double-layer capacitance and does not drive current through the system's resistance (see Fig. 2). This problem was addressed in Ref. 29 by using a large DC voltage, making the portion across the double layer negligible. The bipolar pulsed conductance method solves the problem in another way. With reference to Fig. 7, at $t = 0$ one applies a potential pulse across the solution. Any capacitance parallel (C_p) to the solution impedance charges quickly, and the full applied potential is applied across the solution resistance (R_s) and the double-layer capacitance in series (C_s) with it. We can assume that at $t = 0$ the double-layer charge is 0 without loss of generality. C_s begins to charge in response to the applied voltage while C_p is charging, so that by the time the C_p is charged, C_s has a small charge on it, with an associated voltage across it. One then does not know how much of the voltage applied is actually impressed across R_s and therefore one cannot calculate R_s from the measured current. When the pulse time is less than the time constant, R_sC_s, the charge on C_s grows to a value Q that is much less than the equilibrium charge

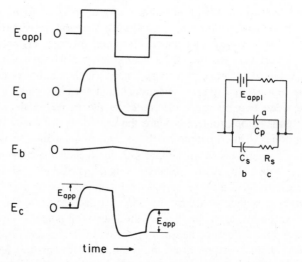

Figure 7. Operation of bipolar pulsed conductance. The applied voltage is E_{app}. Traces of voltages across the other points in the circuit are shown; (a) parallel capacitance, (b) series (double-layer) capacitance, and (c) solution resistance.

for the applied voltage. Now one reverses the polarity of the potential pulse. C_s is now able to discharge. At a time equal to the length of the first pulse, the charge Q will have decreased to 0, and *all* the applied potential is across R_s only. A measurement of the current at this point leads to the accurate measurement of R_s and thus conductance.

This technique has been used successfully in detection work (7,24). The elimination of the complications of the capacitances in the circuit allows detection over a wide linear range (24).

An interesting area is the interaction between the detector and the eluent. Consider once again the equation for the specific conductance of a salt, given by Eq. 33.

$$K = \frac{C}{1000} (\lambda_+ + \lambda_-)\alpha \qquad (33)$$

Recall C is the concentration of the species, λ are the equivalent ionic conductances for the ions in the salt, and α is the degree of dissociation of the salt. From the point of view of the *eluent* one wants the specific conductance to be low, and clearly there are three ways to accomplish this. One can design a system with a low eluent concentration. In order to elute ions in a reasonable time this will require the use of low capacity resin or short columns. One can cause α to be low. This is the idea behind "suppression" (28). For example, anion chromatography can be effectively carried out with sodium phenate ($\alpha \approx 1$) as an eluent, then by passing the eluent through an ion-exchange column or a membrane suppressor in the H^+ form, the phenate is converted to phenol, which has a low α. Finally, one can use eluents that have low equivalent ionic conductances, for example, the use of tetrabutylammonium salicylate has advantages in this regard (6).

Now consider the case of the analyte. Here one wants the maximum conductance. To increase C one should minimize band spreading. This requires low-capacity, monodisperse, small particles. Because of the exchange behavior of ions, their concentration on an ion-exchange precolumn is very simple and advantageous for ppb detection. Generally one has no choice over α, because it is a property of the analyte in the given eluent. In anion chromatography there is some advantage as an indirect result of the suppression step described above. Anions are typically eluted as Na^+ salts, but weakly basic anions are converted to the more highly conducting H^+ salts before detection. The effects of α can be interesting. The influence of pH has been studied in a system employing tartrate as the eluent (35). One can control the elution of anions by controlling the pH because the dianion of tartrate is responsible for most of the eluting "power" of the eluent. But when one lowers the pH, salts of weak acids are protonated, and their conductance decreases. Also, using organic modifiers can suppress ionization leading to decreased analyte conductance (29,31).

3.5. Potential

Potentiometric detectors operate based on the measurement of electrode potential at zero current. Two types, metal electrode and membrane, have been used. In a series of papers Haddad, Alexander, and co-workers have given theoretical and experimental evidence of the utility of a copper electrode (43–46). The electrode works by detecting changes in the activity of Cu^{2+} ion due to the presence of complexing agents. For this type of detector the limiting process is the presence of and fluctuations in "side reactions," reactions of the eluent with the copper ion resulting in a change in its activity. Detection limits, unfortunately usually stated as a concentration or amount, vary widely because the sensitivity of the detector varies depending on the relevant equilibrium constants. In the determination of complexing agents the copper electrode is capable of determining about 1 nmol of a strong complexing agent, but 30 nmol of a weak complexing agent. In use as an indirect detector for alkaline earth ions, a 2-ppm detection limit was obtained. The calibration curve can be linear for very low concentrations of ligand but becomes logarithmic at higher concentrations. Appropriate analytes for primary detection are donor molecules, and for secondary detection, ionic species that can compete with copper(II) for ligands may be detected.

Potentiometric detectors based on membranes have been used as well (47–51). In certain cases the combined selectivity of the chromatography and the detector yields a method that is well suited to repetitive determinations of a species. Of course, such highly selective methods are of limited general utility. Iodide in seawater may be determined using an silver iodide electrode-based detector (47). Fluoride is selectively determined by fluoride ion-selective-electrode detection following preconcentration on a zirconium(IV)-containing column (50). A very clever cell was designed that detected the presence of various anions or cations depending on the type of ion-exchange membrane interposed between two reference electrodes. Ionic activity on one side of the membrane was held constant while the eluent passed through the other side. Noise levels were 5 μV, which allowed small potential changes to be measured (51). Membranes incorporating neutral carrier ligands have been used for the detection of ammonium ion at the 1 μM level (49). An indirect method of general utility is based on the preparation of a silver precipitate-based sensor from oxidizing a silver wire in the presence of a solution of the eluting anion in an anion chromatography procedure. When using this eluting ion in a chromatographic determination, the analyte ions replace the eluting ion. This activity change is detected potentiometrically (52). Cyanide and sulfide have been detected using a silver sulfide membrane electrode (48).

3.6. Current

3.6.1. Discussion of Table 1

Because of the popularity of aperometric detection, the field has been reviewed many times. A review of the detector's general operating principles and useful features appeared in 1977 (53). Later, a broad review, including mercury-based detectors, appeared (54). Criteria for a cell's effectiveness and a discussion of performance evaluation were subjects for another review (55). A fairly comprehensive review has appeared in a chromatographic series (56). The most recent review, team-written, is unique in its breadth of coverage (57).

Amperometric detectors measure the current created by the constant-potential electrolysis of species in a flowing stream. A great deal of selectivity is had by the control of the electrode potential, because it can be set to just oxidize (or reduce) an analyte, obviating interference from species that are more difficult to oxidize (or reduce). The electrode material can have a significant effect on selectivity as well. Analytical electrochemistry got its start with the mercury electrode, and it is therefore appropriate that the first amperometric detector for LC employed a dropping mercury electrode (58). Mercury has the advantage of not catalyzing the reaction

$$2H_{ads} \rightarrow H_2$$

Thus it requires a significantly more reducing potential than that required by the Nernst equation to cause the reaction

$$2H^+ \rightarrow H_2$$

to proceed at a mercury surface. Consequently reductions, many of which occur at potentials more negative than the equilibrium potential for the reduction of protons to hydrogen, can be carried out in protic solutions. Also, many metals form amalgams with mercury. This avoids a kinetically slow crystallization step in the overall redox process of metal ions. Amalgam formation causes many such reactions to be reversible. Nonetheless, because of the practical difficulties involved in making low dead volume, easy to use dropping mercury electrode detectors, they have not been widely employed. A commercial detector solves part of the difficulty by using a static mercury drop that is replaced periodically (EG&G PAR).

Solid electrodes have been more widely used. A gold electrode can be coated with mercury, forming the amalgam. This can then be used for reductive work (59,60) with the convenience of solid electrodes. The chemistry of mercury can be exploited with gold-amalgamated mercury electrodes. For example, thiols can be detected at low potentials because of the following reaction (61):

$$2RSH + Hg \rightarrow Hg(SR)_2 + 2H^+ + 2e^-$$

An analogous concept can be employed to detect poorly ionized species in ion chromatography, for example, cyanide and sulfide with a silver electrode in which the reaction creating the current is the oxidation of the silver metal to the ion as a complex or a precipitate (62). In recent interesting work it has been shown that silver halide chemistry can be used to accomplish reductions of alkyl and aryl bromides and iodides (62,63). Platinum electrodes have been used for their chemistry as well. Johnson (64) has modified a platinum electrode with adsorbed iodine in order to detect chromium(VI). More recently, Johnson and his group have exploited the catalytic and adsorptive properties of clean platinum surfaces for detection (65–71). The difficulty with platinum has been its tendency to adsorb strongly, leading to a surface that is inaccessible to solutes. Johnson's work, recently reviewed (70), solves the problem using pulsed amperometric detection. The pulse sequence consists of two or three pulses. The pulse sequence depends on the chemistry. There are three possible effects of the adsorption of an analyte on the oxidation of the surface to the oxide. These are illustrated in Fig. 8. Figure 8a shows the current as a function of time after the potential on the electrode has been stepped to a value sufficient to carry out the oxidation of a species that is in solution. The technique is called chronoamperometry. In this case, the limiting step in the electrolysis is diffusion, and the process is described by Fick's laws. The current is proportional to $t^{-1/2}$, where t is the time after the application of the step of potential. Figures 8b–d show the current–time plot for an electrode that forms an oxide film. In the absence of an adsorbate the current due to oxide formation is proportional to $1/t$. In the presence of an adsorbate that is *not* displaced by the oxide, the current has the same functional form but is reduced in magnitude because part of the electrode is "blocked" by the adsorbate. In the presence of an adsorbate that *is displaced* by the oxide, the initial current is low because of blockage. Because the oxide eventually covers the entire surface, the total charge needed to form the oxidized surface is the same as in Fig. 8b with $\theta = 0$, where θ is the fraction of the surface that is covered by the adsorbate at the initiation of the step of potential. Thus the current at long times must be large relative to the $\theta = 0$ case. Note that a difference between a "blank" solution and a solution with analyte has been recorded in Figs. 8b and c even though the analyte is not electroactive. For flow analysis one wants the

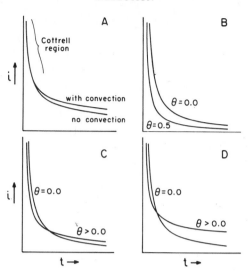

Figure 8. Expected anodic response to a positive potential step. (*a*) Transport-limited faradaic reaction. (*b*) Oxide formation in the absence ($\theta = 0.0$) and presence ($\theta = 0.5$) of an irreversibly adsorbed, electroinactive adsorbate. (*c*) Oxide formation with desorption of electroinactive adsorbate. (*d*) Oxide formation with oxide-catalyzed oxidative desorption of electroactive adsorbate. Reprinted with the permission of Elsevier from Ref. 70.

adsorption to be reversible, so the example shown in Fig. 8*b* is not analytically useful, but the example shown in Fig. 8*c* is. If the adsorbed compound is oxidized, as it is desorbed concomitantly with oxide layer formation, then an added current is obtained at long times as shown in Fig. 8*d*. This approach is a very powerful one for many analytes that are otherwise difficult to detect, such as alcohols, including carbohydrates (65,68), and amino acids (69). Other electrode materials act as catalysts and can be employed for selective detection. Lead oxide (70) has been used to oxidize thiourea, and nickel electrodes, in which the nickel(III) hydroxide is the active species, have been used to oxidize carbohydrates (72–74).

It is clear that one of the powerful aspects of electrochemical detection by amperometric means is the degree of chemistry that is involved. The examples illustrated serve to show how selectivity is obtained from that chemistry. It should also be pointed out that, although the activity in potentiometric detection has been far less than in amperometric detection, the work of Hassad and Alexander with copper electrodes is relevant to the current discussion (43–46). There, the complex chemistry of copper was exploited for detection work.

Although the chemical selectivity just illustrated is useful, there are many circumstances when the selectivity is not required, and one wants the electrode to act as a source or sink of electrons and no more. In these cases the most widely used materials are carbonaceous. Vitreous or glassy carbon and pyrolytic graphite (75) have been used, the former more often. These materials have the advantage of being useful over a wide potential range, one can resurface them, and they are compatible with all chromatographic solvents. Composites of carbon powders with a wide variety of binders have been employed. Carbon paste, a formation of graphite powder and oil, for example, Nujol, (76,77) is useful for chromatography with aqueous mobile phases whereas Kel-graph, a composite of Kel-F and graphite powder, may be used in organic solvents (78,79). These composite electrodes yield better S/N ratios than bulk carbon electrodes under similar conditions.

The limiting process in amperometric detection is the fluctuating background current. When one establishes a potential at an electrode in a flow stream there is a finite current that flows in the absence of analyte. The electrolysis of solvent, supporting electrolyte, and highly retained (chromatographically) compounds occurs to yield this current. The capacitance of the interface is also important in this regard. Recalling that the capacitance is related to the charge on the interface and the potential across the interface, one can derive, for constant potential, Eq. 34.

$$ i = \frac{dQ}{dt} = V\left(\frac{dC}{dt}\right) \tag{34} $$

It is seen that a change in capacitance with time leads to a change in charge with time or a current. Thus changes in capacitance due to adsorption of material on the surface, changes in ionic strength, and changes in the permittivity of the solvent all lead to a current. Of course, temperature affects all these processes. These processes also all depend on the area of the interface; thus smaller electrodes give less noise than larger electrodes. Of course, they give less signal as well. A major contributor to background current during reductions is the reduction of ambient oxygen to hydrogen peroxide. A recent report shows the utility of using a packed column of zinc to reduce the oxygen before the detector (80). Glassy carbon is recommended for work at potentials less negative than -0.8 V (vs. Ag/AgCl) and gold-amalgamated Hg for more negative potentials (81).

In discussion of detection limits and general operating figures of merit the question of the cell design is often raised. Figure 9 shows several designs. Of the three low-efficiency cells (Fig. 9a) one can say that their theoretical performance is approximately equal (55). The channel and "wall-jet" cells have more easily resurfaced electrodes. All three designs are able to accommodate multiple electrodes (see below). The channel design has been the most successful commercially. Of the two high-efficiency cells (Fig. 9b) the reticulated or packed variety is more efficient (82) than the open variety. The positioning of auxiliary and reference electrodes is also important to performance, especially in establishing the linear dynamic range. The ion-containing electrolyte has a resistance, and Ohm's law states that a current passing through that resistance will lead to a decrease in voltage along the current path. That portion of the "voltage drop" that occurs between the reference and working electrodes is measured as being part of the reference electrode-working electrode potential difference. Ideally one wants this entire potential change to occur in the vicinity of the interface (in fact, in the double layer) (see Fig. 2). If this does not occur, then the actual potential change across the interface will be lower in magnitude than the applied potential across the interface. A decrease in the magnitude of the potential across the interface will decrease the current from the electrolysis, with a lowering in

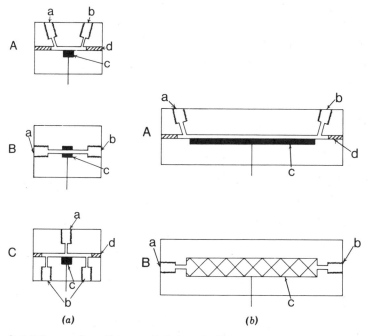

Figure 9. (a) Common low-efficiency cell designs. A, Channel; B, tubular; C, "wall jet." (b) A, High-efficiency channel; B, reticulated. (a) Entrance, (b) exit, (c) working electrode, (d) spacer. Reprinted with permission from S. G. Weber and W. C. Purdy, *I&EC Prod. Res. Dev.*, **20**, 593 (1981). Copyright 1981 American Chemical Society.

sensitivity. The sensitivity can be regained by increasing the potential. Because this lowering in sensitivity occurs when the signal current is largest, the calibration curve becomes nonlinear, becoming less sensitive at high concentrations. Ideally one would have a small reference electrode positioned near the interface [but not closer than about twice the diameter of the reference electrode (83)] with the auxiliary electrode at least as large as the working electrode and parallel to it. This is difficult to do. A convenient and workable solution yielding a linear dynamic range of 10^6 (84) is to put the auxiliary electrode as described with the reference electrode downstream from the cell proper.

The detection limit depends on many factors but mostly on the background current magnitude. The background current depends exponentially on potential, and for reductions, on the efficacy of oxygen removal. For oxidations at less than about 0.8 V (vs. Ag/AgCl 3 M NaCl), a 10 pg detection limit for 20 μL injected is routine. It is perhaps up to 10 times poorer for reductions and for oxidations at more positive potentials. The linear dynamic range also depends on design but, as mentioned above, can be quite large.

This chapter is not primarily intended as an application review. However, a brief look at some recent representative applications will give the reader an idea of the types of compounds well suited to amperometric detection. Many applications are mentioned elsewhere in the chapter. Catecholamines are the class of compound most often determined using amperometric detection. Recent advances have lowered detection limits and simplified analytical procedures (85–88). Caliguri and Mefford have used microbore LC with a carbon-paste-based detector to achieve detection limits of 100 fg of catechols. Most of the improvement is possible because the chromatographic system does not dilute the sample as much as an ordinary chromatographic system (85). Multiple electrodes have been used to assist in identification of neurotransmitters in cerebrospinal fluid (88). A comparison of UV and electrochemical detection demonstrates the lower detection limits of the latter for catecholamines (89).

Other aromatic electron-rich compounds are suitable. Various phenols have been determined in foods (90), human body fluids (hydroxyvanillic acid (91,92), other biogenic amines (92)), and natural water (93). Anilines and their metabolites have been studied with both oxidative and reductive detection (94). Quinones of various sorts have been determined in plants (95), in body fluids as cytostatic drugs (96), and in metabolism studies on pterins (97). The special care needed for microbore-based analysis is discussed by Goto et al. (92) and di Bussolo et al. (98).

Many inorganic species have natural redox activity. Some have been discussed above. An extremely important class of compounds that has been investigated in some detail are the platinum-based antitumor drugs (99).

Indirect determination of enzyme activity can be accomplished if the substrate or product (the latter is preferred) is electroactive. Thus phenylalanine hydroxylase activity could be determined with amperometric detection. Not only the product, but both oxidized and reduced forms of a biopterin cofactor were detectable, allowing the stoichiometry to be unambiguously established (100). Enzymes have been used in immunoassay as labels. Heineman's group has vigorously pursued the avenue of electrochemical detection for the determination of enzyme activity in conjunction with immunoassay. Both phenol and NADH have been determined with excellent results (101–103).

3.6.2. Design Considerations

3.6.2.1 Cell Parameters Influencing Behavior. There are many varieties of amperometric detector. Descriptions of them usually focus on the geometry of the cell, but this can be misleading. It is unfortunate that there are so many important features of the detectors that a convenient nomenclature has not arisen. Some of these features are listed in Table 2. Certainly the number of working electrodes should be specified, as should the electrode material. All workers and commercial outfits do this. The electrode shape(s) (tubular, planar, cylindrical, sperhical, conical, etc.) should be specified. It is important to give a feeling for how the detector works as well. This can be done by specifying the hydrodynamic conditions in the cell. If the fluid path goes by or to the electrode through an open space, then the cell should be so described. Moreover, a distinction should

Table 2. Detector Features Important in Defining Performance

Number of working electrodes
Electrode materials
Electrode shape
Hydrodynamic path
    Open
    Free
    Constrained
    Angle between v and surface
    Reticulated
Flow type
    Laminar
    Turbulent
    Auxiliary (laminar or turbulent)
Electrolytic efficiency (under given conditions)

be made between those electrodes for which the hydrodynamic boundary layer is allowed to grow unrestrained (free) and those for which the walls of the cell impede its growth (constrained). The angle between the velocity vector **v** and the electrode surface should be specified. If the space through which the fluid goes is reticulated, then it should be so stated. The type of flow is important. The natural flow can be laminar or turbulent, or it can be driven by a source other than the chromatographic pumping system, for example, in the case of a rotating disk detector (104) or the vibrating wire electrode (105,106). Finally, the electrolysis efficiency should be stated.

Two examples of potential confusion resulting from incomplete specifications are given. Kissinger correctly points out that "amperometry" is the technique performed by those who measure current from a detector, and coulometry is the technique performed by those measuring charge. But the term "coulometric detector" has come to mean a cell with an efficiency near 1, and the term "amperometric detector" has come to mean a cell with low efficiency. It would be far preferable to have the efficiency and an idea of its flow rate dependence stated.

Another case in which a simple name for a complex system leads to misunderstanding is that of the so-called wall-jet detector. Figure 10 shows the cross section of this type of detector with five different electrodes. This cell with the same geometry, that is, nozzle diameter, cell thickness, is capable of operating in fully *five* different hydrodynamic domains. The smallest electrode is a microelectrode (107), its current independent of flow rate. A larger electrode, but still smaller than the nozzle diameter, is one with uniform current density over its surface. It has a current proportional to the $\frac{1}{2}$ power of flow rate (108). The next electrode is significantly larger than the jet diameter. It operates with a freely growing boundary layer above it. This electrode has a current that depends on flow rate to the $\frac{3}{4}$ power (109). When a significant fraction of the electrode area is under solution with fully developed laminar flow, then the current is proportional to the $\frac{1}{3}$ power of the flow rate. This cell has been shown to be identical in theory, and similar to in practice, the more common low-efficiency cell with an open rectangular conduit with laminar flow parallel to the electrode surface (20). Finally, as the radius becomes larger, the electrolysis empties the solution of analyte, and the cell has unit efficiency, current being directly proportional to mobile phase flow rate. In fact, the third cell mentioned is the one that was coined wall jet by Yamada and Matsuda (109), but the fourth style until recently (110,111) was the one used in practice. Other examples of hydrodynamic misunderstandings have been pointed out (112,113).

3.6.2.2 Effective Dead Volume.   The influence of the detector on band spreading is roughly determined by its physical volume. However, there are cases in which

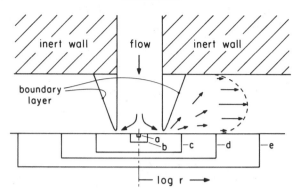

Figure 10. Five different operating environments of a putative wall-jet detector. (a) Microelectrode, (b) uniform accessibility, (c) developing flow, (d) fully developed laminar flow, low efficiency, (e) fully developed laminar flow, high efficiency. Note that the radius scale is logarithmic.

the dead volume can be much smaller or larger than the physical volume. In a high-efficiency reticulated detector based on reticulated vitreous carbon the dead volume of the cell is related to the time for 100% conversion of the analyte (82). The product of this time and the volume flow rate can be less than the physical volume of the detector. In the cell of the wall-jet design it has been shown above that several hydrodynamic regimes obtain. The cell represented by electrode c in Fig. 9 (called the large-volume wall jet by Guinasingham) can have a re-markably large volume, but it has a dead volume that is described by hydro-dynamics (110,114). Figure 11 shows detector responses from systems that were identical except for the distance between the wall and the jet, and thus the physical volume. Note that the band spreading in the cell with a physical volume greater than 1 mL (wall–jet distance 0.318 cm) is no worse than that for the cell with the lower physical volume. Hydrodynamics in enclosed places can lead to unexpected behavior. The dead volume can be a function of flow rate (115); in the case cited it was abnormally large at high flow rates. This may have to do with separated flow, when an initially continuous stream breaks into one or more flowing at different velocities. The hydrodynamics in packed beds, if not perfectly well understood, certainly leads to low band spreading. From this perspective, reticulated detectors have an advantage.

It is also true that the effective dead volume can be greater than the physical volume of the cell. Any sluggishness in the response of the detector to a step input contributes to the effective dead volume. Thus many workers have pointed out that the detector electronics is a source of band-broadening that contributes to "dead volume." The convolution of all linear effects that are detector-related

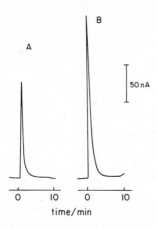

Figure 11. Peaks from an injection of 100 μL of 5×10^{-5} M Ru(CN)$_6^{4-}$ at a potential sufficient to oxidize the compound. Flow rate of 0.1 M NaClO$_4$ carrier stream was 0.5 mL/min. (A) Spacer thickness 0.005 cm; (B) spacer thickness 0.318 cm.

gives rise to the effective dead volume. It has already been pointed out that the mass transfer to the surface results in a steady-state behavior. But certainly, if one injects into the flow stream a perfectly sharp step of concentration of some substance, one does not obtain a perfectly sharp step-function current response. For one thing, the entire electrode surface must be exposed to the analyte. Because the flow stream moves at an average velocity $\bar{v}$ and the electrode has a length L, one can characterize the time taken for this as $L/\bar{v}$. This represents the actual volume contribution to the effective dead volume (really the effective dead time). At the same time, the system is relaxing to the steady state. Digital simulations of the time dependent response of a rectangular channel open cell (116) have shown that for small values of the parameter r the relaxation to the steady state is important.

$$r = \frac{DA}{\overline{U}b} \tag{35}$$

The parameter r is given by Eq. 35, where D is the solute's diffusion coefficient, A is the electrode's area, $\overline{U}$ is the average volume flow rate, and b is the cell thickness. Figure 12 shows the time to 95% of steady-state response (in units of $L/\bar{v}$) as a function of r. The principles of operation of other geometries are sufficiently similar that one should be able to extrapolate to these geometries for which r is perhaps not a good parameter. A second abscissa, in efficiency of the cell, is shown in Fig. 12 for use with other geometries. The dynamic behavior of the cell is a contributing factor in the effective dead volume for cells of low efficiency, say, less than 10%. This encompasses a significant fraction of the detectors in use.

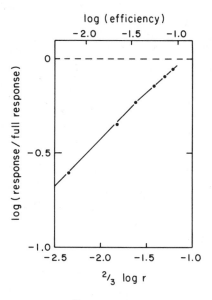

Figure 12. Fractional response at $t = L/\bar{v}$ for cells of various efficiencies. Valid for open-channel cells.

3.6.2.3 Mass Transfer. Mathematical treatments of the mass-transfer problem have been undertaken. A brief overview is given, followed by a general procedure for the estimation of mass-transfer properties without excessive mathematics.

Mass transfer in cells of high efficiency has recently been reviewed by Sioda and Keating (117). A general formula for a cell's efficiency is given by Eq. 36.

$$\text{efficiency} = 1 - \exp\left(\frac{-akL}{\bar{v}}\right) \tag{36}$$

In this equation a is the specific surface area (electrode area/cell volume), k is a mass transfer coefficient, L is the electrode's length, and $\bar{v}$ is the average velocity in the cell. The mass-transfer coefficient may be given by the empirical Eq. 37 with empirically determined parameters j and α.

$$k = j\bar{v}^{\alpha} \tag{37}$$

The latter often falls in the region $0.3 < \alpha < 0.5$ for laminar flow. The relationship has been employed to explain data obtained with a cell based on reticulated vitreous carbon (82). Curran and Tougas obtained values of j = 0.009 and α ≅ 0.42. At a flow rate of 1 mL/min, k ≅ 5 × 10^{-3} cm/s. This is the same order of magnitude as that obtained for low efficiency cells at the same flow rate (81).

For more well-defined geometries less empirical expressions are obtained. For rectangular open channels with one (118,119) or two electrodes (120), and for cylindrical open-channel electrodes (120), the equations for efficiency are (respectively)

$$\text{efficiency} = 1. - 0.90 \exp\left(\frac{-2.44DL}{\bar{v}b^2}\right) \tag{38}$$

$$= 1. - 0.91 \exp\left(\frac{-7.56DL}{\bar{v}b^2}\right) \tag{39}$$

$$= 1. - 0.82 \exp\left(\frac{-19.84DL}{\bar{v}b^2}\right) \tag{40}$$

where D is the solute diffusion coefficient, L is the channel length, $\bar{v}$ is the mean velocity of the fluid, and b is the rectangular channel height or the cylinder diameter. For the rectangular channels it is assumed that the electrode spans the entire width of the channel. The specific surface areas (surface to volume ratio) of each cell are given by (respectively)

$$a = \frac{1}{b} \tag{41}$$

$$= \frac{2}{b} \tag{42}$$

$$= \frac{4}{b} \tag{43}$$

which yields mass-transfer coefficients of (respectively)

$$k = 2.44\frac{D}{b} \tag{44}$$

$$= 3.78\frac{D}{b} \tag{45}$$

$$= 3.65 \frac{D}{b} \qquad (46)$$

Not surprisingly, the two-electrode rectangular channel and the cylindrical electrodes have similar mass transfer coefficients because virtually all of the inner wall of the conduit is electrode. However, the two-electrode rectangular channel rewards diffusion in only one direction with electrolysis, whereas the cylindrical channel rewards diffusion in two directions. This influence appears in the surface-to-volume ratio term.

In cell construction the most significant parameter is b. This is more easily controlled in the rectangular channel case than the cylindrical case because of the ease of manufacture of the sandwich design of the rectangular channel cell as opposed to drilling a hole (or milling a pair of hemicylindrical channels) for a cylindrical channel. In the former case, the flatness of electrodes is critical to good performance. The control of potential is less than perfect in any cell consisting of extended, small cross-section regions of electrolyte. This problem is given coverage by Sioda and Keating (117). Jorgenson and co-workers (121,122) used small carbon fibers (OD 9 and 5 μm) in open capillaries (ID 15 and 8 μm, respectively) to obtain high-efficiency detection. Because the annular space is of micrometer dimensions, the mass transfer coefficient is large. (For example, using Eq. 42 with $D = 5 \times 10^{-6}$ cm^2/s and $b = 1.5 \times 10^{-4}$ cm, k is about 1.8×10^{-1} cm/s.) This is about two orders of magnitude larger than that obtained for other high-efficiency cells. An attendant difficulty is the appearance of irreversible behavior when such large mass transfer coefficients are used. It is not suggested that high mass-transfer coefficients should be avoided; quite the opposite is true. But to take full advantage of high mass transfer each analyte species that strikes the electrode surface should be electrolyzed before it has a chance to diffuse away. Added overpotential, resulting in higher background currents, may be required to accomplish the task. Mathematical solutions to the problem of mass transfer in an annular space yielding high-efficiency cells have not appeared (to my knowledge) in the electrochemical literature, but they have in the heat transfer literature (123,124).

The mathematical description of low-efficiency cells has been widely treated. Many of the solutions are adaptations of the analogous work in heat transfer. Several different mathematical approaches have been used to solve the general convective diffusion equation given as Eq. 47, where ∇ is the differential operator given for cartesian coordinates as Eq. 48. The unit vectors in the x, y, and z directions are i, j, and k, respectively.

$$D\nabla^2 C - \bar{\mathbf{v}} \cdot \nabla C = 0 \qquad (47)$$

$$\nabla \equiv \left(i \frac{\partial}{\partial x} + j \frac{\partial}{\partial y} + k \frac{\partial}{\partial z} \right) \tag{48}$$

This is the steady-state solution that reflects mass balance based on diffusion and convection as modes of solute motion. In this equation, motion due to migration is assumed to be zero. This equation states that, at the steady state, material lost (gained) by diffusion from adjacent regions is gained (lost) by convection. Note that velocity is a vector quantity.

When one determines the concentration, one finds the current by determining the flux (mol/cm^2 · s) of electroactive species to the surface. Because of charge balance at the interface Eq. 49 relates current and flux:

$$i = nFA(J)_{surface} \tag{49}$$

The flux is determined from the concentration by Eq. 50, where it is understood that only motion normal to the electrode surface is of interest.

$$(J)_{surface} \doteq -D(\nabla C)_{surface} + (Cv)_{surface} \tag{50}$$

For most applications one can use a coordinate system in which a single dimension represents the direction normal to the electrode surface, so the vector quantities J, ∇C, and v become scalars.

In flow-through cells the solution is not in an equilibrium; the portion of the solution far from the electrode is relatively unaffected by the electrolysis, whereas the solution at the electrode held at a potential suitable for complete electrolysis is devoid of analyte. The solution can then be broken into two halves, a part affected by the electrolysis and a part unaffected. Of course there is not a clean boundary between the two regions, but one can specify an approximate boundary. The region inside (close to the electrode) is called the diffusion layer, a concept developed by Nernst (see Fig. 4). The diffusion layer has a thickness, the distance between it and the electrode surface. This diffusion layer thickness is not necessarily the same at every point on the electrode. The implication is that $(\nabla C)_{surf}$ and $(Cv)_{surf}$ will also depend on position. Consequently, one may have to solve the mass-transfer problem as a function of position over the electrode and integrate over the entire electrode to obtain the measured current. Alternatively, if the concentration is known, one can calculate the current by mass balance. Current is proportional to the quantity of analyte electrolyzed per unit time. This must be equal to the difference between the quantity flowing into the cell and the quantity flowing out of the cell (Eq. 19).

Many mathematical techniques have been used to solve such problems, including Laplace transformation and separation of variables followed by eigenvalue determination. Approximations are often used to simplify the mathematical approach. The techniques applied to a popular detector geometry, the rectangular open channel, are discussed in a recent paper from Poppe's group (125).

Results for several detectors are given in Table 3 in terms of i/nFC. Note that these results are for steady-state currents, not for peak currents. Chromatographic or flow-injection peaks are easily related to steady-state equations by integration (110,126). The peak area is proportional to the mass injected according to Eqs. 51–55. Thus any of the equations given for high-efficiency or low-efficiency cells can be used to determine peak area.

$$i = fCU^q \tag{51}$$

$$= f(CU)U^{q-1} \tag{52}$$

Table 3. Equations for Current at Low-Efficiency Electrodes

			Ref.
1. Flow Direction Parallel			
Unrestricted			
$\dfrac{i}{nFC}$	$=$	$0.68wD^{2/3}v^{-1/6}(L\bar{v})^{1/2}$	14
Restricted			
Rectangular conduit	$=$	$1.47\dfrac{(DA)^{2/3}}{b^{2/3}\bar{U}^{1/3}}$	
Cylindrical conduit			127
—electrode on outer wall	$=$	$1.61\dfrac{(DA)^{2/3}}{r^{2/3}\bar{U}^{1/3}}$	
Annular conduit*			127
—electrode on inner wall	$=$	$10.15[\phi(a)]^{1/3}\left(\dfrac{D^2L^2\bar{v}}{d_h}\right)^{1/3}r_1$	
			128
2. Flow Direction Perpendicular			
Unrestricted			
$r_{jet} \lessapprox r_{electr}$	$=$	$0.898\,D^{2/3}v^{-5/12}a^{-1/2}A^{3/8}\bar{U}^{3/4}$	109
$r_{jet} \gtrapprox r_{electr}$	$=$	$0.903D^{2/3}v^{-1/6}A^{3/4}\bar{v}^{1/2}$	129
Restricted			
$r_{jet} \lessapprox r_{electr}$	$=$	$1.47\left(\dfrac{DA}{b}\right)^{2/3}\bar{U}^{1/3}$	110

Symbols: w = electrode width; v = kinematic viscosity; r_1 = inner radius of annulus; r_2 = outer radius of annulus; a = jet diameter. All other symbols are defined in the text.

*$\phi(a) = \dfrac{a-1}{a}\left[0.5 + \left(\dfrac{a^2}{1-a^2}\right)\ln\left(a\right)\right] \bigg/ \left[\left(\dfrac{1+a^2}{1-a^2}\right)\ln(a) + 1\right]$; $a = \dfrac{r_1}{r_2}$; $d_h = \dfrac{2(r_2^2 - r_1^2)}{(r_2 - r_1)}$.

$$= f\frac{d(\text{moles})}{dt}U^{q-1} \tag{53}$$

$$\int i\,dt = fU^{q-1}\int \frac{d(\text{moles})}{dt}dt \tag{54}$$

$$Q = fU^{q-1}\,(\text{moles}) \tag{55}$$

Often a semiquantitative or qualitative picture of the cell's operation is sufficient for understanding or determination of design strategy. This can be done by considering the various times of importance in the system. For example, the two competing processes in the establishment of a cell's efficiency are diffusion to the electrode surface and flow out of the cell. The time required for diffusion is given by the Einstein equation (Eq. 56).

$$l^2 = 2nDt \tag{56}$$

In this equation l^2 is the mean square distance traveled in time t in an n-dimensional system. The average distance a molecule must travel depends on the system under scrutiny. However, a generally useful approximation for the average distance that must be traveled by diffusion is $V/2A$, where V is the interstitial, or fluid-filled, volume of the cell and A is the surface area of the electrode. On the other hand, the average time required for a molecule to pass through the system is the interstitial volume divided by the average volume flow rate $V/\overline{U}$. Finally, defining the diffusional time as t_D and the residence time as t_R, one has the cell efficiency as shown in Eqs. 57 and 58.

$$\phi = \frac{t_R}{t_R + t_D} \tag{57}$$

$$= \frac{1}{1 + (UV/8A^2nD)} \tag{58}$$

The route to high efficiency is thus shown to be increasing the electrode area *at constant cell volume* and increasing the dimensionality of the electrode. Of course low flow rate can be used as well, but flow rate is usually dictated by the chromatography. The highest efficiency is then expected from electrodes made of reticulated material with a small cell or particle size. This hardly comes as a surprise to a chromatographer.

3.6.2.4 Selectivity. Chromatography does not always provide the selectivity required for an accurate determination. The amperometric detector provides selectivity through potential control. Approximate potentials for oxidation of various functional groups are shown in Fig. 13 and, for reduction, in Fig. 14. The electrode material can aid selectivity. This has been discussed above. Three other strategies for increasing selectivity are the use of multiple electrodes, voltammetry, and the use of modified electrodes.

Because the electrolysis at an electrode creates a chemical change in the solution, in some cases it ought to be possible to detect the changes brought about by one electrode with a second electrode. This has been done in what has been called the dual-electrode series configuration (see Fig. 15) (130). This configuration is most useful when the analyte has reversible electrochemistry and a potential interferent does not. Then at the first (generator) electrode one can, for example, reduce both the analyte and the interferent, and at the second electrode (collector) attempt to oxidize the products of that reaction. If the interfering species does not have reversible electrochemistry, then it will not be oxidizable; only the analyte will be detected. This can be particularly useful for the avoidance of the common interferent in reductive detection, oxygen (131).

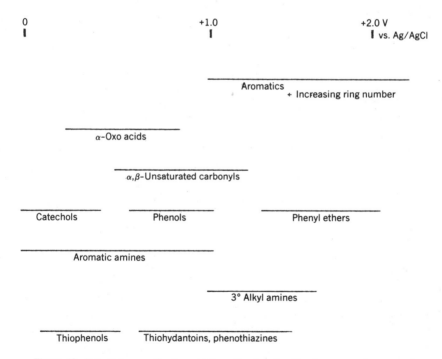

Figure 13. Potential ranges for the oxidation of various functional groups and compounds.

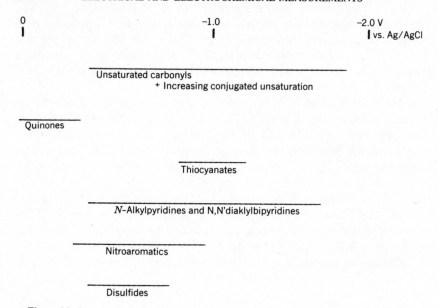

Figure 14. Potential ranges for the reduction of various functional groups and compounds.

One sacrifices some sensitivity, because only a fraction [near 40% for closely spaced, equal-sized electrodes (132)] of the material generated is collected. However, the detection limit may be improved because of the decrease in the background signal.

Single-electrode amperometry cannot, in a single determination, yield any qualitative information. Single- and multiple-electrode voltammetry can provide such information allowing the possibility of more specific detection. The voltammetry can be carried out with single electrodes by measuring current as a function of potential. This requires that one change the potential in a time that is well-suited to chromatography. Recall that the electrode-solution interface is a capacitor and that differentiation of Eq. 1 at constant capacitance yields

$$i_{cap} = \frac{CdV}{dt} \tag{59}$$

where i_{cap} is the capacitive current and dV/dt is the sweep rate. The rapid sweep rates required for voltammetry yield a background current of significant magnitude. For this and other reasons (133), the detection limit in voltammetry is poorer than in amperometry. However, a modest decrease in precision (poorer

Front View   Side View

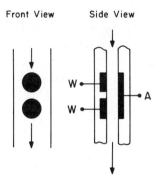

Figure 15. Dual-electrode series configuration. The upstream electrode can act as a generator and the downstream electrode acts as a collector. Figure reprinted with the permission of Bioanalytical Systems, Inc.

detection limit) is accompanied by an increase in selectivity. The voltage–time wave forms used have been the triangle wave (134–135), staircase (136–138), and square wave (139,140). A coulostatic approach has proved very useful (133).

One can also obtain voltammetric information by measuring current from electrodes poised at different potentials. This has been done most often with two electrodes (130) but can be extended to n electrodes (88,141). The selectivity gained can be beneficial in analysis (88). Because the potential is constant on each electrode the background current is the same (or lower) as in the amperometric case, so detection limit need not be sacrificed in this mode of voltammetry.

Electrodes may interact chemically with analytes. This has been demonstrated above for the pulsed amperometric detector. Control over that chemistry is, in some cases, also possible, leading to increased selectivity due to the chemical reactions occurring at the electrode. The control is by some form of electrode treatment, either an electrolytic treatment or a synthetic treatment. Hydrazines can be detected with excellent (<pmol) detection limits following electrochemical treatment of glassy carbon electrodes (142) or by incorporating cobalt phthalocyanine into a carbon paste electrode (143). Note that carbon paste is dissolved by organic solvents; thus it can be used only in mostly (>90%) aqueous solvents. A robust and easy to prepare modification, (probably) consisting of a ruthenium analog of Prussian Blue, has been employed to catalyze the oxidation of arsenic(III) (144). Although its use in a flow-through detector is only discussed, the robustness of the modification is worthy of note. The catalytic activity remained intact for 10 weeks. The adsorption of iodide onto a platinum electrode to yield a mixed H_{ads}, I_{ads} surface has been used to catalyze the oxidation of antimony(III) (145) and the reduction of chromium(VI) (64). This modification has also been used to catalyze the oxidation of nitrite (146). Although not employed in a flow stream, the modification discussed in this paper points out an interesting idea for the protection of the modification. A second robust polymer modification was used to protect the I modification. This protection procedure may have some general utility in flow-through detectors. Carbon electrodes have been modified by leaving embedded in them particles of alumina used in polishing (147). This modification decreased the overpotential required for oxidation of ascorbate, epinephrine, and oxalate. From data provided (147), the modification appears to have about a 5-h half-life. Although the alumina "modification" has shown unarguably good analytical results, the response may have to do more with electrode cleanliness than with the presence of alumina (148).

Molecular weight selectivity can be gained by modifying the electrode surface with cellulose acetate (149,150). This technique has been used to detect H_2O_2 selectively (149). The effective pore size can be modified by base hydrolysis of the film. Species with molecular weights in the 100–300-MW region can be selectively detected in the presence of interfering species of higher molecular weight (150).

3.6.2.5 Noise, S/N Ratio, and Optimization.

All determinations are accomplished with some degree of imprecision. The understanding of the causes of imprecision, or noise, is an important part of the optimization of an analytical system. There have been relatively few studies of interest to analytical chemists that have focused on the noise in analytical electrochemistry. For this reason few statements can be made with any certainty. Nonetheless, an appreciation of what has been done will help in the design and use of these detectors.

Much of the noise problem can be due to slow leakage of accumulated material on the column. Column flushing or using a new column can reduce the lower frequency noise. Environmental electrical noise can be a problem as well. Grounding the flow stream near the cell and maintaining connecting tubing (including tubing to waste) stationary can help avoid this (151). Using a Faraday cage around the cell is necessary for low-current measurements (122,152). It is a cardinal rule to remove noise at its source. Thus the best way to avoid "flow noise" is to use appropriate pumps/pulse dampeners. However, this may not be sufficient. For ordinary HPLC work the pump fluctuations can easily be filtered out electronically using a high-order filter (153).

A scheme of the cell and its noise sources is shown in Fig. 16. In this figure a voltage ΔE drives current through the cell's impedance z_c to be measured as current in the ammeter A. The cell and ammeter together have a voltage noise e_n that is in series with z_c, and there is a noise current in parallel with the ammeter. Note that the noise current goes through the (zero impedance) ammeter irrespective of the magnitude of z_c. There is also an impedance noise z_n. The latter is a departure from traditional electronic treatment of noise (136) in which all impedances are assumed to be noiseless, and any variation in the impedance

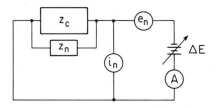

Figure 16. The cell with noise sources. The variable voltage is impressed across the cell's impedance, z_c, resulting in a current being measured through the ammeter A. The cell has a voltage noise e_n, a current noise, i_n, and an impedance noise, z_n.

ends up as a noise voltage or noise current term. The use of an impedance noise term has two beneficial effects. It is the most straightforward way to build into the theory a dependence of the noise on the background current (an experimentally observed dependence), and it is more physically realistic to have a varying impedance than not.

Inspection of Fig. 16 shows that the applied potential ΔE impressed over z_c will yield the measured current i. The noise in the overall measurement can be understood by using the propagation of (independent) errors treatment shown in Eq. 60.

$$\sigma_F^2 = \sum_i \left(\frac{\partial F}{\partial x_i}\right)^2 \sigma_{x_i}^2 \tag{60}$$

This treatment assumes that the variances of all x_i are independent. Applying this treatment to a function F that represents the measured current yields

$$\sigma_{i,T}^2 = \frac{1}{z_c^2} \{\sigma_e^2 + i^2\, \sigma_z^2\} + \sigma_i^2 \tag{61}$$

where F is taken as the sum of the Ohm's law current and the constant current from the current source (i_c):

$$F = \frac{E}{z_c} + i_c \tag{62}$$

The latter has a mean value of nearly zero, but has a nonzero variance (i_n in Fig. 16). The variances of the noises are given as

$$\sigma_{i,T}^2 = \text{total measured noise in current}$$

$$\sigma_e^2 = \text{voltage noise}$$

$$\sigma_i^2 = \text{current noise in constant current source}$$

$$\sigma_z^2 = \text{impedance noise}$$

The units of electrical noise can be confusing. The variances listed above have the units (quantity)^2/Hz, where (quantity) = volts, amperes, or ohms, respectively. Each of the variances may be frequently dependent. One is interested in the determination of the noise in a particular band of frequencies, say, from f_1 to f_2 Hz. Then the rms noise is determined by integrating the variance over the bandwidth and taking the square root, as shown in Eq. 63.

$$\sigma_{i,T} \text{ (amperes)} = \sqrt{\int_{f_1}^{f_2} \sigma_{i,T}^2 \, df} \qquad (63)$$

Let us now consider Eq. 61 term by term. The first term, σ_e^2/z_c^2 says that voltage noise creates a current by being passed through the cell's impedance. Possible sources of voltage noise are the operational amplifiers in the potentiostat (151,154), particularly the current-to-voltage converter and the auxiliary electrode driver (151), Johnson noise and fluctuations in liquid junction potentials (154). The latter is very temperature dependent. Most detection work is done with a silver/silver chloride-based reference electrode that is relatively temperature sensitive. The calomel-based electrodes are less temperature sensitive but less convenient as well. The cell's impedance is frequency dependent. For typical amperometric detectors in the frequency range relevant to "DC" detection, the z_c is the impedance of the double-layer capacitance, $1/2\pi fC$. Therefore, the current noise from voltage fluctuations, $\sigma_{i:E}^2$ is given by Eq. 64.

$$\sigma_{i:E}^2 = (2\pi f)^2 C^2 \sigma_e^2 \qquad (64)$$

Because C is proportional to electrode area, decreasing the area decreases the influence of σ_e^2 on the measured current noise. Experimental measurements have shown that σ_e^2 is proportional to $1/f$. Hence $\sigma_{i:E}^2$ *increases* as the first power of f. This means that effective filtering of this noise can only be done with low-pass filters having more than one pole (155).

The next term is due to impedance fluctuations. Note that the *relative* fluctuations are of concern:

$$\sigma_{i:z}^2 = \left(\frac{\sigma_z^2}{z_c^2}\right) i \qquad (65)$$

Because the impedance is largely capacitive, fluctuations in capacitance yield noise. Fluctuations in capacitance may arise from fluctuations in temperature, ionic strength, solvent permittivity, electrode area, and rate of corrosion of the electrode. The magnitude of this noise also depends on the current being passed at the time. Thus low background currents, meaning operating potentials that are mildly oxidizing or reducing, absence of oxygen for reductive work, and small electrode areas, yield low noise from this source.

Finally we consider the current noise. It derives from the electronics, specifically the input to the current-to-voltage converter, and also perhaps from corrosion at metal–metal junctions (152). In modern systems the noise from this source is small and independent of electrode area, applied potential, and other parameters.

Several routes to noise reduction are possible. It has already been mentioned that detectors based on composite materials, part electrode and part insulator, display higher S/N ratios than a solid electrode of the same total area. Part of this can be attributed to the decrease in the noise based on the decrease in the electrode area. Some of the enhancement is due to "depletion layer recharging" (156). The effect is so-named because as solution that has been depleted of analyte near an electrode surface traverses a region of insulating surface, the layer depleted is "recharged" by diffusion from the bulk.

It is possible to switch the electrode in and out of the potentiostatic circuit in such a way that an increased S/N ratio is obtained (157). The root of the improvement is the decrease in average noise power because of the decreased "duty cycle" of the measurement. For example, using a 2% duty cycle (measure current 92 μs, rest 4.508 ms) resulted in an order of magnitude improvement in S/N ratio.

Many of the noise sources are environmental in nature, for example, temperature fluctuation and variation in eluent composition. Strategies that attempt to neutralize such environmental effects have been beneficial. A high-efficiency cell can be used to clean the eluent before the stream passes to the detector cell (158). In a similar vein, two low-efficiency electrodes in series can be used. The advantage of this approach is that the electrodes can be well-matched; thus a differential signal is obtainable. If each electrode is held at the same potential, then the second electrode yields a smaller signal than the first electrode because of the growth of the diffusion layer at the first electrode. The difference in signals between the two electrodes is, therefore, nonzero and significant (e.g., 70% of the single-electrode signal). However, many environmental noises are eliminated by this procedure; thus improvements in S/N ratio occur (159).

There are two ways to consider optimization. One is to establish rather broad criteria suitable for many situations in order to guide the experimenter in design and implementation. The other is to establish operational criteria for evaluation of existing systems. Let us consider the latter case first.

Roe has discussed a figure of merit (FOM) for detectors that combines band spreading, minimum detectable signal, and sensitivity (115). The equation for the FOM is given as

$$FOM = \frac{V_{INJ}}{V_{PEAK}} \frac{I_{PEAK}}{W_{INJ}} \frac{1}{I_N} \qquad (66)$$

where the first term represents dilution due to band spreading, V_{INJ} is the injected volume, and V_{PEAK} is the peak volume measured in some consistent fashion (e.g., full width at half-maximum height). The next fraction is the sensitivity, peak current per weight injected, and the last term is the inverse of the noise current. Although this FOM is useful and was instrumental in focusing attention on the band spreading in commercial thin-layer cells, it does overemphasize band broadening. If two cells are compared that have equal mass-transfer coefficients and noise levels but whose effective dead volumes differ by a factor of 2, then the cell that dilutes the sample more will have a FOM that is four times poorer than the other. A factor of 2 arises in the V_{PEAK} term, and another factor of 2 arises in the I_{PEAK} term. It would be appropriate to consider a purely sensitivity term, for example, the mass-transfer coefficient or perhaps the quantity Q_{PEAK}/nF (moles injected), the electrolytic efficiency, in place of the I_{PEAK}/W_{INJ} term. Then the FOM gives appropriate weight to band broadening.

Theoretical work has been done in which optima are sought (160,161,162). Such studies can be valuable but only if the correct theories are used. The hydrodynamic complexities of amperometric cells have been discussed. Owing to the changing nature of a cell's hydrodynamics as one changes the cell's size or other conditions, prediction of behavior can be difficult. These difficulties for the open rectangular channel, low-efficiency cell and for two types of cell with the wall-jet geometry have been discussed (110).

Less work has been done in the area of the establishment of broad criteria for electrochemical detectors. Certainly in the broader field of analysis, though, these criteria have been successfully applied.

First, it must be understood that there is no single best cell to be used. The best cell depends on the objectives of the user. An outline of the steps to be taken in optimization follows (163).

1. Establish objective.
2. Determine bandwidth of measurement and low-frequency limit required to accomplish objective.
3. Determine electrode geometry (especially the parameter a, area to perimeter ratio) that can benefit the measurement given the frequency range established.

4. Determine the noise as a function of electrode area in the frequency range established.

5. Using steps 3 and 4, determine best electrode size and shape.

6. Consider chosen electrode from the point of view of band broadening, IR drop, and practicalities of construction, and make necessary modifications.

The first step, then, is to define one's objective. This objective will then define a range of frequencies that need to be measured. For example, detection in an industrial flow stream (without chromatography) may require only one datum every 100 s, and one wants the response time to be $^1/_5$ of that to ensure accuracy. In chromatography one wants at least 20 points to define a peak, and a peak has a certain width in time. The inverse of the peak width gives a low frequency, and 20 times that gives a high frequency, and together they yield a frequency band. If one wants to do voltammetry at each point, then the bandwidth might need to be increased by another factor of 35 or so. These relationships are shown in Fig. 17.

One would like to make use of the edge effect. This effect is pronounced for microelectrodes (107). In the broadest sense, it is the surface area to perimeter ratio, a, that is important in determining the extent of this behavior. This behavior is also time dependent (164). Simple dimensional arguments lead to the conclusion that the edge effect becomes significant only after a time $t \approx a^2/2D$ has passed. Thus the parameter a is determined by earlier considerations of the time scale of the experiment. For example, if one wants to enjoy the benefits of the edge effect in a detector that will be used for peaks that have a 1 min width, then the maximum value of a must correspond to the time required to take one datum. It has previously been suggested that $^1/_{20}$ of the peak width is adequate for one datum. For $t = 3$ s and a diffusion coefficient 10^{-5} cm^2/s a maximum a of 77 μm is called for.

In most cases, many small electrodes would be used in a detector. The regions from which individual electrodes (in an ensemble of electrodes) draw solute overlap after a time $t \approx l^2/2D$, where l is the distance separating the edges of

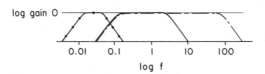

Figure 17. Log gain versus log f (Hz) for three situations: ●–●–●, process analysis; ——, liquid chromatography; – – –, liquid chromatography with voltammetry. See text for details.

the electrodes. One would like that distance to be large so that overlapping due to purely diffusional transport does not occur. Using the same chromatographic example, one wants signal independence over the entire peak, thus $t = 60$ s. This requires that the minimum value of l be about 350 μm. In this and the preceding discussion of a the arguments are approximate but useful and applicable to both geometrically regular and irregular ensembles.

The parameter a does not specify the electrode area. One must first know the characteristics of the noise *for the bandwidth* under consideration and *in the environment* where the analyses will occur. The signal and the noise are related to the electrode area. In the few studies that have been done noise seems to be proportional to electrode area (116,151,154,163) or to the amount of graphite in a graphite/insulator composite such as Kelgraf (165,166). Thus one needs to keep the area small to avoid excessive noise. There is a limit to the reduction in size that will yield an improvement, however. A detailed treatment has been given of this (151). The one noise source that is constant σ_i^2, will play a large roll for small electrodes. The S/N ratio in electrochemical detectors has a maximum at some area. It decreases as one increases the electrode area from the optimum because increasing area yields a less than proportional increase in signal (Table 3, Eqs. 36, 38–40) but a proportional increase in noise. It decreases as one decreases the electrode area from the optimum because in this region the noise is dominated by the area-independent σ_i^2, but the signal decreases. In this regard it should be noted that band electrodes with a sufficiently small a are preferable to circular electrodes because the rate of decrease of signal with decreasing area is lower for bands (167).

Further considerations of area must be made if voltammetry is planned. The electrode's capacitance, proportional to its (microscopic) area, is also of concern in voltammetry. The capacitance in conjunction with the solution resistance yield a time constant, RC, that should not be greater than about 1% of the sweep time (for linear sweep voltammetry) or about 5–10% of a pulse or step width (for pulse, staircase, square wave, etc. voltammetry).

Given the area and recognizing that the distance l is given by the slow frequency (long time) limit of the analysis, one must now ensure that the final cell design is acceptable from the point of view of dead volume, IR drop, and other practical considerations.

3.6.2.6 The Future. The beginnings of several important directions in electrochemical detector design have been recorded above. This final section discusses future directions.

Electrode modification will play a larger role in the future (168–170). Increased selectivity is a major driving force. In fact, doing away with the chromatography in those applications where the chromatography is used only to separate a single analyte from the matrix would be ideal. The resulting flow-injection determinations would be faster than the corresponding chromatographic determinations. All this is well known. The point is that modified electrodes will be crucial in implementing such tactics. A problem with such approaches is that the filtering effect of the column is removed. Thus flow-injection approaches seem more prone to electrode fouling than the corresponding chromatographic approach (170). A more insidious problem may be the higher demands on short response times in the flow injection determination. Roe (171) has seen evidence that adsorption is required before the electrolysis can occur. This has been demonstrated by Johnson in works cited above. In the flow-injection approach it has meant low sensitivity for initial injections while building up to a certain electrode coverage.

There will be continued development of detectors capable of yielding more information per unit time than current detectors. Both multielectrode and single-electrode approaches will continue to be developed. It is likely that the large-volume wall-jet detector will enjoy prominence in both areas. Because the dead volume is defined hydrodynamically and can be small, whereas the physical volume is large, one has room for good positioning of auxiliary and reference electrodes without sacrificing peak shape (172). Multiple-electrode voltammetry is well suited to the large-volume wall jet. The number of electrodes that can be put in front of the jet is limited, but it seems possible to put enough there to describe a set of voltammograms (141,171).

It has been pointed out here and elsewhere that electrostatic potential is not the only means of selectivity control in the detector. Electrode modifications, electrode materials, and pulse sequences can render the detector selective. It is very likely that efforts to obtain as much information as one can about a sample will culminate in cells with numerous electrodes of various sorts at any number of different potentials and undergoing a panoply of potential perturbations.

ACKNOWLEDGMENTS

I would like to thank my many associates, particularly Professors R. M. Wightman, P. T. Kissinger, and D. E. Tallman, for helpful discussions, and Drs. J. A. Cox, W. R. Heineman, D. C. Johnson, J. W. Jorgenson, P. T. Kissinger, I. S. Krull, D. K. Roe, R. E. Shoup, P. E. Sturrock, D. E. Tallman, J. T. Wang, and R. M. Wightman for releasing data and thoughts prior to publication.

I am also very grateful to the National Institutes of Health (GM 28112) for supporting me during the course of the preparation of this chapter.

REFERENCES

1. R. C. Weast, Ed., *CRC Handbook of Chemistry and Physics, 59th ed.*, CRC Press, Boca Raton, FL, 1979, p. F305.

2. B. R. P. Scaife, *Complex Permittivity*, The English Universities Press, London, 1971.

3. D. A. MacInnes, *The Principles of Electrochemistry*, Dover, New York, 1961.

4. F. J. Holler and C. G. Enke, in P. T. Kissinger and W. R. Heineman, Eds., *Laboratory Techniques in Electroanalytical Chemistry*, Dekker, New York, 1984 pp. 235–266.

5. S. A. Wilson, E. S. Yeung, and D. R. Bobbitt, *Anal. Chem.*, **56**, 1457 (1984).

6. R. M. Cassidy and S. Elchuck, *J. Chromatogr.*, **262**, 311 (1983).

7. J. M. Keller, *Anal. Chem.*, **53**, 344 (1981).

8. J. S. Fritz, D. T. Gjerde, and R. M. Becker, *Anal. Chem.*, **52**, 1519 (1980).

9. D. T. Gjerde, G. Schmuckler, and J. S. Fritz, *J. Chromatogr.*, **187**, 35 (1980).

10. R. P. Buck, *CRC Crit. Rev. Anal. Chem.*, **5**, 323 (1976).

11. J. R. Reitz and F. J. Milford, *Foundations of Electromagnetic Theory*, 2nd ed., Addison-Wesley, Reading, MA, 1967.

12. J. O. M. Bockris and A. K. N. Reddy, *Modern Electrochemistry*, Vols. 1 and 2, Plenum, New York, 1973.

13. D. C. Johnson and J. Larochelle, *Talanta*, **20**, 953 (1973).

14. V. G. Levich, *Physicochemical Hydrodynamics*, Prentice-Hall, Englewood Cliffs, NJ, 1962.

15. F. J. Bayley, J. M. Owen, and A. B. Turner, *Heat Transfer*, Nelson & Sons, London, 1972.

16. A. J. Bard and L. R. Faulkner, *Electrochemical Methods*, Wiley, New York, 1980.

17. P. A. St. John, W. J. McCarthy, and J. D. Winefordner, *Anal. Chem.*, **39**, 1495 (1967).

18. M. L. Parsons, *J. Chem. Ed.*, **46**, 290 (1969).

19. S. Haderka, *J. Chromatogr.*, **91**, 167 (1974).

20. V. Slavik, *J. Chromatogr.*, **148**, 117 (1978).

21. J. F. Alder, P. K. P. Drew, and P. R. Fielden, *J. Chromatogr.*, **212**, 167 (1981).

22. J. F. Alder, P. K. P. Drew, and P. R. Fielden, *Anal. Chem.*, **55**, 256 (1983).

23. J. F. Alder, P. R. Fielden, and A. J. Clark, *Anal. Chem.*, **56**, 985 (1984).

24. M. Goebl, *GIT Fachz. Lab.*, **27**, 373 (1983).

25. S. Matsushita, *J. Chromatogr.* **312**, 327 (1984).

26. R. M. Cassidy and S. Elchuk, *J. Chromatogr. Sci.*, **21**, 454 (1983).

27. J. B. Nair and M. F. Delaney, *J. Chromatogr.*, **283**, 361 (1984).

28. H. Small, T. S. Stevens, and W. C. Bauman, *Anal. Chem.*, **47**, 1801 (1975).

29. D. Kourilova, K. Slais, and M. Krejci, *Collect. Czech. Chem. Commun.*, **48**, 1129 (1983).

30. K. Tesarik and P. Kalab, *J. Chromatogr.*, **78**, 357 (1973).

31. V. T. Wee and J. M. Kennedy, *Anal. Chem.*, **54**, 1631 (1982).

32. M. Hutta, D. Kaniansky, P. Havasi, and V. Lipsky, *Acta Fac. Rerum Nat. Univ. Comenianae, Chim.*, **32**, 101 (1984).

33. V. Svoboda and J. Marsal, *J. Chromatogr.*, **148**, 111 (1978).

34. D. Heisz, *GIT Fachz. Lab.*, **29**, 113 (1985).

35. D. L. Manning and M. P. Maskarinec, *J. Liq. Chromatogr.*, **6**, 705 (1983).

36. M. J. Shepherd, M. A. Wallwork, and J. Gilbert, *J. Chromatogr.*, **261**, 213 (1983).

37. H. Malissa, J. Rendl, and W. Buchberger, *Anal. Chim. Acta*, **90** 137 (1977).

38. J. W. Dolan and J. N. Seiber, *Anal. Chem.*, **49**, 326 (1977).

39. P. Ciccioli, R. Tappa, and A. Guiducci, *Anal. Chem.*, **53**, 1309 (1981).

40. D. J. Popovich, J. B. Dixon, and B. J. Ehrlich, *J. Chromatogr. Sci.*, **17**, 643 (1979).

41. Y. Hashimoto, Y. Asai, M. Moriyasu, and A. Uji, *Anal. Lett.*, **14**(B17-18), 1483 (1981).

42. D. E. Johnson and C. G. Enke, *Anal. Chem.*, **42**, 329 (1970).

43. P. R. Haddad, P. W. Alexander, and M. Trojanowicz, *J. Chromatogr.*, **321**, 363 (1985).

44. P. R. Haddad, P. W. Alexander, and M. Trojanowicz, *J. Chromatogr.*, **315**, 261 (1984).

45. P. R. Haddad, P. W. Alexander, and M. Trojanowicz, *J. Chromatogr.*, **294**, 397 (1984).

46. P. W. Alexander, P. R. Haddad, G. K. C. Low, and C. Maitra, *J. Chromatogr.*, **209**, 29 (1981).

47. E. C. V. Butler and R. M. Gershey, *Anal. Chim. Acta*, **164**, 153 (1984).

48. W. N. Wang, Y. J. Chen, and M. T. Wu, *Analyst* (London), **109**, 281 (1984).

49. K. Suzuki, H. Aruga, and T. Shirai, *Anal. Chem.*, **55**, 2011 (1983).

50. N. Kokubu, Y. Hayasida, T. Kobayasi, A. Yamasaki, *Denki Tsushin Daigaku Gakuho*, **31**, 113 (1980).

51. R. S. Deelder, H. A. J. Linssen, J. G. Koen, and A. J. B. Beeren, *J. Chromatogr.*, **203**, 153 (1981).

52. H. Hershcovitz, C. Yarnitzky, and G. Schmuckler, *J. Chromatogr.*, **252**, 113 (1982).

53. P. T. Kissinger, *Anal. Chem.*, **49**, 447A (1977).

54. R. J. Rucki, *Talanta*, **27**, 147 (1980).

55. S. G. Weber and W. C. Purdy, *I&EC Prod. Res. Dev.*, **20**, 593 (1981).

56. A. M. Krstalovic, H. Colin, G. A. Guiochon in J. C. Giddings, E. Grushka, J. Cazes, and P. R. Brown, Eds., *Advances in Chromatography,* Vol. 24, Dekker, New York, 1984.

57. D. C. Johnson, S. G. Weber, A. M. Bond, R. M. Wightman, R. E. Shoup, and I. S. Krull, *Anal. Chim. Acta,* **180,** 187 (1986).

58. W. Kemula, *Roca. Chem.,* **26,** 281 (1952).

59. W. A. MacCrehan and R. A. Durst, *Anal. Chem.,* **14,** 2108 (1978).

60. M. Funk, M. Keller, and B. Levinson, *Anal. Chem.,* **52,** 771 (1980).

61. L. Allison and R. E. Shoup, *Anal. Chem.,* **55,** 8 (1983).

62. R. D. Rocklin and E. L. Johnson, *Anal. Chem.,* **55,** 4 (1983).

63. I. Mizuho, A. Webber, and R. J. Osteryoung, *Anal. Chem.* **56,** 1202 (1984).

64. J. H. Larochelle and D. C. Johnson, *Anal. Chem.,* **50,** 240 (1978).

65. S. Hughes, P. L. Meschi, and D. C. Johnson, *Anal. Chim. Acta,* **132,** 1 (1981).

66. S. Hughes and D. C. Johnson, *Anal. Chim. Acta,* **132,** 11 (1981).

67. S. Hughes and D. C. Johnson, *J. Agric. Food Chem.,* **30,** 712 (1982).

68. S. Hughes and D. C. Johnson, *Anal. Chim. Acta,* **149,** 1 (1983).

69. J. S. Polta and D. C. Johnson, *J. Liq. Chromatogr.,* **6,** 1727 (1983).

70. D. S. Austin, J. A. Polta, T. Z. Polta, A. D.-C. Tang, T. D. Cabelka, and D. C. Johnson, *J. Electroanal. Chem.,* **168,** 227 (1984).

71. J. A. Polta and D. C. Johnson, *Anal. Chem.,* **57,** 1373 (1985).

72. T. N. Morrison, K. G. Schick, and C. O. Huber, *Anal. Chim. Acta,* **120,** 75 (1980).

73. B. S. Hui and C. O. Huber, *Anal. Chim. Acta,* **134,** 211 (1982).

74. J. B. Kafil and C. O. Huber, *Anal. Chim. Acta,* **139,** 347 (1982).

75. R. M. Wightman, E. C. Paik, S. Borman, M. A. Dayton, *Anal. Chem.,* **50,** 1410 (1978).

76. R. N. Adams, *Anal. Chem.,* **30,** 1576 (1958).

77. G. Dryhurst and D. L. McAllister in P. T. Kissinger and W. R. Heineman, Eds., *Laboratory Techniques in Electroanalytical Chemistry,* Dekker, New York, 1984, pp. 289–319.

78. D. E. Tallman and D. E. Weisshaar, *J. Liq. Chromatogr.,* **6,** 2157 (1983).

79. J. L. Anderson, K. K. Whiten, J. D. Brewster, T.-Y. Ou, and W. K. Nonidez, *Anal. Chem.,* **57,** 1366 (1985).

80. W. A. MacCrehan and W. E. May, *Anal. Chem.,* **56,** 625 (1984).

81. K. Bratin and P. T. Kissinger, *Talanta,* **29,** 365 (1982).

82. D. J. Curran and T. P. Tougas, *Anal. Chem.,* **56,** 672 (1984).

83. D. Britz, *J. Electroanal. Chem.,* **88,** 309 (1978).

84. P. T. Kissinger in P. T. Kissinger and W. R. Heineman, Eds., *Laboratory Techniques in Electroanalytical Chemistry,* Dekker, New York, 1984, p. 611.

85. E. J. Caliguri and I. N. Mefford, *Brain Res.,* **296,** 156 (1984).

86. C. Kim, C. Campanelli, and J. M. Khanna, *J. Chromatogr.*, **282**, 151 (1983).

87. M. Goto, G. Zou, and D. Ishii, *J. Chromatogr.*, **273**(2), 271 (1983).

88. W. R. Matson, P. Langlais, L. Volicer, P. H. Gamache, E. Bird, and K. A. Mark, *Clin. Chem.*, (Winston-Salem), **30**, 1477 (1984).

89. P. Grossman, *Chimia*, **37**, 91 (1983).

90. Y. Kitada, Y. Ueda, M. Yamamota, K. Shinomiya, and H. Nakazawa, *J. Liq. Chromatogr.*, **8**, 47 (1985).

91. J. I. Javaid, T. S. Liu, J. W. Maas, and J. M. Davis, *Anal. Biochem.*, **135**, 326 (1983).

92. M. Goto, E. Sakurai, and D. Ishii, *J. Liq. Chromatogr.*, **6**, 1907 (1983).

93. J. V. Geil, J. Schaefer, and H. Kraensler, *Gewaesserschutz, Wasser, Abwasser*, **67**, 229 (1983).

94. D. M. Radzik and P. T. Kissinger, *Anal. Biochem.*, **140**, 74 (1984).

95. S. Okayama, *Plant Cell Physiol.*, **25**, 1445 (1984).

96. W. J. Van Oort, J. Den Hartigh, and J. J. M. Holthius, *Mod. Trends Anal. Chem.*, *Pt. A*, **1984**, 365.

97. C. E. Lunte and P. T. Kissinger, *J. Chromatogr.*, **317**, 407 (1984).

98. J. M. Di Bussolo, M. W. Dong, and J. R. Gant, *J. Liq. Chromatogr.*, **6**, 2353 (1983).

99. W. J. F. Van der Vijgh, H. B. J. Van der Lee, G. J. Postma, and H. M. Pinedo, *Chromatographia*, **17**(6), 333 (1983).

100. C. E. Lunte and P. T. Kissinger, *Anal. Biochem.*, **139**, 468 (1984).

101. H. M. Eggers, H. B. Halsall, and W. R. Heineman, *Clin. Chem.* (Winston-Salem), **28**, 1848 (1982).

102. K. R. Wehmeyer, M. J. Doyle, D. S. Wright, H. M. Eggers, H. B. Halsall, and W. R. Heineman, *J. Liq. Chromatogr.*, **6**, 2141 (1983).

103. M. J. Doyle, H. B. Halsall, and W. R. Heineman, *Anal. Chem.*, **56**, 2355 (1984).

104. B. Oosterhuis, K. Brunt, B. H. C. Westerink, and D. A. Doornbos, *Anal. Chem.*, **52**, 203 (1980).

105. J. A. Lown, R. Koile, and D. C. Johnson, *Anal. Chim. Acta*, **116**, 33 (1980).

106. K. W. Pratt and D. C. Johnson, *Anal. Chim. Acta*, **148**, 87 (1983).

107. R. M. Wightman, *Anal. Chem.*, **53**, 1125A (1981).

108. H. Matsuda, *J. Electroanal. Chem.*, **15**, 109 (1967).

109. J. Yamada and H. Matsuda, *J. Electroanal. Chem.*, **44**, 189 (1973).

110. J. M. Elbicki, D. M. Morgan, and S. G. Weber, *Anal. Chem.*, **56**, 978 (1984).

111. H. Gunasingham, *Anal. Chim. Acta*, **159**, 139 (1984).

112. H. Gunasingham and B. Fleet, *Anal. Chem.*, **55**, 1409 (1983).

113. S. G. Weber, *J. Electroanal. Chem.*, **145**, 1 (1983).

114. B. Fleet and H. Gunasingham, *Anal. Chem. Symp. Ser.*, **17** (*Chem. Sens.*), 556 (1983).

115. D. K. Roe, *Anal. Lett.*, **16**, 613 (1983).
116. S. G. Weber, unpublished data.
117. R. E. Sioda and K. B. Keating, in A. J. Bard, Ed., *Electroanalytical Chemistry*, Vol. 12, Dekker, New York, 1982, p. 1.
118. J. Lankelma and H. Poppe, *J. Chromatogr.*, **125**, 375 (1976).
119. E. Ya. Klimenkov, B. M. Grafov, V. G. Levich, and I. V. Strizherskii, *Elektrokhimiya*, **6**, 1028 (1970).
120. E. Ya. Klimenkov, B. M. Grafov, V. G. Levich, and I. V. Strizherskii, *Elektrokhimiya*, **5**, 202 (1969).
121. L. A. Knecht, E. J. Guthrie, and J. W. Jorgenson, *Anal. Chem.*, **56**, 497 (1984).
122. R. L. St. Claire, III, and J. W. Jorgenson, *J. Chromatogr. Sci.*, **23**, 186 (1985).
123. S. N. Singh, *Appl. Sci. Res.*, **A7**, 237 (1958).
124. R. E. Lundberg, N. C. Reynolds, and W. M. Kays, *NASA TN, D-1972* (1963).
125. E. M. M. Roosendaal and H. Poppe, *Anal. Chim. Acta*, **158**, 323 (1984).
126. P. L. Meschi, D. C. Johnson, and G. R. Leuke, *Anal. Chim. Acta*, **124**, 315 (1981).
127. H. Matsuda, *J. Electroanal. Chem.*, **15**, 325 (1967).
128. T. K. Ross and A. A. Wragg, *Electrochim. Acta*, **10**, 1093 (1969).
129. H. Matsuda, *J. Electroanal. Chem.*, **15**, 109 (1967).
130. D. A. Roston, R. E. Shoup, and P. T. Kissinger, *Anal. Chem.*, **54**, 1417A (1982).
131. K. B. Bratin and P. T. Kissinger, *J. Liq. Chromatogr. Suppl.*, **4**, 321 (1981).
132. H. Matsuda, *J. Electroanal. Chem.*, **16**, 153 (1968).
133. T. A. Last, *Anal. Chem.*, **55**, 1509 (1983).
134. N. Thorgersen, J. Janata, and R. Ruzicka, *Anal. Chem.* **55**, 1986 (1983).
135. Z. Kowalski and W. Kubiak, *Anal. Chim. Acta*, **159**, 129 (1984).
136. W. L. Caudill, A. G. Ewing, S. Jones, and R. M. Wightman, *Anal. Chem.*, **55**, 1877 (1983).
137. J. J. Scanlon, P. A. Flaquer, G. W. Robinson, G. E. O'Brien, and P. E. Sturrock, *Anal. Chim. Acta*, **158**, 169 (1984).
138. P. A. Reardon, G. E. O'Brien, and P. E. Sturrock, *Anal. Chim. Acta*, **162**, 175 (1984).
139. J. Wang, E. Ouziel, Ch. Yarnitsky, and M. Ariel, *Anal. Chim. Acta*, **102**, 99 (1978).
140. R. Samuelsson, J. O'Dea, and J. Osteryoung, *Anal. Chem.*, **52**, 2215 (1980).
141. D. K. Roe and I. P. Ho, Paper 532, Pittsburgh Conference, 1984.
142. K. Ravichandran and R. P. Baldwin, *J. Liq. Chromatogr.*, **7**, 2031 (1984).
143. K. M. Korfhage, K. Ravichandran, and P. Richard, *Anal. Chem.*, **56**(8), 1514 (1984).
144. J. A. Cox and P. J. Kulesza, *Anal. Chem.*, **56**, 1021 (1984).
145. L. R. Taylor and D. C. Johnson, *Anal. Chem.*, **46**, 262 (1974).

146. J. A. Cox, and P. J. Kulesza, *J. Electroanal. Chem.*, **175**, 105 (1984).

147. J. Wang and B. Freiha, *Anal. Chem.*, **56**, 2266 (1984).

148. T. Kuwana, Paper 38, 1985 Electroanalytical Symposium, Chicago, IL, May 1985.

149. G. Sittampalam and G. S. Wilson, *Anal. Chem.*, **55**, 1608 (1983).

150. J. Wang and L. D. Hutchins, *Anal. Chem.*, **57**, 1536 (1985).

151. D. M. Morgan and S. G. Weber, *Anal. Chem.*, **56**, 2560 (1984).

152. H. W. Ott *Noise Reduction Techniques in Electronic Systems*, Wiley, New York, 1976.

153. S. G. Weber, *Anal. Chem.*, **54**, 2126 (1982).

154. H. W. van Rooijen and H. Poppe, *J. Liq. Chromatogr.*, **6**, 2157 (1983).

155. R. M. Wightman, personal communication.

156. D. K. Cope and D. E. Tallman, *J. Electroanal. Chem.*, **21**, 188 (1985).

157. S. G. Weber and W. C. Purdy, *Anal. Chem.*, **54**, 1757 (1982).

158. G. N. Schieffer, *Anal. Chem.*, **52**, 1994 (1980).

159. C. E. Lunte, P. T. Kissinger, and R. E. Shoup, *Anal. Chem.*, **57**, 1541 (1985).

160. H. B. Hanekamp and H. J. van Nieuwkerk, *Anal. Chim. Acta*, **121**, 13 (1980).

161. H. B. Hanekamp and H. G. de Jong, *Anal. Chim. Acta*, **135**, 351 (1982).

162. S. G. Weber and W. C. Purdy, *Anal. Chim. Acta*, **100**, 531 (1978).

163. S. G. Weber, Paper 27, 1985 Electroanalytical Symposium, Chicago, IL, May 1985.

164. H. Reller, E. K. Eisner and E. Gileadi, *J. Electroanal. Chem.*, **138**, 65 (1982).

165. J. L. Anderson, K. K. Whiten, J. D. Brewster, T.-Y. Ou, and W. K. Nonidez, *Anal. Chem.*, **57**, 1366 (1985).

166. D. E. Weisshaar, D. E. Tallman, and J. L. Anderson, *Anal. Chem.*, **53**, 1809 (1981).

167. K. R. Wehmeyer, M. R. Deakin, and R. M. Wightman, *Anal. Chem.*, **57**, 1913 (1985).

168. P. T. Kissinger, personal communication, April 1985.

169. J. T. Wang, personal communication, April 1985.

170. J. A. Cox, personal communication, April 1985.

171. D. K. Roe, personal communication, April 1985.

172. P. E. Sturrock, personal communication, April 1985.

CHAPTER

8

MASS SPECTROMETRY AS AN ON-LINE DETECTOR FOR HPLC

JONATHAN B. CROWTHER, THOMAS R. COVEY, and
JACK D. HENION

Equine Drug Testing and Research Program Cornell University Ithaca, New York

1. INTRODUCTION

In discussing the attributes of a molecular specific detector such as the mass spectrometer, Desiderio describes the inverse relationship between chromatographic resolution and detector specificity. "This inverse relationship theoretically means that if chromatographic resolution is infinitely high no detector is needed because an analyst can unambiguously state that a unique, single structure elutes at a specific retention time whereas, on the other hand, if molecular specificity of the detector were infinite, then chromatographic resolution would not be required" (1).

Unfortunately neither chromatography nor MS individually fully possess the unique ability to establish unknown identities in complex mixtures. However, the combined techniques offer the chromatographer a detector that is capable of unequivocally determining peak identity and the mass spectroscopist a continuous inlet system that routinely "injects" purified components into the mass spectrometer source.

Such potential has been fully realized in coupling GC to the mass spectrometer, but practical MS interfaces are just now being introduced to couple this powerful detection technique to HPLC. The difficulties in coupling HPLC to the mass spectrometer are well documented (2–5), with most emphasis placed on the process of efficiently converting the solute to the gas phase with subsequent removal of the chromatographic eluent either prior to the source inlet or by supplementing the pumping ability of the existing mass spectrometer. Whenever different techniques such as LC and MS are interfaced, the combined analytical capabilities are greatly enhanced but not without some compromise in the performance of each instrumental method. The most recent advances in LC–MS have been related to reducing the number of concessions that must be made for both HPLC and MS when they are coupled in an on-line manner.

The current trend in LC–MS interface design has been in developing specialized interfaces for specific separation problems (reverse-phase or normal-phase separations, routine or trace analysis, volatile or involatile buffers) as well as specialized MS requirements [high- or low-resolution mass spectra, high or low molecular weight, characteristic electron impact ionization (EI) type spectra, or simplied chemical ionization (CI) type spectra]. Thus LC–MS interfaces differ quite drastically depending upon the specific application requirements and instrumentation that is utilized. Certainly there are different precautions for a liquid chromatograph interfaced to a mass spectrometer with an ion source near ground potential than for one interfaced to a high-resolution sector instrument with an 8-kV accelerating potential. Therefore, this brief review only highlights the interfaces and interface designs that have been proved to provide routine operation and reliable performance in several laboratories, as well as those offered commercially.

2. PRINCIPLES OF LC/MS

2.1. Mass Spectrometry Considerations

At present, the mass spectrometer is unique in that it offers a combination of three very important analytical features. These include sensitivity, selectivity, and near universal detection for organic analytes. There is currently no other detector for GC or HPLC that can match these combined benefits of specificity and sensitivity.

From the standpoint of sensitivity MS can routinely provide full-scan mass spectra for low-nanogram quantities of many organic compounds. If instead of scanning the entire mass range, certain selected ions are monitored in what is called selected ion monitoring (SIM), then detection limits may be increased up to 100-fold and sometimes to as much as 1000-fold. This capability provides detection limits with very good selectivity down to low or sub-part-per-billion (ppb) levels. The choice of ionization mode may vary from conventional EI to a host of CI experiments with a diverse choice of CI reagent gases that may provide certain desired characteristic data. This flexibility of experimental conditions for mass spectrometry makes the technique extremely versatile in the hands of experienced personnel.

In addition to the sensitivity performance of mass spectrometry, the technique provides unequaled universality to most organic compounds. Although there will always be certain compounds that give a disappointing or unexpected response, EI mass spectrometry provides a fairly consistent response to most analytes. In addition, for purposes of EI quantitative analyses, the technique demonstrates linearity over three to six orders of magnitude and thus is well suited for such experiments. MS is able to quantify an analyte quite reliably because it can distinguish between the analyte of interest and interference by a coeluting component that cannot be distinguished by nonspecific detectors. A good example of this is when stable isotope-incorporated internal standards are utilized that usually coelute with the unlabeled analyte but can be distinguished because of their differing molecular weight and fragment ions.

Because of these and many other desirable features the mass spectrometer is a natural choice as a detector for HPLC. None of the 20 or so HPLC detectors commonly used today can provide all the features described above. On the other hand, no HPLC detector suffers from the complexity and cost of the mass spectrometer. In spite of this, however, the development of a means of combining HPLC with mass spectrometry has been underway for more than 10 years. The focus of the difficulty for this desirable goal is the fact that a mass spectrometer operates under high vacuum and is not suitable for use with some of the agents that are common eluents for HPLC. The latter include acidic and alkaline modifiers, inorganic agents, and polar, involatile eluents normally used in HPLC separations. The direct introduction of these HPLC eluents into the mass spectrometer at first seemed unlikely. However, the removal of the HPLC eluent and its modifiers prior to introduction into the mass spectrometer turns out to be rather difficult as well. It is now possible to accomplish LC–MS by either of these approaches provided certan precautions are taken. These are described in more detail below.

The successful LC–MS experiment requires an interface between the HPLC and the mass spectrometer. Although it is generally accepted that the best interface is no interface, no one has been successful at this endeavor as yet. Because an interface is currently a necessary component of an LC–MS system, let us look at the features and characteristics required.

The accepted LC–MS interface should be simple, rugged, and routine. It should not significantly compromise either the HPLC or the mass spectrometer and it should not unnecessarily complicate the LC–MS experiment beyond what would be required for operation of either of the instruments by themselves. This is a tall order and has not yet been totally achieved.

The chromatographer would prefer that the LC–MS interface be simply a means of "attaching" the mass spectrometer to the HPLC instrument and that the same inorganic salts be used, including potassium hydrogen phosphate, with the mass spectrometer simply providing a mass spectrum that "identifies" the HPLC peak(s) obtained from the UV or other HPLC detector. The mass spectroscopist, on the other hand, would prefer to keep the instrument under high vacuum with only low levels of clean samples being introduced into it. It appears from our experience that neither person may enjoy the LC–MS combination without some departure from normal operation. The chromatogrpaher must appreciate that an inorganic modifier will simply deposit inside the mass spectrometer vacuum system and quickly shut the system down. Instead, a suitable volatile eluent modifier such as ammonium acetate or acetic acid may be required to aid the chromatographic process. These materials will not deposit inside the mass spectrometer, but simply be pumped away. Thus the chromatographer makes a compromise by altering the modifier to something else which, it is hoped, still accomplishes the chromatographic separation, but does not harm the mass spectrometer. On the other hand, the mass spectroscopist allows the continuous introduction of a low percentage by weight modifier that in time may require cleaning or maintenance of the instrument. However, by a modest compromise a successful LC–MS outcome may be accomplished that can help solve problems not previously solvable.

A key feature of LC–MS today that should be appreciated is that many of the approaches to LC–MS do not provide the classical EI ionization that has been the cornerstone of interpretative MS (6). EI mass spectra are often rich in fragmentation information that facilitates structural characterization. Thus one can often deduce the structure of an unknown compound from its EI mass spectrum alone using established interpretation procedures. The chromatographer may be familiar with this capability and naturally expect the same information from the LC–MS system. To date only the moving belt LC–MS system has routinely provided EI mass spectral data and thus allowed the opportunity for classical mass spectral interpretation. All other proven LC–MS systems utilize some form of direct liquid introduction (DLI) that provides chemical ionization-like mass spectra. These data usually are rich in molecular weight information, but weak in fragmentation data. Thus DLI LC–MS data may be obtained, but the limited structural information does not allow identification of the HPLC peak based upon mass spectral information alone. This may come as a disappointment to many who are new to LC–MS, but one should not despair. Usually considerable history is available for the sample of interest and this information combined with the LC–MS information often provides a substantial increase of knowledge concerning the unknown's identity that may contribute significantly towards the actual identification.

To complement the sometimes insufficient mass spectral information from a DLI LC–MS experiment it is often beneficial to perform the LC–MS experiment in different ionization modes. Usually a mass spectrometer used for LC–MS has the capability to provide either positive ions or negative ions. These data may be easily obtained in sequential LC–MS experiments. Thus the mass spectroscopist may be able to discern more about the identity of an unknown analyte when both positive chemical ionization (PCI) and negative chemical ionization (NCI) mass spectral data are available.

Finally, it has often been suggested that LC–MS really is not worth the effort when all you really need to do is preparatively collect the HPLC peak(s) of interest and submit them for conventional solid probe mass spectral analysis. In principle, this is true, but in practice not practical. In the first place preparative HPLC collection of peaks requires removal of all the excess solvent, such as water, and all traces of any organic or inorganic modifiers that may have been used in the chromatographic process. This may require several washings or manipulations, each of which contributes to possible contamination or loss of the analyte. In addition, if there is more than one HPLC peak of interest the task is further complicated and may require unreasonable time with discouraging results. It seems fair to admit that off-line GC–MS was not acceptable. If GC–MS had not been a better way we might all still be preparatively collecting GC effluents. It should go without saying that if a viable, routine LC–MS interface was available, on-line LC–MS would also be a welcome analytical tool to many laboratories.

2.2. Chromatographic Detector Considerations

An interface used to combine the separation power of LC with the unique detection capabilities of the mass spectrometer must be designed to retain the chromatographic fidelity of a separated band, maintain the sensitive detection capabilities of the mass spectrometer (1 ng), and respond linearly in a concentration range of several orders of magnitude. In many regards the mass spectrometer is competitive with the best of HPLC detectors in addressing these parameters. Nonetheless, a discussion of the "limitations" of the mass spectrometer as an HPLC detector is merited.

When the support size and column diameter are reduced, extracolumn variances must necessarily be reduced to minimize instrumental band broadening. Ideally, extracolumn dispersion should not reduce the effective efficiency of the column by 10%. In other words, for a column of 10,000 theoretical plates, a maximum loss of 1000 plates due to extracolumn peak broadening is assumed to be acceptable. This extracolumn contribution to total variance (0.1σ total) is distributed among the injector, tubing, and connections, and the detector cell geometry and time constant. One goal in the design of any LC–MS interface is to minimize the variance contributions due to the detector rise time and cell geometry. Several comprehensive articles that detail the extracolumn variance contribution by the mass spectrometer have recently been published (3,7,8).

Coupling the mass spectrometer to the HPLC is best done with short lengths of stainless steel tubing of sufficiently small diameter (0.006–0.004 ID). In-line filters are often used to prevent disruption in interface operation resulting from impurities in the mobile phase and particulates passed from the column. These filters, as well as tubing connectors, should be of the low dead volume type. These precautions are especially important if the mass spectrometer is expected to couple with a microbore HPLC (9,10).

In addition to band broadening due to connecting tubing, the variance contribution of the interface itself must be considered. The interface must not expose the solutes to any adsorptive surfaces that may degrade the sample profile. Not only would this affect peak shape, but recoveries of trace quantities would be low. Transport interfaces should not permit the solute to "flow" on the belt once the sample is placed on it because detrimental broadening will result. Likewise, spray and pneumatic based interfaces should provide smooth, tapered paths for the vapor to travel into the ion source. Heating, although necessary in most interfaces for transition of the solute or sample to a gaseous state, must be minimized to prevent decomposition of the analyte peak.

An often unaddressed issue in LC–MS design is the geometry of the MS source. Similar to other LC detectors, such a "cell" should be cleanly swept so no residual solute remains in the source after the band passes. Quite often, an excessively long tail on a concentrated peak is diagnostic of poor ion-source design particularly when the same band is symmetrical in a well-designed UV detector cell.

The time constant associated with modern computer-based detection/data acquisition system for mass spectrometry is suited for most chromatographic peaks and should not exceed 1/20 of the peak width in seconds. However, this contribution could indeed be significant when analyzing early eluting peaks or when operating at high mobile phase linear velocities. Detector time constants have been shown to be on the order of 0.08 (9) when monitoring a single ion which is quite comparable with most LC detectors.

In addition to limitations imposed by the mass spectrometer on chromatographic efficiency, the chromatographer must be aware of the difficulties obtaining suitable mass spectra of a rapidly eluting peak. In practice, one hopes to obtain a minimum of 15 complete spectra for accurate determination of a peak's purity and concentration. For a quadrupole mass spectrometer a complete scan over a mass range of 1000 may require 1 s. Thus to retain reasonable chromatographic integrity, a peak should have a minimum peak width of about 20 s, whereas early eluting peaks in modern high-speed HPLC may have peak widths of 1 or less. One solution to this problem is either to minimize the mass scan range or to use SIM to reduce the scan rate to the millisecond time frame. Such a procedure used in targeted analysis or when characteristic ions of the solute are known has the added advantage of increasing the sensitivity of the experiment (depending upon the number of ions scanned) by a factor of up to 1000 (3). The limitations imposed by slow mass spectrometer scanning rates should be addressed in future MS advances. An attractive solution is to couple HPLC interfaces to FTMS instrumentation where scan cycle time is substantially reduced (11). These scanning rates would more closely ally the full-scan MS data acquisition with the speed of the chromatographic experiment.

Sensitivity of the LC–MS system varies according to the frequency of the

Table 1. HPLC-MS Interface

Feature	Interface				
	Moving Belt	Split-effluent DLI	Micro DLI	Thermospray	APCI
Optium flow rate (mL/min)	1–2 (NP) 0.1–1.0 (RP)	0.2–2.0	0.01–0.05	1–2	1–2
Chromatographic integrity	Good	Good	Good	Good	Good
Detection limit (full scan) (ng)	10–20	100–200	1–10	5–10	1–10
Modes of ionization	EI/CI/FAB	CI	CI	Thermospray/ CI/discharge	APCI
Operation with inorganic buffers	Possibly	No	No	No	Possibly
Mass range (minimum) (daltons)	50	130	130	120	150
Optimized in what HPLC mode[a]	NP	NP/RP	NP/RP	RP	RP

[a]NP = Normal phase; RP = reverse phase.

scan and the total number of data points. Furthermore, because the mass spectrometer is a mass flow sensitive detector, its sensitivity is increased at higher flow rates. Unfortunately, each interface design limits the flow rate the mass spectrometer can accept typically because of vacuum considerations. Even with these restrictions, detection limits often reach low nanogram limits for favorable analytes in the full-scan mode. Sensitivity varies between interfaces and is often compound dependent. Ultimate sensitivities for several designs are shown in Table 1. ·

In an effort to achieve lower detection limits, "noise" in the form of solvent impurities must be removed or reduced. All solvents must be filtered and of "HPLC Grade." The volatile buffer salts should be of low molecular weight and of reagent-grade quality. Ionization of solvent produces adduct ions which limits the mass spectrometer scan range in LC–MS to above 125 daltons.

Along with sensitivity, the chromatographer wants a detector that yields a linear calibration curve over several orders of magnitude. A linear response is difficult to achieve in some HPLC–MS interfaces. Because detector response is a function of several variables including temperature, reagent ion concentration, pressure in the ion source, stability of the overall system, and solvent ion concentration, it is not surprising that accurate quantitative analysis was difficult in early LC–MS systems. Recently, reliable quantitative determinations have become more routine, which further attests to the advances realized by improvements in LC–MS interfaces. The best analytical results are obtained when using an internal standard (preferably a stable isotope or chemical analog of the analyte of interest) and diluting the sample if necessary to operate in a specific linear concentration regime.

It is helpful to keep these parameters in mind when comparing interfaces or proposing new interface designs. As mentioned, the HPLC detector should not substantially alter the chromatographic performance, and as will be shown, has been quite successfully adapted to coupling with modern LC instrumentation.

3. COMPARISON OF HPLC–MS INTERFACES

There is often some confusion among chromatographers concerning the array of HPLC–MS available, and some confusion related to understanding the abilities and potential of each "technique." Unfortunately, unlike the advances made in coupling capillary GC–MS where there is essentially no longer an interface, such simplification of the LC–MS experiment is unlikely in the immediate future. Thus the chromatographer will have to distinguish which interface will suit his or her needs best (types of samples, instrumentation available, sensitivity requirements). This will persist at least until a truly novel LC–MS interface is developed that is suited for all types of LC mobile phase conditions and suited for several forms of ionization (EI, CI, etc.).

Each of the popular interface designs is now discussed along with some examples. Furthermore, a brief review of more novel interfaces that are not yet commercially available is presented. A listing of these techniques is presented in Table 1 to serve as a quick reference when comparing these techniques and predicting which approach would be most relevant to a specific application problem.

3.1. Moving Belt or Transport LC–MS Interface

One of the first and most developed interfaces for LC–MS is the moving belt interface pioneered by Scott et al. (12). This transport interface required a stainless steel wire to carry the solute and mobile phase through a series of vacuum locks where the solvent was removed. The solute was then vaporized by directly heating the wire upon passing into the ion source region. Although this system has low sensitivity (the stainless wire could accept only 10-μL/min mobile phase), it remains the model even for today's advanced moving belt system.

The basic components for an updated version of the moving belt system are illustrated in Fig. 1 (13). The moving wire is replaced by a Kaptan belt which yields low background but suffers from poor wetting by aqueous solvents. Several modifications to improve the wetting characteristics of the belt include coating the surface with Carbowax (14) and the use of a spray deposition device that nebulizes the sample onto the belt (15). An example of the use of spray deposition on a moving belt interface is shown in Fig. 2.

Of the several desirable features of the moving belt interface, the most noteworthy is its ability to yield EI spectra readily. This is highly desirable in structure

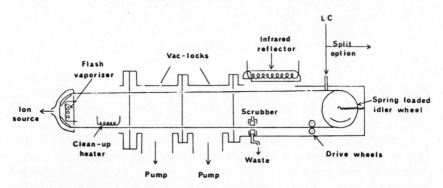

FINNIGAN MOVING BELT LC/MS INTERFACE

Figure 1. Schematic of a moving belt interface, Finnigan MAT Reproduced with permission from Ref. 26.

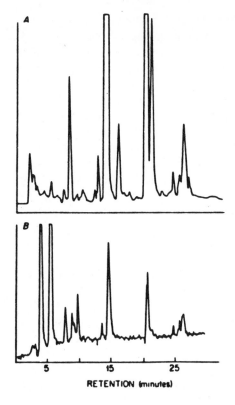

Figure 2. Separation of a coal gasifier condensate sample. The column was a 1-mm × 350-mm Zorbax BP-ODS, 7.5 μm. Solvent A, 5% acetonitrile, 95% water 0.1% TFA; solvent B, 95% acetonitrile, 5% water, 0.1% TFA; gradient, 5–70% B in 40 min; flow rate 100 μL/min. (a) LC–UV detection wavelength 280 nm, 0.01 AUFS; (b) LC–EI–MS (5× magnification) selected ion chromatogram. Reproduced with permission from M. J. Hayes, H. E. Schwartz, P. Vouros, B. L. Karger, A. D. Thruston, Jr., and J. M. McGuire, *Anal. Chem.*, **56,** 1229 (1984). Copyright 1984 American Chemical Society.

RETENTION (minutes)

elucidation work because EI mass spectra can be compared to data stored in a computer-based mass spectral library.

The moving belt interface is also suited to CI spectra upon the addition of appropriate reagent gas. The flexibility of this interface is further enhanced by several recent applications of LC–MS using FAB (fast atom bombardment), a soft ionization technique that yields molecular weight information from very labile compounds (16).

An additional highlight of this interface is that the ion source of the mass spectrometer is usually unchanged because its conventional ionization modes are available when using the moving belt LC–MS interface. Thus without much difficulty LC–MS results could be obtained in the morning and the mass spectrometer used for other projects during the rest of the day. Finally, the moving belt interface is quite compatible with both a quadrupole mass spectrometer as well as operating at high mass/resolution conditions found in magnetic sector instruments.

Commercial versions of the belt interface are moderately expensive and require additional pumping to facilitate removal of the chromatographic solvent. Although advances have been made for operation in the reverse-phase mode, operation of the moving belt interface appears to be easiest when a nonpolar, highly volatile mobile phase of low molecular weight is used. Such conditions permit the rapid and efficient removal of solvent and provide even deposition of the mobile phase on the Kaptan belt.

Because the belt is continuous, care must be taken to vaporize the sample completely to preclude memory effects. For this reason scrubbers and more intense heaters are applied to the belt as it is readied again to accept mobile phase. Although it is hoped that these measures are 100% efficient many involatile compounds present some difficulty. Because of considerable chemical background and the tendency toward surface adsorption/decomposition effects, the moving belt has been shown to suffer in certain trace analyses.

3.2. Direct Liquid Introduction (DLI)

The direct introduction of HPLC effluent into the ion source of a mass spectrometer was another early approach tried for LC–MS. The earliest report of this was that of Talrose et al. (17), wherein a very small fraction of HPLC effluent was introduced into an EI source while trying to achieve the very low pressures required of EI MS. Preliminary results were not impressive from an applied standpoint, but when McLafferty and co-workers (18) tried the same approach using a CI mass spectrometer ion source, the feasibility of this approach was demonstrated. The development of DLI LC–MS continued, albeit slowly, from the mid-1970s to the present (19–27).

The two main problems with DLI LC–MS centered around the routineness of the technique and the actual sensitivity of the method. The former obtained from the fact that the HPLC solvent delivery rate must be extremely stable in order to maintain the proper CI pressure conditions in the mass spectrometer. The sensitivity problems stemmed from the fact that at 1-mL/min HPLC flow rates only about 5% of the total HPLC effluent was actually introduced into the mass spectrometer. This meant that if one injected 10 ng of a drug or metabolite onto the HPLC column only about 500 pg actually went into the mass spectrometer. Because this level is below the usual full-scan mass spectrometric detection limit, no signal could be seen from the DLI LC–MS experiment.

One possible solution to this problem appeared to be the implementation of micro HPLC flow rates, and this has been reported (27). In DLI micro LC–MS 1-mm ID packed microbore columns are typically utilized at eluent flow rates ranging from 10 to 50-µL/min flow rates. At these reduced flow rates the entire micro HPLC effluent may be introduced into the CI mass spectrometer ion source. In this instance no sample splitting results and considerably improved detection limits may be realized. In practice, when the split effluent DLI LC–MS experiment is utilized with a flow rate of 1 mL/min and a 5% split to the mass spectrometer, approximately 200 ng of analyte is required for a good full-scan LC–MS mass spectrum. If 2 mm ID columns are used, operating at 200 μL/min, a 20% split to the mass spectrometer can be achieved, resulting in approximately 50-ng full-scan detection limits. When DLI micro LC–MS is utilized at 40-μL/min eluent flow rate, for example, one can obtain 5–10 ng full-scan detection limits. This capability nicely meets the demands of real-world analytical problems (23,27), but the technique of micro HPLC made that of DLI micro LC–MS less than routine. Early micro HPLC experiments used modified pumps and accessories that were not well suited for reliable, easy operation. Often the success of a DLI micro LC–MS experiment hinged upon the reliability of the technique of micro HPLC alone.

Finally, one important potential weakness of the DLI LC–MS technique is the fact that the HPLC effluent from either a conventional or micro HPLC system must enter the mass spectrometer through a small, laser-drilled pinhole whose diameter is usually 4–5 μm. This is potentially the "weak link" in a DLI LC–MS experiment, because if the pinhole becomes plugged the experiment must be interrupted. However, this is rarely a problem in practice if all solvents and samples are scrupulously filtered. The split effluent technique is considerably more forgiving in this regard because the entire effluent is not forced through the pinhole and particles tend to pass out the split line. Continuous operation for up to 18 h has been reported (28) with quantitative DLI micro LC–MS. All that routine operation requires is clean laboratory practices and good maintenance of all associated equipment.

In order to understand better how DLI LC–MS occurs let us take a look at what is actually happening inside the mass spectrometer ion source. When the HPLC effluent passes through the DLI probe and into the mass spectrometer ion source via the 5-μm pinhole a desolvation process commences. Tiny droplets of HPLC effluent ranging anywhere from low micrometer to perhaps 100 μm in diameter enter a heated desolvation chamber intimately attached to the heated CI ion source of the mass spectrometer. When a chromatographic peak is eluting each tiny droplet contains a small amount of the dissolved analyte in "solution" as it passes rapidly toward the ionization chamber of the mass spectrometer. If the droplet solution is too large in diameter when it reaches the ionization chamber it will pass right through that region, not be ionized, and thus not be detected. However, if it is small enough when it reaches the ionization chamber, the high-energy electrons from the mass spectrometer filament will induce CI of the excess solvent and its dissolved analyte. In this instance the analyte is analyzed and passed through the mass filter, and a DLI LC–MS mass spectrum is obtained. The maintenance of the droplet size into the ion source, the heat of the desolvation chamber and ion source, and the actual pressure within the CI ion source are all therefore very important parameters to control (24). These features are not difficult to establish, however, so long as certain important experimental conditions are met. For example, all that is required is the correct pinhole diameter, proper effluent flow rate through the pinhole, and an ion source temperature of 200–300°C.

A problem associated with the DLI concept is the excessive volume of HPLC solvent gases introduced into the conventional CI mass spectrometer vacuum system. Normally the mass spectrometer is not expected to handle such gas volumes, so the pumping system of the instrument is assisted by means of a liquid nitrogen cryopump. This inexpensive device simply freezes much of the excess solvent on a liquid nitrogen-cooled surface situated directly over the CI ion source and thus assists the standard vacuum system during the course of the LC–MS experiment. The cryopump may be "defrosted" during the night and the vacuum system returned to normal operating order automatically for the next day.

3.2.1. Examples of DLI LC–MS

In an effort to highlight the inherent capabilities of DLI LC–MS a few examples have been selected from our own work. It is hoped that these applications will emphasize the analytical utility of the technique by means of real examples. It is our belief that although DLI LC–MS will by no means solve everyone's problems, it is one of the easiest and least expensive means of getting started in LC–MS.

Previous split-effluent DLI LC–MS results have shown that chiral stationary phase (CSP) LC–MS analysis of a derivatized TLC scrape could detect and identify ibuprofen isolated from an equine urine (29). The same analysis has also been performed on the crude urine extract. The crude urine extract was derivatized with benzylamine as described previously to form the ibuprofen benzylamine. CSP LC–MS analysis of this sample provided the ion current profile shown in Fig. 3b. The lower total ion intensity (TI) trace clearly revealed several component peaks in addition to the peak for (S)-(+)-ibuprofen bensylamide at a retention time of 4.6 min (Fig. 3a).

Significantly, the shoulder component that appears at 4.9-min retention time is not all due to the (R)-(−)-ibuprofen benzylamide enantiomer. Inspection of a mass spectrum (not shown) in the 4.9-min component peak observed in Fig. 3b shows the presence of a different component whose mass spectrum is different from that observed for the (R)-(−)-ibuprofen benzylamide enantiomer. This mass spectral information allows one to distinguish between samples of interest and coeluting components undetected by conventional UV or fluorescence chromatographic detectors. Thus it is clear that the small chromatographic shoulder at 4.9-min retention time in the TI profile of Fig. 3b is not all due to the (R)-(−)-ibuprofen benzylamide enantiomer. Thus as can be seen in the m/z 296

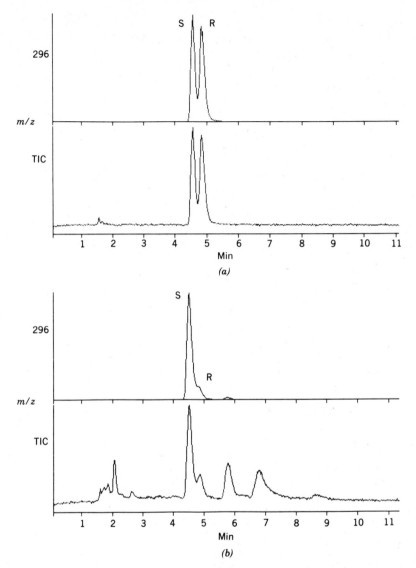

Figure 3. (a) CSP–LC–MS TI and (M + 1) extracted ion current profile of an authentic (R,S)-ibuprofen benzylamide mixture (5 μg each) obtained on a J. T. Baker packed Pirkle covalent bound (R)-*N*-(3,5-dinitrobenzoyl) phenylglycine 4.6 mm × 25 cm column operated at 2.0 mL/min hexane–isopropyl alcohol (92:8). (b) CSP–LC–MS TI and (M + 1) extracted ion current profile of the benzylamide derivative resulting from a crude equine urine extract collected 2 h postadministration of ibuprofen. Reproduced with permission from J. B. Crowther, T. R. Covey, E. A. Dewey, and J. D. Henion, *Anal. Chem.*, **56,** 2921 (1984). Copyright 1984 American Chemical Society.

extracted ion current profile in the upper portion of Fig. 3*b,* a small portion of the 4.9-min component is the requisite enantiomer. Only the unique specificity provided by MS allows one to readily distinguish coeluting organic interferences as observed in this example.

The extracted ion current profiles shown in Fig. 4 clearly show the presence of different components in the crude urine extract. The data system software allows one to plot the different ions observed for each component. Thus Fig. 4 shows the sequential elution of selected components possessing ions of *m/z* 239, 279, 296, 294, 270, and 212, respectively. The mass spectrum for each component may be reviewed to facilitate further characterization or identification of components of interest. These data demonstrate that the 4.9-min retention time component has both a small *m/z* 296 ion and a significant *m/z* 294 ion characteristic of an unknown coeluting interferent. These data enhance our ability to determine the composition of an unknown complex sample containing targeted optical isomers of drugs.

As discussed above, the original reason for implementing micro LC–MS was to achieve trace analysis capability comparable to that of modern GC–MS.

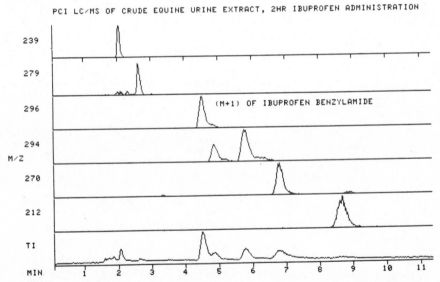

Figure 4. TI and extracted ion current profiles for several components observed from the on-line CSP–LC–MS analysis of the benzylamide derivative mixture resulting from a crude equine urine extract collected 2 h postadministration of ibuprofen. Reproduced with permission from J. B. Crowther, T. R. Covey, E. A. Dewey, and J. D. Henion, *Anal. Chem.,* **56,** 2921 (1984). Copyright 1984 American Chemical Society.

Although DLI micro LC–MS of 50-ng levels of most analytes is easy to accomplish, the ability to achieve low nanogram detection limits requires careful attention to experimental details just as is required for high-sensitivity capillary GC–MS.

The determination of illegal estrogenic residues such as diethylstilbestrol (DES) and its metabolites is a good example of problems associated with trace analysis studies. DES residues in edible meats are not allowed at any level. Thus analytical methods capable of detecting and confirming this compound in tissue matrices may involve levels of DES at low parts per billion or even lower. Studies have been initiated to evaluate the potential for DLI micro LC–MS in the determination of DES and its metabolites in bovine tissues (30).

Figure 5 (lower) shows the micro HPLC UV chromatogram for the separation of 50 ng standards of *cis/trans*-DES from its major metabolite, dienestrol. The micro HPLC used in this work was a Partisil, glass-lined, 1-mm ID × 25-cm microbore column generously provided by Whatman. The eluent of hexane–isopropanol (96:4) was maintained at 40 μL/min by a Brownlee Laboratories Micropump operated in the isocratic mode. The lower UV trace in Fig. 5 shows base-line separation of *trans*-DES from dienestrol. This separation is very difficult to accomplish in the reverse-phase mode (31).

The upper panel of Fig. 5 shows the DLI micro LC–MS total ion intensity chromatograph (TIC) and extracted ion current profile for the sample described above. The only HPLC equipment change was the removal of the micro-HPLC UV detector from the system to preclude its extracolumn void contribution. Careful comparison of the UV and mass spectral data shown in Fig. 5 shows that good chromatographic fidelity is maintained in the DLI micro LC–MS interface using the normal-phase system described here. The DLI micro LC–MS mass spectra for these compounds (not shown here) allowed identification of these important carcinogenic compounds.

It is usually easy to identify standards of compounds, but it is often a more difficult challenge to identify successfully such compounds found at low levels in complex matrices. We have continued our efforts to determine DES in bovine tissues in order to demonstrate the utility of DLI micro LC–MS for such purposes. Figure 6 shows the full-scan DLI micro LC–MS ion current profiles for 50-ppb levels of *cis/trans*-DES and dienestrol spiked into a bovine liver extract. The liver homogenate was worked up by solid-phase extraction described previously (30) and a portion of the 3-g liver extract was injected onto the normal-phase silica column for DLI micro LC–MS analysis. A 50-ppb spike of octadeutero (D8) *cis/trans*-DES was also spiked into the liver for use as an internal standard. The extracted ion current profile for m/z 277 shown in Fig. 6 shows the presence and coelution of the internal standard with the undeuterated DES. These positive ion chemical ionization (PCI) DLI micro LC–MS data show that the 50-ppb

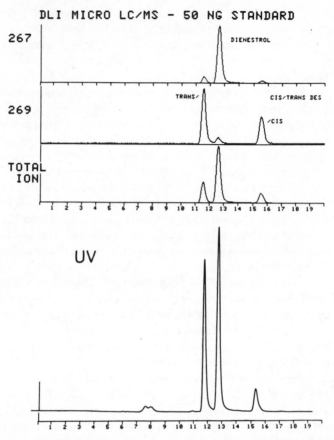

Figure 5. Micro HPLC UV chromatogram, TIC, and extracted ion current profiles for the separation of *cis/trans*-DES (*m/z* 269) and its metabolite dienestrol (*m/z* 267) standards. Reproduced with permission from Ref. 30.

levels of DES and its analogs are readily detected in the liver extract under these experimental conditions.

To further test the utility of the DLI micro LC–MS technique, the liver extract obtained from a bovine calf that had received a 10-mg dose of DES 10 days prior to its sacrifice was analyzed. The DLI micro LC–MS data shown in Fig. 7 show the comparison data from this experiment. The presence of *cis/trans*-DES at about 40 ppb is clearly evident along with the 50-ppb spike of D8 DES internal standard. The absence of any detectable dienestrol in Fig. 7 suggests that very little metabolism of DES had occurred in the rather young calf used in this experiment. However, the data offer convincing evidence that the DLI

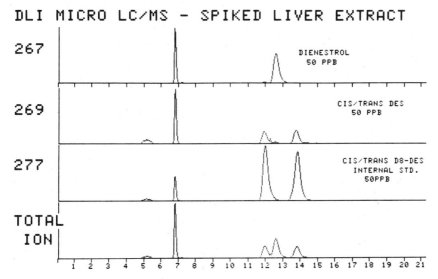

Figure 6. TIC and extracted ion current profiles for the separation of 50-ppb levels of internal standard, *cis/trans*-DES (*m/z* 277), *cis/trans*-DES (*m/z* 269), and metabolite dienestrol (*m/z* 267) spiked into a bovine liver extract. Reproduced with permission from Ref. 30.

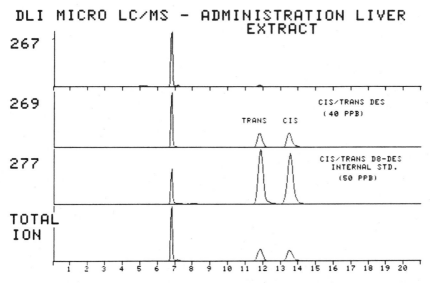

Figure 7. TIC and extracted ion current profiles for the separation of internal standard, *cis/trans*-D8-DES (*m/z* 277), *cis/trans*-DES (*m/z* 269), and metabolite, dienstrol (*m/z* 267), from a 10-day postadministration (10-mg DES) bovine liver extract. Reproduced with permission from Ref. 30.

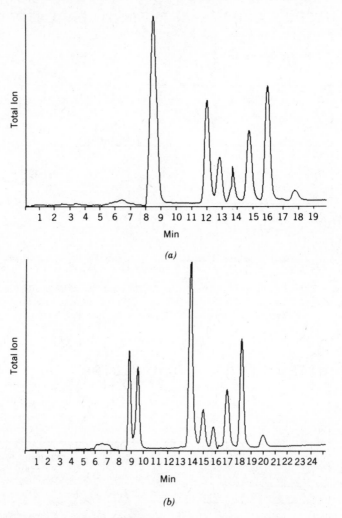

Figure 8. PCI DLI gradient micro LC–MS TIC profiles for a standard mixture of eight pesticides: Flow rate 40 μL/min over 20 min. (*a*) gradient of 55:45 CH₃CN–H₂O to 100% CH₃CN; (*b*) gradient of 45:55 CH₃CN–H₂O to 100% CH₃CN. Reproduced with permission from Ref. 27.

micro LC–MS system described provides the capability for trace analysis of biological samples without undue limitations.

A final example of DLI micro LC–MS is shown in Figs. 8 and 9. Figure 8*a* shows the PCI DLI gradient micro LC–MS TIC for a standard mixture of eight pesticides at 24-ng levels injected on column. The gradient was from 55:45 CH₃CN–H₂O to 100% CH₃CN over 20 min at 40 μL/min on an ODS-3, 1-mm

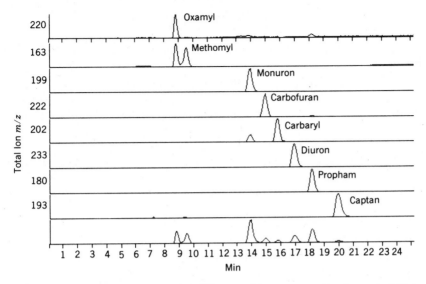

Figure 9. PCI DLI gradient micro LC–MS TIC extracted ion current profiles for standard mixture of eight pesticides. Experimental conditions were the same as in Fig. 8*b*. Reproduced with permission from Ref. 27.

ID × 25-cm microbore column (Whatman). All HPLC, LC–MS interface, and MS hardware were the same as described elsewhere (27).

The TIC in Fig. 8 shows only seven distinct chromatographic peaks because oxamyl and methomyl were unresolved in the first peak centered at 8.6-min retention time. When the micro HPLC gradient was changed to an initial eluent composition of 45:55 CH_3CN-H_2O, the DLI micro LC–MS TIC shown in Fig. 9 was obtained. Although the total run time was increased by approximately 3 min, the first two components, oxamyl and methomyl, were now resolved and all eight pesticides were readily detected.

The ability to perform gradient elution is an important capability for HPLC. Micro HPLC should naturally provide gradient performance, but this places a demand of unusually low flow rates on the pumping system in such instances. The micropump used in this work is capable of providing reliable micro HPLC gradients under these micro LC–MS conditions. The extracted ion current profiles in Fig. 9 show how each component of the pesticide mixture described above may be identified by extracting the ion characteristic of each component. Of course, the CI mass spectrum of each component may be viewed for confirmation purposes. The ability to reliably utilize gradient micro HPLC conditions makes it possible to easily obtain data as shown in Figs. 8 and 9.

3.3. Thermospray

Perhaps the most exciting recent development in LC–MS is the development of the thermospray interface by Vestal and co-workers (33,34). This research, developed out of initial attempts with the "crossed-beam" interface, has evolved into a popular interface for water-soluble component mixtures analyzed by reverse-phase HPLC. Thermospray's popularity is attributed to the fact that the total aqueous eluent is introduced to the chromatograph system with a minimum of band broadening because the interface is directly coupled to the mass spectrometer source. The unique aspect of thermospray is that ionization of the sample may be obtained without the use of a filament. Thermospray has developed into not only a means for coupling the liquid chromatograph to the mass spectrometer, but has evolved into a form of ionization itself.

Ionization in thermospray much like the ion-evaporation model proposed by Thomson and Iribane (35) occurs when a solute in a salt solution is vaporized, producing a supersonic jet of charged vapor. As the solvent evaporates and the droplet radius diminishes the surface field on a droplet increases until the ultimate evaporation of an ion occurs. The mechanism has many similarities to field desorption ionization. The solute ions formed (positive or negative) by this process are typically either molecular species or clusters with solvent ions. The ions are subsequently focused and accelerated through the mass analyzing region of the instrument.

Thermospray, which appears to require a mobile phase rich in volatile buffer (ammonium acetate), has been found quite suited for the analysis of highly functionalized polar solutes typically analyzed by reverse-phase HPLC. Interestingly, thermospray's appeal is founded upon its ability to operate in high concentrations of water, where other interfaces begin to fail.

Thermospray spectra are quite simple, yielding primarily protonated $(M + 1)^+$ adducts in the positive-ion mode and $(M - 1)^-$ ions in the negative ion mode. With this in mind, one expects increased sensitivity with the thermospray technique (over EI) but the technique often does not provide the molecular specificity desired from the mass spectrometer. To circumvent this shortcoming, tandem mass spectrometers are available that are capable of fragmenting molecular ions into structurally significant daughter fragments upon collision with an inert gas.

Several commercial and homemade thermospray interfaces have been constructed. As opposed to other LC–MS interfaces the source region of the mass spectrometer must be removed and a dedicated thermospray source installed (36,37). The thermospray source differs in that it is tighter to maintain low pressure in the analyzer region, and it has additional heaters to provide reliable temperature control, as well as additional pumping to facilitate removal of the vaporized solvent. The inlet probe to the mass spectrometer is heated either directly (38) or indirectly (37) to provide a means of vaporizing the chromatographic eluent. Vapor temperature must be accurately controlled to optimize ionization of solute and ultimately to achieve optimum sensitivity. The optimum probe temperature must be adjusted for large changes in mobile phase composition and velocity. Alternatively, a thermocouple connected in a feedback circuit can be positioned in the vapor steam to monitor vapor temperature. Probe temperature can be adjusted to maintain constant vapor temperature (38).

3.3.1. Thermospray Applications

The thermospray source is rugged and along with APCI LC–MS (next section) perhaps the most adaptable to complete automation. The thermospray source used in our laboratory has provided routine consistent analysis for more than 2 years despite several modifications. The schematic for our homemade thermos-

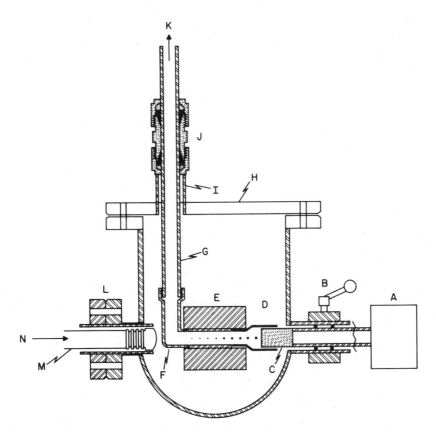

Figure 10. Schematic of a homemade thermospray inlet, ion source, and pump-out region for a Hewlett-Packard 5985B GC–MS. Reproduced with permission from T. R. Covey, J. B. Crowther, E. A. Dewey, and J. D. Henion, *Anal. Chem.*, **57,** 474 (1985). Copyright 1985 American Chemical Society.

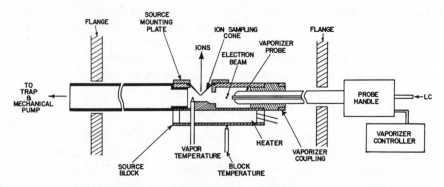

Figure 11. Schematic of the Vestec thermospray LC–MS inlet, ion source, and pump-out region. Reproduced with permission from M. L. Vestal and G. L. Fergusson, *Anal. Chem.*, **57**, 2372 (1985). Copyright 1985 American Chemical Society.

pray system is shown in Fig. 10. The original thermospray system (Fig. 11) is commercially available to retrofit several existing mass spectrometers (39), and most commercial instrument companies are now offering their own version of a thermospray LC–MS interface. Figure 12a shows the thermospray LC–MS analysis of several carbamate pesticides using our homemade thermospray LC–MS interspace. Sensitivities for the compounds varied from no detection to about 5 ng using SIM. As witnessed from the ion trace, chromatographic fidelity is maintained.

Because sensitivities vary among closely related species and over a wide range, "filament on" thermospray has become a practical necessity (39) to mimic the ionization process that occurs in DLI (chemical ionization using solvent reagent ions). Figure 13 shows the NCI thermospray LC–MS analysis of an acid drug mixture using a Vestec interface. These data indicate that the sensitivity for furosemide is greatly enhanced relative to the series of compounds analyzed with the "filament on." These data presumably superimpose chemical ionization conditions upon the "classical" thermospray ionization. It is our experience that greatest sensitivity is obtained in the "filament off" mode, provided the compound of interest is well suited to such direct ionization.

Thus "filament off" thermospray LC–MS is suited primarily for reverse-phase separation of certain analytes, and yields spectra that are rich in molecular weight information because of the abundant molecular ions being formed. When "filament off" thermospray LC–MS provides little or no ion current signal from a particular analyte it is necessary to apply an external source of ionization. This is easily done either by a conventional electron emitting filament or by discharge ionization as reported by Vestal (39). These alternative sources of ionization appear both necessary and complementary for universal thermospray LC–MS applications.

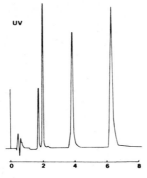

THERMOSPRAY LC/MS OF 500NG PESTICIDES

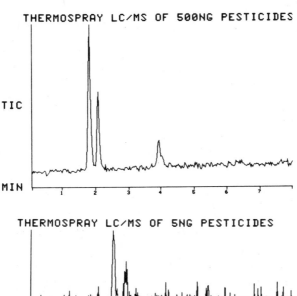

THERMOSPRAY LC/MS OF 5NG PESTICIDES

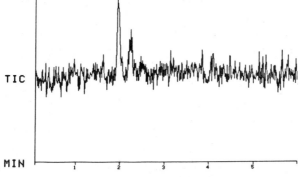

Figure 12. Top, HPLC UV chromatogram for the separation of 500-ng levels of carbaryl, carbofuran, warfarin, and zearalenone. Chromatographic conditions included an Altex 4.6-mm × 10-cm RP-18 Ultrasphere 3-μm column with an eluent of 72:28 [CH₃OH–CH₃CN (50:50)]/0.1 M NH₄OAc maintained at a flow rate of 1.4 mL/min. Middle, thermospray ionization LC–MS TIC for the pesticide mixture described above. Bottom, thermospray ionization LC–MS TIC for 5-ng levels of the four pesticides described. Reproduced with permission from T. R. Covey, J. B. Crowther, E. A. Dewey, and J. D. Henion, *Anal. Chem.*, **57**, 474 (1985). Copyright 1985 American Chemical Society.

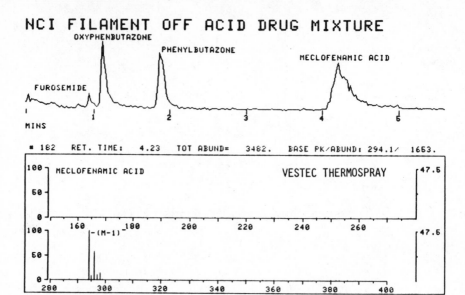

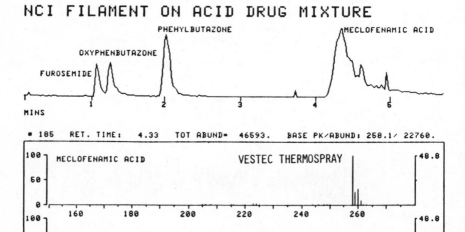

Figure 13. Top, "filament off" thermospray LC–MS TIC analysis of a four-component acid drug mixture using a Vestec system. Bottom, "filament on" thermospray LC–MS TIC analysis of the same four-component acid drug mixture. The mass spectrum shown in both parts was obtained from the fourth component, meclofenamic acid, eluting in the chromatogram. The sample mixtures consisted of 1 μg each of the acid drugs and were separated using 40:60 MeOH/0.2 M NH₄OAc at 1.4 mL/min on a homemade 4.6-mm × 25-cm 5-μm C8 column.

3.4. Heated Nebulizer Interface with APCI

The direct introduction of liquids into the vacuum system of a mass spectrometer is an unnatural procedure in spite of the fact that DLI LC–MS is an established technique. It would appear more straightforward if the HPLC effluent could be introduced into a region at atmospheric pressure whereupon ions could be formed and these ions extracted into the vacuum system for mass analysis. Such an atmospheric pressure ionization system actually exists and works quite well. Because the ionization of analyte in such a system occurs in the presence of excess solvent vapor, the process is called atmospheric pressure chemical ionization (APCI).

The earliest report of APCI LC–MS was by Carroll et al. (40) and represents one of the best detection limits to date. Unfortunately, other reports of APCI LC–MS did not occur until the early 1980s with the advent of the heated nebulizer LC–MS interface from Sciex. This interface (Fig. 14) consists of three concentric tubes that effectively nebulize the HPLC effluent with a combination of heat and high gas flow into the APCI ion source of a quadrupole mass spectrometer. The process generates tiny droplets of liquid that, if they contain analyte due to an eluting HPLC peak, are ionized by a high-voltage corona discharge. The ions thus formed are extracted through a small (100-μm) orifice that separates the APCI ion source from the high-vacuum region typical of mass analyzers. Once these ions enter the mass analyzer they are separated and detected in the same manner as any other quadrupole mass spectrometer.

The advantage of this heated nebulizer APCI LC–MS system is its ruggedness. There are no fragile components to the APCI ion source and no excessive exposure of high-vacuum components to solvent vapors. The pumping speed of the high-vacuum mass analyzer region is very high and effectively compensates for the differential pumping between the source and analyzer regions. The process of APCI is very mild ionization and produces abundant molecular ions of unstable analytes. Unfortunately, just as described above for thermospray ionization, this

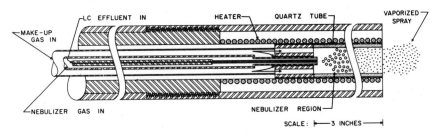

Figure 14. Sciex heated nebulizer LC–MS interface. Reproduced with permission from Ref. 50.

mild ionization effectively precludes mass spectral fragmentation for many compounds.

A solution to the problem of too little fragmentation in such a situation is the implementation of tandem MS, or MS–MS (41). A description of the principles of MS–MS is beyond the scope of this text, but the benefits include increased mass spectral structural information and mixture analysis capability. In the latter, if a HPLC peak is not homogeneous the so-called LC–MS–MS experiment could in principle provide a clean, full-scan mass spectrum of each component of the unresolved HPLC peak in spite of the fact that they are not chromatographically resolved. In conventional GC–MS or LC–MS this would not be possible because the resulting mass spectra would be a combination or mixture of each component. Thus LC–MS–MS offers two valuable solutions of practical importance to LC–MS. The additional fragmentation information complements the very mild ionization resulting from DLI, thermospray, and APCI LC–MS, and the mixture analysis capability facilitates dealing with the unresolved components often found in HPLC.

3.4.1. Examples of APCI LC–MS–MS

The benefits afforded by LC–MS–MS may be demonstrated from some examples of current interest in our laboratory. Because the tandem mass spectrometer is typically very expensive, it must be capable of high sample throughput to be cost effective. Pursuant to these needs we have investigated the potential of very fast sample analysis (one sample per minute) by LC–MS–MS (42). These experiments are based upon establishing HPLC experimental conditions that effectively "resolve" the analyte(s) of interest from the unretained endogenous material of a biological extract. Of course analyte(s) of interest may or may not be resolved from each other, but the tandem mass spectrometer's mixture analysis capability can accomplish that. The short, 3-μm HPLC column in effect is used as a final sample cleanup step that essentially separates the high levels of endogenous materials from the analyte. This subtle but important step is very important because it minimizes problems associated with matrix effects when such crude samples are analyzed by tandem mass spectrometry without adequate cleanup.

Figure 15 shows the UV trace from the fast HPLC elution of a standard mixture of 50 ng each diethylstilbestrol (DES) and zeranol. The single chromatographic peak in Fig. 15a clearly shows that the two analytes are unresolved in the 1-min time period utilized for this experiment.

Figures 15b and c show the corresponding total ion current traces (TIC) for the APCI LC–MS determination of 50-ng standard DES and zeranol. The ion currents for each of these analytes appear at the same chromatographic retention

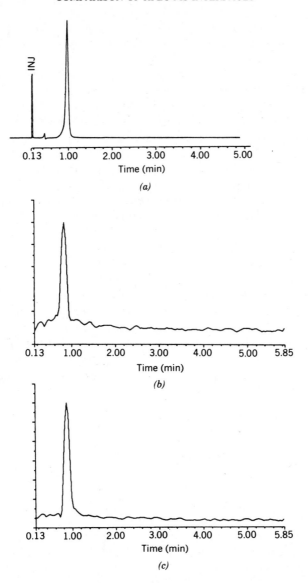

Figure 15. (*a*) HPLC–UV determination of a mixed standard of DES and zeranol (50 ng each). (*b*) APCI LC–MS TIC of 50-ng standard DES. (*c*) APCI LC–MS TIC of 50-ng standard zeranol. Reproduced with permission of Ref. 42.

time and thus are indistinguishable chromatographically. However, the ability to identify a component by LC–MS–MS while that component is in the presence of another analyte is shown in Fig. 16; in Fig. 16a LC–MS–MS TIC of standard DES and zeranol shows them to be chromatographically unresolved as before (cf. Fig. 15). The collisionally activated dissociation (CAD) experiment of the corresponding (M + 1) ions at m/z 269 and m/z 323 of DES and zeranol respectively produces the full-scan CAD mass spectra shown in Figs. 16b and 16c, respectively (43). These mass spectra correspond exactly to those of pure standards and allow facile identification of these compounds even though they elute unresolved from the HPLC column. This mixture analysis capability can be extremely useful in routine, real-world LC–MS–MS.

Figure 17 shows a final example of the rapid sample analysis capabilities of LC–MS–MS with APCI. The UV chromatogram shown in Fig. 17a was obtained from repetitive 1-min injections of exracts from spiked liver homogenates containing 0.0, 0.25, 0.375, 0.5, and 1.0 ppb of zeranol. The short HPLC column effectively resolved the components eluting with low k' from zeranol. However, there is still considerable interference from coeluting material with zeranol that would render LC–MS or related techniques ineffective. Only the mixture analysis capability of tandem mass spectrometry combined with the sensitivity and specificity of SIM LC–MS–MS (41) can effectively and rapidly provide analytical data such as those shown in Fig. 17. It should be noted that Fig. 17b shows the analysis and identification of zeranol in four samples plus one control within 5 min. Thus the technique of LC–MS–MS, albeit expensive and complex, can provide rapid sample throughput.

4. INTERFACES "ON THE HORIZON"

4.1. MAGIC

The Monodispersed Aerosol Generation Interface (MAGIC) developed by Willoughby and Browner (44) is a more recent approach to LC–MS interfacing. The interface relies on the formation of a well-defined aerosol (45) formed by pneumatic nebulization. The solvent evaporates and is removed by two stages of differential pumping. The inlet flow rate into the interface was optimized between 0.1 and 0.5 mL/min. Both EI and CI spectra have been generated using this technique. The schematic representation of this technique is shown in Fig. 18a. The ion current for a normal phase separation of retinal acetate isomers is shown in Fig. 18b. These rather nice chromatographic profiles demonstrate potential feasibility for this MAGIC LC–MS interface at least for relatively volatile analytes such as these.

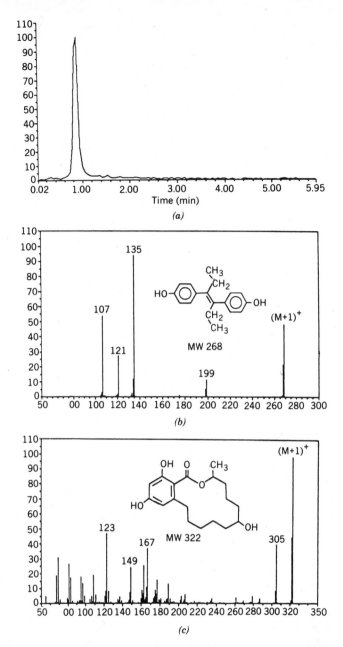

Figure 16. (a) APCI LC–MS–MS TIC of the unresolved standard mixture of DES and zeranol described in Fig. 15a. (b) APCI LC–MS–MS full-scan CAD mass spectrum of the (M + 1) ion of standard DES present in the TIC shown in a. (c) APCI LC–MS–MS full-scan CAD mass spectrum of the (M + 1) ion of standard zeranol present in the TIC shown in a. Reproduced with permission from Ref. 42.

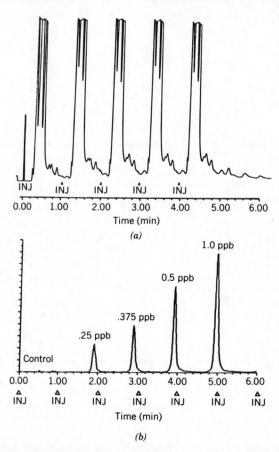

Figure 17. (*a*) HPLC–UV chromatogram from successive 1 min injections from extracts of bovine liver homogenate spiked with 0.0, 0.25, 0.375, 0.5, and 1.0 ppb zeranol, 0.2 AUFS. (*b*) APCI LC–MS–MS selected reaction monitoring ion current chromatograms from the same sequence of zeranol injections shown in *a*. Reproduced with permission from Ref. 42.

4.2. Electrospray

Another spray technique, somewhat a cross between APCI and thermospray, is "electrospray" where the droplets are formed by charging (46) as opposed to the APCI and thermospray techniques of forming charged solutes in liquid droplets. Electrospray spectra are similar to those observed by thermospray and this is a very mild ionization technique. The ions are formed in electrospray by passing a few microliters per minute of eluent through a high-voltage gradient (5 kV) and the charged ions formed are accelerated through a countercurrent of nitrogen

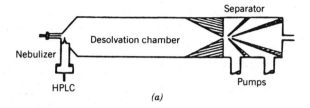

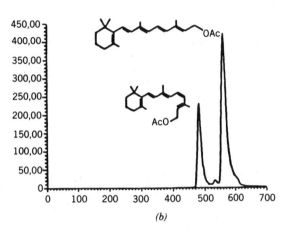

Figure 18. (*a*) Schematic of MAGIC LC–MS interface. (*b*) MAGIC LC–MS ion current profile for the separation of retinol acetate isomers using silica gel and 5% ether–hexane at 1 mL/min. Reproduced from Ref. 46*a*.

gas and into a differentially pumped skimmer/molecular beam apparatus (Fig. 19). Sensitivities are extremely good, although some restrictions are placed upon the conductivity of the mobile phase. Electrospray theory predicts that it will be quite suitable for analysis of high-molecular-weight polymers.

4.3. Supercritical Fluid Chromatography (SFC)

Any in-depth discussion of LC–MS interfaces should include a brief description of SFC–MS. SFC is suited for the analysis of nonpolar as well as moderately polar compounds. The technique has been particularly suited for the determination of polynuclear aromatic compounds and more recently for relatively polar drugs. SFC interfaces must have a restriction to prevent the expansion of the mobile phase and a subsequent loss in solvent power as the supercritical fluid leaves the interface. Typical SFC interfaces resemble those of the DLI type using

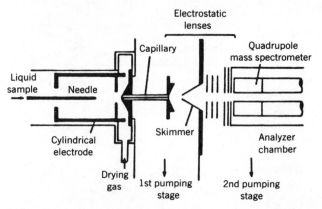

Figure 19. Schematic of the apparatus for mass spectrometry with an electrospray ion source. Reproduced with permission from C. M. Whitehouse, R. N. Dreyer, M. Yamashita, and J. B. Fenn, *Anal. Chem.*, **57**, 675 (1985).

either a diaphragm or a small piece of small diameter (5 μm) fused silica or crimped tubing to provide this restriction (47–49). SFC interfaces have the added advantage that the mobile phase readily converts to a gas upon expansion in the ion source and can be efficiently removed by the mass spectrometer vacuum system. In addition to many impressive applications of SFC with open tubular capillaries we have applied SFC separations using a microbore column. Figure 20 shows the micro SFC–MS ion current profiles for several polynuclear aromatic hydrocarbons in a synthetic mixture. The upper (Fig. 20*a*) three-dimensional total ion current profile and the lower extracted ion current profiles (Fig. 20*b*) readily show the capability of providing micro SFC–MS data from such mixtures.

5. CONCLUSIONS

It should be clear from this chapter that LC–MS is a useful technique that is here to stay. Whether we have any significant advancements in the ease of carrying out LC–MS remains to be seen. We expect that 5 years from this writing we shall see progress that makes today's approach to LC–MS archaic. We need simplified mass spectrometers that are easier to use and are significantly less expensive. We also need more specificity from the LC–MS experiment so that unknown analytes may be "identified" by LC–MS rather than just confirmation of a suspected identity. A significant improvement in the ease of operation of a LC–MS experiment would make the technique more acceptable to the chromatographer. At present chromatographers far outnumber mass spectroscopists, so it should be intuitively clear that the market for LC–MS could be as large as

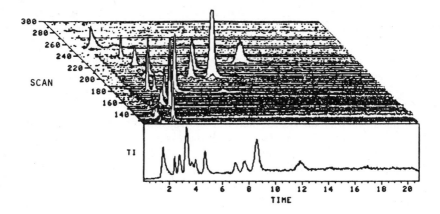

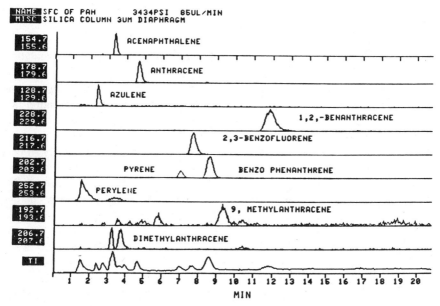

Figure 20. Top, TIC and its three-dimensional view for the packed column SFC–MS analysis of a mixture of polynuclear aromatic hydrocarbons using a homemade 1 mm × 25 cm silica column. The eluent was 3% methanol in carbon dioxide at 200 bar maintained at a flow rate of 85 μL/min. Bottom, TIC and extracted ion current profile for the characteristic ions observed under the same conditions.

for HPLC itself if the mass spectrometer and the LC–MS interface were as simple and inexpensive as today's UV detector. Whether this goal is ever realized remains to be seen, but it should be remembered that the potential for widespread application of LC–MS is very large.

Perhaps one of the most frequently asked questions is "Which LC–MS interface should I purchase?" Unfortunately this question has no universal answer unless it is "the one that appears to handle best your kinds of problems." It is a fact of life, unfortunately, that no existing LC–MS interface will handle all problems. The moving belt interface is unique in that it can provide electron impact and thus structural information, but is mechanically complex and appears to have higher detection limits for thermally unstable compounds. Thermospray LC–MS is very good for involatile compounds that are difficult to ionize by other ionization methods, but thermospray LC–MS often provides little or no fragmentation information. DLI LC–MS is relatively inexpensive and handles compounds of medium volatility and thermal lability, but it only provides CI spectra. APCI LC–MS provides molecular-weight information for even involatile, thermally unstable compounds, but provides essentially no sructural information, similar to the thermospray and DLI techniques. This can be overcome by using tandem MS, but this currently requires very complex mass spectrometric equipment and unreasonable expense. In addition, APCI is not widely available from commercial manufacturers, so it will probably be slow to be implemented and accepted in many laboratories.

For these reasons the individual analyst must assess his or her analytical problems and determine which approach to LC–MS appears best suited to handle the samples and compounds that are of interest. This is best done by sending legitimate samples of interest to the respective manufacturer's application laboratory, followed by a personal visit to witness the actual LC–MS analysis of the samples. In this way the application chemist can have some time to "practice" on the samples, and the individual will have the opportunity to personally witness just what the successful LC–MS experiment entails. When this has been done with two or three different commercial systems, a decision should be possible as to which LC–MS system best handles the problem. Only in this way can one realistically determine which LC–MS interface will best handle the samples of interest.

At the time of this writing the cost of implementing LC–MS into a laboratory varies from about $8000 to over $40,000. This assumes that the laboratory already has a mass spectrometer capable of performing LC–MS, and available HPLC equipment. The least expensive way to "try" LC–MS is by implementing a DLI LC–MS probe onto an existing CI quadrupole mass spectrometer. This is reasonably straightforward because the standard $\frac{1}{2}$-in. direct insertion probe inlet will accommodate the DLI LC–MS interface, and the low voltages of the quadrupole mass spectrometer make operation simpler. The addition of a liquid nitrogen-cooled cryopump will facilitate pumping the excess gases resulting from the DLI LC–MS experiment. Of course the interested reader should be aware of potential difficulties of plugging the pinhole orifice, but given that all samples and solvents are prefiltered, plugging the orifice should not be a problem. Use of the split DLI probe reduces pinhole plugging to a negligible level. Implementation of the DLI LC–MS interface should enable the procurement of results similar to those which have been published (27), and could allow an initial feasibility study within a given laboratory towards determining whether LC–MS offers problem solving ability for that laboratory.

The next easiest way to "get into" LC–MS is to purchase a thermospray LC–MS interface, either from the manufacturer of a particular mass spectrometer (if available as a retrofit) or from an independent supplier (30), that will provide an interface suitable for a particular mass spectrometer. This route is viable in that one can obtain a complete package suited in principle for the instrument along with professional help toward its successful implementation. This commercial route toward LC–MS should be satisfactory for implementing thermospray LC–MS on existing equipment, but will cost $30,000 or more. Alternatively one might build one's own thermospray LC–MS interface (37). It should also be remembered that a mass spectrometer will be limited to LC–MS capability while the thermospray LC–MS ion source is installed. That is, to return to conventional CI or EI capability the thermospray ion source must be removed and replaced by the standard ion source. This exchange may require anywhere from 1 h to overnight.

Of course, for those laboratories that are in such a position, it is quite possible to purchase an entirely new LC–MS system. As of this writing, a new, dedicated thermospray LC–MS system may be purchased at a price range of $125,000–$250,000 (50–55). These systems may be updated with GC–MS options along with a wide range of other state-of-the-art options for mass spectrometry. There is a trend toward simplification and automation of these systems, and we will surely see reductions of cost and increases in capabilities in the future.

The moving belt interface has clearly withstood the test of time, and is commercially available from more than one manufacturer (50,55). The cost of this LC–MS interface is rather high, but it does have the capability of ionization in three different modes (55). These include EI, CI, and FAB (fast atom bombardment). This capability makes the moving belt transport LC–MS interface a very versatile system, and should be considered if these capabilities are of interest.

In addition to the cost and varying complexity of the LC–MS interfaces described herein, the reader should be aware of the fact that the successful outcome to any experiment in part depends on the capability and attitude of the analyst(s). The LC–MS experiment, at least initially, has several different, important features that require that the liquid chromatographer and the mass spectroscopist be very familiar and capable with their respective instruments. At present the mass spectrometer is in no way as simple as a UV detector, and thus requires careful control and monitoring of its "pulse" during the course of an LC–MS experiment. The liquid chromatograph, too, must be operated in a careful, deliberate manner. In all LC–MS experiments the mass spectrometer depends upon the HPLC pumps to deliver a reliable, pulse-free supply of HPLC eluent. Thus the HPLC pumps must be well maintained and operating in "fine-tuned" condition.

Finally, the chromatographer and the mass spectroscopist must work together toward a common goal of LC–MS success. Each specialist should be aware of the important criteria necessary for both HPLC and mass spectrometry. This includes that the chromatographer recognizes it is necessary to exchange inorganic modifiers for volatile alternatives and that the mass spectroscopist recognizes there is no need to bother with LC–MS unless the chromatographic process is allowed to take its course.

It is our sincere belief that LC–MS offers the analytical community the potential for solving more problems than has ever been available previously. This is to say that LC–MS has a potential market that is probably far in excess of GC–MS simply because HPLC can handle more compounds (in particular those that are involatile, polar, and thermally labile) than GC. In addition, HPLC separation can often be accomplished without derivatization which simplifies characterization of the analytes by LC–MS. Thus when the mass spectrometer becomes of comparable cost and simplicity as a conventional UV or similar detector, LC–MS will be welcomed by an army of chromatographers who are "sitting on the fence" waiting for the technique to mature. Although this cost/simplicity factor is a tough goal to achieve for mass spectrometry, we should not discount the possibility of this goal given the major advances seen in the recent past by benchtop GC–MS. With these possibilities in mind the future of LC–MS is very exciting indeed.

REFERENCES

1. D. Desiderio, *Adv. Chromatogr.*, **22**, 1 (1983).
2. P. J. Arpino and G. Guiochon, *Anal. Chem.*, **51**, 682A (1979).
3. G. Guiochon and P. J. Arpino, *J. Chromatogr.*, **271**, 13 (1983).
4. S. P. Markey, *Anal. Chem.*, **42**, 306 (1970).
5. D. E. Games, *Adv. Chromatogr.*, **21**, 1 (1983).
6. J. H. Beynon, R. A. Saunders, and A. E. Williams, *The Mass Spectra of Organic Molecules*, Elsevier, Amsterdam,. 1968.

7. B. L. Karger and P. Vouros, *J. Chromatogr.*, **323**, 13 (1985).

8. K-P. Hupe, R. J. Jonker, and G. Ronjing, *J. Chromatogr.*, **285**, 253 (1984).

9. D. E. Games, M. J. Hewlins, S. A. Westwood, and D. J. Morgan, *J. Chromatogr.*, **250**, 62 (1982).

10. G. Guiochon and H. Colin, *J. Chromatogr. Libr.*, **28**, 1 (1984).

11. C. L. Wilkins and M. L. Gross, *Anal. Chem.*, **53**, 1661A (1981).

12. R. P. W. Scott, C. G. Scott, M. Munroe, and J. Hess, Jr., *J. Chromatogr.*, **99**, 394 (1974).

13. W. H. McFadden, *J. Chromatogr. Sci.*, **18**, 97 (1980).

14. J. Van Der Greef, A. C. Tas, M. C. Ten Noever De Brahm, M. Hohn, G. Meijerhoff, and U. Rapp, *J. Chromatogr.*, **323**, 81 (1985).

15. M. J. Hayes, H. E. Schwartz, P. Vouros, B. L. Karger, A. D. Thruston, Jr., and J. M. McGuire, *Anal. Chem.*, **56**, 1229 (1984).

16. P. Dobberstein, E. Korte, G. Meyerhoff, and R. Pesch, *Int. J. Mass Spectrom. Ion Phys.*, **46**, 185 (1983).

17. V. L. Talrose, V. E. Skurat, and G. V. Karpov, *J. Phys. Chem.* (Moscow), **43**, 241 (1969).

18. F. W. McLafferty, R. Knutti, R. Venkataraghavan, P. J. Arpino, and B. G. Dawkins, *Anal. Chem.*, **47**, 1503 (1975).

19. J. D. Henion, *Anal. Chem.*, **50**, 1687 (1978).

20. A. Melera, *Adv. Mass Spectrom*, **8B**, 1597 (1980).

21. C. N. Kenyon, A. Melera, and F. Erni, *J. Anal. Tox.*, **5**, 216, (1981).

22. P. J. Arpino, P. Krien, S. Vajta, and G. Devant, *J. Chromatogr.*, **203**, 117 (1981).

23. C. Eckers, D. S. Skrabalak, and J. D. Henion, *Clin. Chem.*, **28**, 1281 (1982).

24. F. R. Sugnaux, D. S. Skrabalak, and J. D. Henion, *J. Chromatogr.*, **264**, 357 (1983).

25. J. D. Henion, presented at the Eastern Analytical Symposium, Nov. 13–16, 1984, paper no. 47.

26. C. Eckers and J. D. Henion, "Combined Liquid Chromatography-Mass Spectrometry of Drugs," in S. H. Y. Wong, Ed., *Therapeutic Drug Monitoring and Toxicology by Liquid Chromatography*, Dekker, New York, 1985, Chapter 6, pp. 115–149.

27. E. D. Lee and J. D. Henion, *J. Chromatogr. Sci.*, **23**, 253 (1985).

28. J. D. Henion and D. S. Skrabalak, "Quantitative Analysis of Betamethasone in Equine Plasma and Urine by DLI Micro LC–MS," in *Proceedings of Fifth International Conference on Control of the Use of Drugs in Racehorses*, Toronto, Canada, June, 1983, pp. 91–95.

29. J. B. Crowther, T. R. Covey, E. A. Dewey, and J. D. Henion, *Anal. Chem.*, **56**, 2921 (1984).

30. T. R. Covey, G. A. Maylin, and J. D. Henion, *Biomed. Mass Spectrom.*, **12**, 274 (1985).

31. C. E. Parker, personal communication.

32. T. R. Covey, J. D. Henion, and G. A. Maylin, "Determination of Dithylstilbestrol (DES) in Meat," presented at the Pittsburg Conference and Exposition on Analytical Chemistry and Applied Spectroscopy, Atlantic City, NJ, March 5–9, 1984, paper no. 932.

33. C. R. Blakley, M. J. McAdams, and M. L. Vestal, *J. Chromatogr.*, **158**, 264 (1978).

34. C. R. Blakley and M. L. Vestal, *Anal. Chem.*, **55**, 750 (1983).

35. B. A. Thomson and J. B. Iribane, *J. Chem. Phys.*, **71**, 4459 (1979).

36. T. R. Covey and J. D. Henion, *Anal. Chem.*, **55**, 2275 (1983).

37. T. R. Covey, J. B. Crowther, E. A. Dewey, and J. D. Henion, *Anal. Chem.*, **57**, 474 (1985).

38. M. L. Vestal and G. L. Fergusson, *Anal. Chem.*, **57**, 2372 (1985).

39. C. H. Vestal, D. A. Garteiz, R. Smit, and C. R. Blakley, "Discharge Ionization as a Secondary Ionization Method for Use with Thermospray," presented at the 33rd Conference on Mass Spectrometry and Applied Topics, San Diego, CA, May 1985.

40. D. I. Carroll, I. Dzidic, R. N. Stillwell, M. G. Horning, and G. C. Horning, *Anal. Chem.*, **46**, 706 (1974).

41. F. W. McLafferty, *Tandem Mass Spectrometry*, Wiley, New York, 1983.

42. J. D. Henion, T. R. Covey, D. R. Silvestre, and K. K. Cuddy, "Chemical Analysis and Characterization of Estrogens," in *Estrogens in the Environment, II: Influence on Development*, J. A. McLachlan Ed., pp. 116–138. Elsevier, 1985.

43. T. R. Covey, G. A. Maylin, and J. D. Henion, "Heated Nebulizer APCI LC–MS and LC–MS–MS Analysis of Environmental Estrogens," presented at the 10th International Mass Spectrometry Conference, Swansea, UK 9–13 September, 1985, paper no. 36.

44. R. C. Willoughby and R. F. Browner, *Anal. Chem.*, **56**, 2626 (1984).

45. R. F. Browner and A. W. Boorn, *Anal. Chem.*, **56**, 786A (1984).

46. (*a*) L. E. Abbey, D. E. Bastwick, P. C. Winkler, and R. F. Browner, presented at the 33rd Annual Conference on Mass Spectrometry and Allied Topics, San Diego, CA, May 26–31, 1985. (*b*) C. M. Whitehouse, R. N. Dreyer, M. Yamashita, and J. B. Fenn, *Anal. Chem.*, **57**, 675 (1985).

47. B. W. Wright and R. D. Smith, *J. High Res. Chromatogr. Chromatogr. Comm.*, **8**, 11 (1985).

48. J. C. Fjelsted and M. L. Lee, *Anal. Chem.*, **56**, 619A (1984).

49. J. B. Crowther and J. D. Henion, *Anal. Chem.*, **57**, 2711 (1985).

50. J. B. Crowther, T. R. Covey, D. R. Silvestre, and J. D. Henion, *LC Mag.*, **3**, 240 (1985).

CHAPTER

9

MISCELLANEOUS METHODS

EDWARD S. YEUNG

Ames Laboratory-USDOE and Department of Chemistry,
Iowa State University, Ames, Iowa

In this chapter we briefly discuss several detection schemes in LC. Some of these simply do not fit into one of the categories represented by the preceding chapters. Others are very recent developments that show excellent potential. Because they have not been streamlined for routine use, the ultimate applicability of these schemes may still be unclear. It is naturally not possible to include every known detection scheme. The following are chosen to illustrate the broad spectrum of physical phenomena that have been adapted for LC detection.

1. CONDUCTIVITY DETECTOR

In most types of chromatography, the analytes are neutral (uncharged) species. However, recent interest in ion chromatography (IC) has led to many situations where the eluted analytes are charged species (1,2). It is then very convenient to monitor changes in conductivity as a universal detection scheme.

The principles underlying conductance measurements can be found in many textbooks (3). Briefly, the conductance of a solution, G, is the reciprocal of the resistance. If the two electrodes used for the measurement have area A (cm^2) and are separated by a distance l (cm), then

$$G = k\frac{A}{l} \qquad (1)$$

where k is the specific conductance (units $\Omega^{-1}cm^{-1}$). The characteristic parameter for an ionic species is the equivalent conductance, Λ, which is defined to be the conductance when 1 gram-equivalent of the ion is contained between two electrodes spaced 1-cm apart. Thus

$$\Lambda = \frac{1000k}{C} \qquad (2)$$

where C is the concentration in equivalents per cubic centimeter. Unfortunately, because of interaction among ions, Λ depends on concentration. It is thus proper to use the equivalent conductance at infinite dilution, Λ_0, which is the sum of contributions from the cations, λ_+^0, and the anions, λ_-^0:

$$\Lambda_0 = \lambda_+^0 + \lambda_-^0 \tag{3}$$

For IC applications, the ionic concentrations are quite low, and Λ_0 can be assumed for all situations. In practical detector cells, A and l are held constant. Thus the background conductance of a chromatographic eluent (one anion and one cation) can be predicted to be

$$G = \frac{C(\lambda_+^0 + \lambda_-^0)}{1000K} \tag{4}$$

where K is the cell constant $(cm^{-1}) = l/A$. The conductance of a chromatographic effluent consisting of sample ions being eluted by an eluent can be described similarly. For example, in the case of a single analyte (anion) being eluted through a column, the conductance is predicted by

$$G_s = \frac{[C_E\lambda_{E^+} + (C_E - C_S)\lambda_{E^-} + C_S\lambda_{S^-}]}{1000K} \tag{5}$$

where λ_{E^+} and λ_{E^-} are the equivalent conductances of the eluent (cation and anion, respectively) and λ_{S^-} is the equivalent conductance of the sample anion. The eluent and sample concentrations (normalities) are C_E and C_S. In Eq. 5, the principles of electroneutrality and equivalence of exchange require that the total number of equivalents of cations equal the total number of anions. Regardless of the particular cation that was associated with the analyte anion at injection, only the eluent cation is relevant at the chromatographic peak. However, associated with the analyte peak is a system peak. This is because at injection, the analyte ions exchange with the eluting ions already attached to the stationary phase. There will thus be a higher instantaneous C_E due to displacement. There will also be increased conductance due to the unretained associated cation of the analyte ion. Counteracting the increase is a dilution effect due to the solvent (water) introduced with the sample. Therefore, the system peak depends on injection volume as well as injected amount. Quantitation of the analyte peak is possible only if it is separated from the system peak, in which case Eq. 5 holds.

A differential conductivity detector does not measure the actual conductance, but rather a change in the conductance of the effluent stream. To derive an equation for the change in conductance, ΔG, we need only subtract the conductance of the eluent above, as predicted by Eq. 4, from the conductance of the eluent and sample ions, as predicted by Eq. 5. Thus we obtain

$$\Delta G = \frac{C_S (\lambda_{S^-} - \lambda_{E^-})}{1000K} \tag{6}$$

For a conductivity detector to work in IC, the volume must be small compared to the peak volume. This does not present much technical difficulty. It is also necessary for the detector to be able to function in the presence of a large background conductance, because of the finite concentration of eluting ions. In principle a dual reference and sample cell arrangement can be used to compensate for the large background, or an electronic null can be used. However, the temperature dependence of conductance is large, typically being 2% per degree Celsius. It is the fluctuations associated with this background conductance that limit detectability. In fact, the reason the conductivity detector was not successful in early ion-exchange chromatography (4) is that relatively high concentrations of eluting ions are needed to elute various ions from the stationary phase, and the conductance background severely limits the detectability. There are two separate solutions to the conductance background, generally known as suppressed IC and nonsuppressed IC. Detectabilities in the low nanogram range are thus possible.

1.1. Suppressed IC

To illustrate the scheme (1), one can use the example of eluting Cl^- (analyte) by OH^- (eluting ion) in a strong base anion-exchange column. If R is symbolic of the resin on the column,

$$RCl + OH^- \rightarrow ROH + Cl^-$$

There will naturally be an equivalent amount of a cation (e.g., Na^+) for charge balance at all times. One is then trying to distinguish the small conductance change from the replacement of OH^- by Cl^- in a large background of Na^+ and OH^-. If, however, the eluted material is then passed through a second column containing a strong acid cation-exchange material in the hydrogen form, R', the NaOH eluent is converted into H_2O:

$$R'H + Na^+ + OH^- \rightarrow R'Na + H_2O$$

This "suppressor" column therefore reduces the background conductance to a negligible level—that of deionized water. When Cl^- ions elute,

$$R'H + Na^+ + Cl^- \rightarrow R'Na + H^+ + Cl^-$$

Because ions are still formed, there is a detectable increase in conductance. Moreover, the analyte peak now contains H^+ rather than Na^+ as the counterion, greatly increasing the signal because of the large λ^0_+ of the hydrogen ion. An analogous system can be devised for studying cations as analytes. The important point about suppressed IC is that the column must be eventually regenerated to be in the correct form. This technical problem has been successfully worked out in the commercial version of the system.

1.2. Nonsuppressed IC

Another approach to reducing the background conductance is by using very low concentrations (10^{-4} M) of eluting ions (2) together with eluting ions of low equivalent conductance. The latter also aids in enhanced sensitivity, according to Eq. 6. This is possible if an ion-exchange resin of very low capacity (10^{-2} meq/g) is used. Table 1 shows the limiting equivalent ionic conductances for ions (5). It can be seen that sodium benzoate is a particularly good eluent based

Table 1. Limiting Equivalent Ionic Conductances in Aqueous Solution at 25° C (5)

Anions	λ^0_-	Cations	λ^0_+
OH^-	198	H^+	350
F^-	54	Li^+	39
Cl^-	76	Na^+	50
Br^-	78	K^+	74
I^-	77	NH_4^+	73
NO_3^-	71	Mg^{2+}	53
HCO_3^-	45	Ca^{2+}	60
Formate	55	Sr^{2+}	59
Acetate	41	Ba^{2+}	64
Propionate	36	Zn^{2+}	53
Benzoate	32	Hg^{2+}	53
Phthalate	48	Cu^{2+}	55
SO_4^{2-}	80	Pb^{2+}	71
$S_2O_3^{2-}$	85	Co^{2+}	53
CO_3^{2-}	72	Fe^{3+}	68
$C_2O_4^{2-}$	84	La^{3+}	70
PO_4^{3-}	69	Ce^{3+}	70
$Fe(CN)_6^{3-}$	101	$CH_3NH_3^+$	58
$Fe(CN)_6^{4-}$	111	$N(Et)_3^+$	33

on background alone. It also turns out that benzoate is a moderately strong eluting ion, thus allowing its use at low concentrations without serious problems. An example of nonsuppressed IC (6) is shown in Fig. 1.

We can compare the two types of IC with respect to detectability. If a commercial conductivity cell with a cell constant K of 33.0 cm^{-1} is used, several eluents can be compared directly (7). A $2 \times 10^{-4} M$ solution of sodium benzoate at pH 7.0 (completely ionized) has a measured conductance of 0.50 μmho at 22.0°C, which is identical to the value calculated from Table 1. A solution of 3.0-mM NaHCO$_3$ plus 2.4-mM Na$_2$CO$_3$, a typical eluent in suppressed IC, has a conductance of 20.9 μmho before the suppressor column and 0.63 μmho after the suppressor column. Thus the second column is necessary to provide a small background contribution. It is interesting that the benzoate solution actually has a lower background conductance at practical concentrations. There is a loss of sensitivity in nonsuppressed IC because of the replacement of Na$^+$ for H$^+$ as the counterion when the analyte ion elutes, according to Table 1. But the convenience and the lack of additional band broadening from the second column makes nonsuppressed IC a practical alternative.

Other fine points in conductivity detection can be explained by Eq. 6. Because the difference in equivalent conductance dictates the magnitude of the signal, one should choose the eluting ion with care. It is possible to obtain a zero signal or a negative signal depending on the choice of the eluent. Phthalate is a stronger IC eluent and can be used at lower concentrations than benzoate, resulting in a lower background conductance and lower noise. For the inorganic anions (Table 1), the loss of sensitivity is not significant. Phthalate is then a clear choice. However, if certain organic anions or HCO$_3^-$ are of interest, one must use benzoate to maintain reasonable sensitivity. From Table 1, the best anion for elution is OH$^-$ in terms of sensitivity. This has been demonstrated in the separation of cyanide and halide ions with a $10^{-3} M$ solution of NaOH (8). The interesting feature is that a suppressor column cannot be used for weak acids such as cyanide because they will be converted to the undissociated form to give negligible conductance.

Another application of the conductivity detector is the quantitation of ions without standards (9). The general scheme for this is explained in Section 7. Using two eluting ions with different equivalent conductances to elute the same sample, one obtains two relationships of the form of Eq. 6, which can be solved to determine the two unknowns, the concentration and the equivalent conductance of the analyte ion. Table 2 shows the results of such an investigation (9). The predictions from solving the set of simultaneous equations agree well with the actual concentrations of the analyte ions. The two eluting ions used were benzoate and S$_2$O$_3^{2-}$. The latter is chosen because of its very high equivalent conductance. It turns out that S$_2$O$_3^{2-}$ is a very strong eluent and a $6 \times 10^{-4} M$ solution is sufficient. At this concentration, the background conductance was 5.4 μmhos,

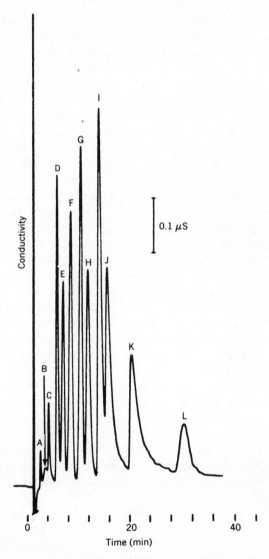

Figure 1. Separation of 12 early-eluting anions. Column, trimethyl quaternary ammonium resin; eluent; 2 m*M* nicotinic acid; detection, conductivity. Peaks: A = acetate, B = propionate, C = azide, D = formate, E = fluoride, F = phosphate, G = chloride, H = bromate, I = bromide, J = nitrate, K = chlorate, and L = propylsulfonate. Reproduced with permission from Ref. 6. Copyright 1984 *LC Magazine*.

Table 2. Quantitation of Ions Without Standards Using the Conductivity Detector

Ion	Benzoate	$S_2O_3^{2-}$	$(Cl)_3Ac^-$	SO_4^{2-}
True C^a	2.72	4.90	1.98	4.02
Calc. C	2.64 ± 0.04	4.85 ± 0.17	2.04 ± 0.02	4.08 ± 0.10
True λ^b	32.4	85.0	36.6	80.0
Calc. λ	34.9 ± 1.1	86.7 ± 0.1	34.0 ± 0.2	78.1 ± 0.2

^{a}Concentrations are in 10^{-3} N units at injection.
^{b}True λ from Ref. 5.

which is actually lower than that (14 μmhos) of the benzoate eluent at 5×10^{-3} M. Good detectability is thus preserved.

2. PHOTOCONDUCTIVITY AND PHOTOIONIZATION

When the analytes themselves do not show a substantial conductivity difference compared to the eluent, one can induce a conductivity change by the interaction of light. This is the basis for the photoconductivity (10) and the photoionization (11) detector. The difference between the two is essentially in the mode of signal acquisition, with the photoconductivity detector still being able to function even if the photochemical products are not ions.

The commercial photoconductivity detector is based on a differential conductivity cell, with the splitting of the column effluent so that one-half goes to the reference cell and one-half goes to a quartz reaction coil before entering the sample cell. The block diagram is shown in Fig. 2. A mercury lamp, primarily acting through its 254 nm output, is used for irradiation. Alternatively, a zinc lamp at 214 nm can be used. Several types of reactions are possible to affect the conductivity. Acids, for example, can be formed from halogenated organics:

$$RX + h\nu \rightarrow R\cdot + X\cdot \rightarrow P + HX$$

where P represents other products. For nitro compounds,

$$C_6H_5NO_2 + h\nu \rightarrow C_6H_5NO\cdot + O$$

and

$$C_6H_5NO_2 + h\nu \rightarrow C_6H_5\cdot + NO_2$$

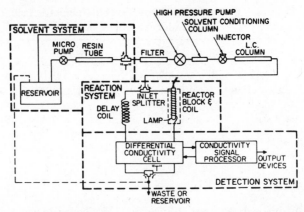

Figure 2. Block diagram showing the relationship of the photoconductivity detector to the other components of the HPLC system (10). Reproduced from the *Journal of Chromatographic Science* by permission of Preston Publications, Inc.

NO_2 is a relatively conductive species, and a good response is obtained. For other species, the products show only a small difference in conductivity, but useful detectabilities are still obtained:

$$R_3N + h\nu \rightarrow R\cdot + R_2N$$

Analogs of these reactions are also found for sulfur compounds and, for example, sulfur–phosphorus insecticides. An example of a chromatogram due to photoconductivity is shown in Fig. 3. Because of the difference in reactivity and in the products, the response is quite different from that of an absorption detector at the same wavelength. The detectabilities for nitrosamines are in the low picogram range.

The main drawback in the commercial photoconductivity detector is the added volume of the reaction coil. This leads to extracolumn band broadening common to all postcolumn derivatization schemes. For high-speed LC or microbore LC, the problem is magnified. An alternative to this is to use a laser as an excitation source. Some extra volume is still introduced because one needs a finite path-length for light absorption, but the collimation of the laser allows a small cross-sectional area to be achieved and the higher photon flux can compensate for a shorter path length. It is also possible to use a pulsed operation to allow discrimination against background conductivity. There are several convenient lasers for such measurements. If dissociation is at 254 nm, a krypton fluoride laser, a frequency-quadrupled neodymium/yttrium–aluminum–garnet laser, a frequency-doubled copper vapor laser, or a frequency-doubled argon ion laser may be used.

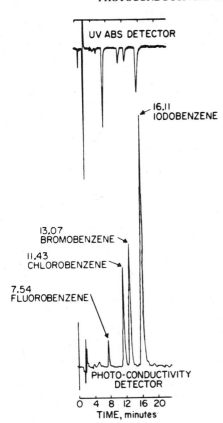

Figure 3. A comparison of response for the monohalogenated benzenes for the photoconductivity detector and the UV absorbance detector. Five nanomoles of each compound were injected. Retention times (min) are indicated for each peak. A Zorbax ODS column was used with a 70:30 methanol–water mobile phase at a flow rate of 1 mL/min (10). Reproduced from the *Journal of Chromatographic Science* by permission of Preston Publications, Inc.

For more stable species, the argon fluoride laser at 193 nm can be used, with the possibility of Raman shifting of the frequency to even shorter wavelengths.

The detection scheme based on photoionization (12) is shown in Fig. 4. To produce ions from organic compounds, shorter wavelengths of light are needed compared to photoconductivity methods. However, at extremely short wavelengths the solvent may contribute. For example, ionization energies range from 9.34 eV for tetrahydrofuran to 12.20 eV for acetonitrile and 12.59 eV for water. Stabilization by solvation of the ionization products, such as in water, reduces the ionization threshold for liquid water to 6.05 eV. This essentially rules out the use of water in the chromatographic solvent. If a microwave-excited xenon lamp is used, the short-wavelength limit of the emission is 7.81 eV. This then allows the photoionization of many organic species without background contributions from the solvent in normal-phase LC. Figure 5 shows a chromatogram

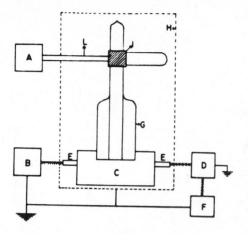

Figure 4. Schematic diagram of PID cell and electrical circuit: (A) microwave generator; (B) collector electrode power supply; (C) Teflon PID cell; (D) Kelthley 610C electrometer amplifier; (E) gold-plated brass electrodes; (F) recorder; (G) Ophthos Xenon lamp; (H) sheet aluminum cage; (I) microwave cavity; and (L) microwave cable. Reproduced with permission from D. C. Locke, B. S. Dhingra, and A. D. Baker, *Anal. Chem.*, **54**, 447 (1982). Copyright 1982 American Chemical Society.

derived from the photoionization detector (12). The detectabilities range from 1 pg for morpholine to 350 pg for fluorene. Halogenated compounds, although known to ionize at these wavelengths, do not show any response because of recombination and electron capture processes. Photoionization in liquids unfortunately cannot approach the sensitivities achieved in the gas phase because of the lack of processes such as space-charge amplification and the lack of a sharp

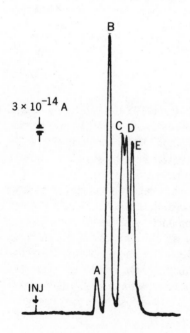

Figure 5. Chromatogram of polycyclic aromatic hydrocarbon mixture detected by photoionization: (A) acenaphthene, 6 ng; (B) pyrene, 12 ng; (C) benz[a]anthracene, 20 ng; (D) benzo[a]pyrene, 4.7 ng; (E) perylene, 1.4 ng. Chromatographic conditions: 25-cm × 4.6-mm ID Whatman Partisil PXS 10 μm; 0.66-mL/min hexane/2% 2-propanol; temperature, ambient; sample size injected, 1 μL; electrometer, 0.3×10^{-11} A f.s.; collector voltage, −3000 V; microwave power, 30% maximum. Reproduced with permission from D. C. Locke, B. S. Dhingra, and A. D. Baker, *Anal. Chem.*, **54**, 447 (1982). Copyright 1982 American Chemical Society.

threshold for solvent ionization. Therefore, one idea is to vaporize the LC effluent and to use a gas-phase detection scheme (13).

Replacing the conventional light source in Fig. 4 with a laser allows more efficient coupling of the excitation into a small volume. Because the background current seems to be independent of irradiation, a high power laser, for example, ArF at 193 nm, may improve the detectability. The pulsed operation will also allow time-gating to discriminate further against the background. A different excitation scheme also becomes feasible at high laser intensities—that of two-photon ionization (14). In this case, the molecule absorbs (simultaneously) two photons of light to reach its ionization threshold. It is necessary only to use photons equivalent to one-half the ionization energy. If there is a real molecular state present at the single photon energy, resonant enhancement occurs and very efficient excitation is possible. A flow cell for photoionization has been demonstrated based on suspending a droplet of the flowing stream between two electrodes and a nitrogen laser operating at 337 nm (10 ns, 20 Hz, 1.3 mJ per pulse). Even though the cell geometry has not been specifically optimized for ionization detection, a detectability of 9 ng/mL was found for pyrene, degrading to 100 ng/mL for naphthacene. An improved version (15) provides a detectability of 2 ng/mL for the drug clonazepam with only 300 μJ per pulse of excitation.

Further improvements in the concentration detectability were reported (16) by using larger electrodes. The detection limit of pyrene was found to be 6 ng/L. When this was converted to a 5 μL cell for LC (17), the detectability was slightly worse. The cell geometry used is shown in Fig. 6. Using a laser at 337 nm (10 ns, 10 Hz, 2.5 mJ per pulse), a detectability of 5 pg of anthracene is

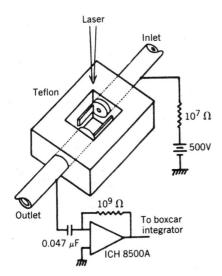

Figure 6. Schematic diagram of the photoionization cell for liquid chromatography. Reproduced with permission from Ref. 17.

achieved. With better matching of the absorption wavelength and the laser output, and with higher laser intensities, it is likely that even better results can be obtained.

3. INDUCTIVELY COUPLED PLASMA ATOMIC EMISSION

Elemental selectivity has already been discussed in Chapter 2 for atomic absorption (AA) methods. Atomic emission (AE) has several distinctive features when coupled to LC. First, simultaneous multielement monitoring is possible with the appropriate multichannel detector because of the inherent optical configuration of atomic emission. Second, for elements that have the first excited state at high energies, absorption measurements may not even be feasible for lack of good light sources, for example, from elemental lines. Third, for most elements, AE can provide better detectabilities than AA. Lastly, the high temperatures available in certain AE sources can help to reduce interferences and to dissociate species that normally would not exist in the atomic form in AA.

In addition to the flame as a source for AE, the microwave-induced plasma (18), DC plasma (19), and the inductively coupled plasma (ICP) (20) have been used for LC. In all the systems, the LC solvent presents a problem. The microwave-induced plasma requires a low solvent flow rate to be stable. The ICP and the DC plasma can handle aqueous-based solvents quite well, for these are commonly used for conventional AE measurements. Organic solvents, however, tend to extinguish the plasma. The normal ceramic nebulizer chamber also does not function properly with hydrocarbon solvents (21). Many of these problems are avoided by the use of microcolumns, which generally have much slower solvent flow rates. The discussion here is restricted to LC–ICP interfaces.

An excellent example of the multielement advantage of LC–ICP is shown in Fig. 7 (22). Gel-permeation chromatography allows the use of aqueous mobile phase for a protein separation. At a normal flow rate of 1 mL/min, the ICP operated properly. In fact, the interface is a simple connection between the column effluent and a commercial ICP system. Because of the various binding metals in the proteins, the individual elemental lines can aid in the identification. It is interesting to note that emission from nonmetals P and C are observed, although at lower sensitivities. One can in principle obtain the ratios of H, C, N, O, P, S, and halogens in a particular analyte in place of the conventional elemental methods for qualitative analysis. The limitations are the precision that can be achieved, which is typically 2% for spectrophotometric methods, and the elemental composition of the solvent.

The interface of microbore column (0.3–1-mm ID) to the ICP requires some modifications. The flow rates of the mobile phase are typically too low for conventional nebulizers to work properly. A modification (23) for using a pneu-

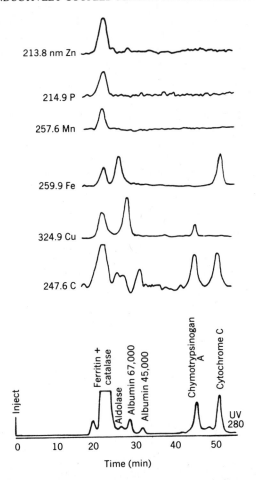

Figure 7. HPLC–ICP of proteins. 100 μg each protein. Steric exclusion separation on two TSK 3000 SW columns, 7.5 mm × 60 cm; 1 mL/min; 0.9% NaCl in water. Reproduced with permission from M. Morita, T. Uehiro, and K. Fuwa, *Anal. Chem.*, **52**, 349 (1980). Copyright 1980 American Chemical Society.

matic nebulizer is shown in Fig. 8. The column effluent is simply mixed with another stream of the solvent as a makeup liquid. There is a dilution effect, however, and the concentration detectability of the ICP is not preserved. The use of low flow rates (16 μL/min) allows an organic solvent—toluene—to be used as the mobile phase, as shown in Fig. 9. The nebulizer has some memory effect, so that some broadening of the chromatographic peak can be expected. An improved interface (24) that does not rely on a makeup liquid is shown in

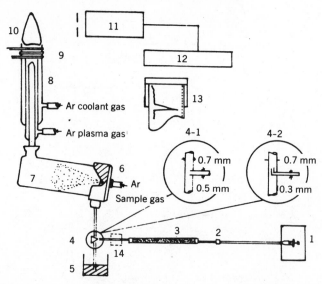

Figure 8. Schematic diagram of micro HPLC–ICP system. 1, micro HPLC pump, syringe type; 2, sample injection point; 3, micro HPLC column; 4, interfacing device (4-1, type 1; 4-2 type 2); 5, reservoir of carrier solvent for ICP—water for reverse-phase mode and methyl isobutyl ketone for normal-phase mode; 6, cross-flow type pneumatic nebulizer; 7, spray chamber; 8, plasma torch; 9, coil; 10, plasma flame; 11, Ebert mounting monochromator with a grating blazed for 300 nm; 12, R-456 photomultiplier; 13, chart recorder; 14, UV detector when used. Reproduced with permission from Ref. 23.

Fig. 10. The entire LC effluent is nebulized in a cross-flow geometry. Only the tubing between the column and the nebulization region contributes to extra volume in the detector. A detectability of 800 ng of carbon was obtained. Applications should be quite broad if one can further improve on the detectability.

4. FLAME PHOTOMETRIC DETECTOR

The flame photometric detector has been widely used for gas chromatography (25). Unlike AE detectors, it is based on molecular emission of primarily non-metallic species in the flame. In LC, the effluent must be separately introduced into the flame by a nebulizer. Emission is observed mainly from HPO and S_2 species, so that phosphorus or sulfur-specific detection is achieved. Usually only optical filters are needed before the phototube, because the emission is spectrally broad. In a crossed-nebulization geometry (26), about 25% of the effluent reaches the flame, and detectabilities are in the 10^{-8}–10^{-7} g/mL range in the column

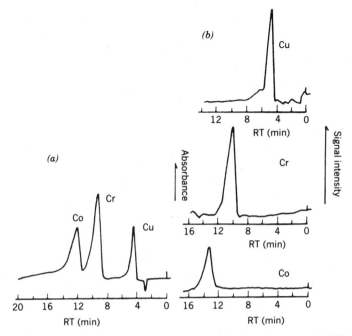

Figure 9. Chromatograms detected by UV and ICP. (*a*) UV detection 300 nm. (*b*) ICP detection (emission lines—Cu, 324.7 nm, 320 ng/μL; Cr, 267.7 nm, 425 ng/μL; Co, 228.6 nm, 425 ng/μL); mobile phase, toluene, 16 μL/min. Reproduced with permission from Ref. 23.

effluent. The main problem is the large background generated when organic solvents are used. A total consumption burner can be used instead (27), as shown in Fig. 11. This has been miniaturized so that packed microcapillary columns (70-μm ID) can be used. Total consumption burners produce larger solution droplets and possibly incomplete vaporization. The diffusion flame is also less stable than premixed flames. However, the small volume that is possible in the total consumption arrangement is necessary to interface to microcolumns. In Fig. 11, the fuel gas is a mixture of H_2 (55 mL/min) and N_2 (90 mL/min) and air is introduced separately (75 mL/min). The fuel ratio is thus 3.7, and is comparable to that for GC applictions. Because of the smaller flow rates, it was found that methanol and acetone up to 50% do not reduce the emission intensity of HPO. There is even a slight enhancement when 10% methanol is used, probably due to a higher flame temperature. The minimum detectable amount was 2 ng of phosphorus injected, which is equivalent to 71 pg/s of phosphorus at the flame. The utility of such a detector is shown in Fig. 12, where selective detection of organophosphorus compounds is demonstrated.

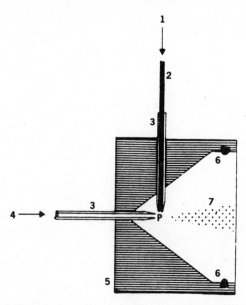

Figure 10. Cross-sectional view of cross-flow nebulizer: (1) eluent inlet from a microcolumn; (2) stainless steel capillary tubing, 130-μm ID (310-μm OD) × 40-mm length; (3) glass tubing, 320-μm ID × 30 mm; (4) sample Ar gas inlet; (5) Teflon nebulizer body; (6) O-ring; (7) nebulized eluent; (P) nebulizing point. Reproduced with permission from K. Jinno, S. Nakanishi, and T. Nagoshi, *Anal. Chem.*, **56**, 1977 (1984). Copyright 1984 American Chemical Society.

5. LIGHT SCATTERING

The scattering of light is related to molecular properties of many different types (28); hence a special kind of selectivity can be obtained for LC detection. The two properties that are of interest in LC are nephelometry and quasi-elastic light scattering (LS).

The optical arrangement for nephelometry is not very different from those for fluorescence, except that no wavelength selection is necessary to analyze the scattering. However, it is unlikely that any solute in the LC effluent can directly contribute to a nephelometric signal beyond the Rayleigh scattering from the eluent, because the LC column acts as a fine filter for particulate matter. Because of this, postcolumn derivatization or precipitation is needed to produce particles or colloids that will scatter light. A flow cell with a volume of 17 μL and collection based on fiber optics (29) is shown in Fig. 13. Only 0.5 mW of power from a laser at 633 nm was needed. It is interesting to note that for nephelometry the beam size of the laser should be substantially larger than the diameters of the particles, or refraction rather than scattering will be measured. Also, the

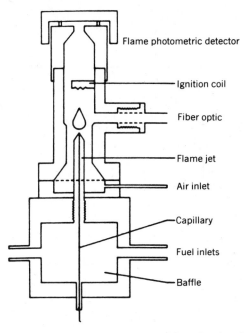

Figure 11. Schematic diagram of the FPD burner base and flame housing. Reproduced with permission from V. L. McGuffin and M. Novotny, *Anal. Chem.*, **53**, 946 (1981). Copyright 1981 American Chemical Society.

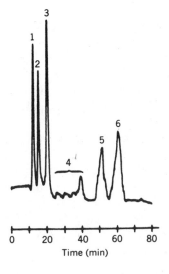

Figure 12. Chromatogram of organophosphorus pesticides: column 10-m octylsilane microcolumn; mobile phase, 42% aqueous methanol; solutes, (1) solvent and phosphorus-containing impurity, (2) cygon, (3) DDVP, (4) phosphorus-containing impurities, (5) malathion, (6) guthion. Reproduced with permission from V. L. McGuffin and M. Novotny, *Anal. Chem.*, **53**, 946 (1981). Copyright 1981 American Chemical Society.

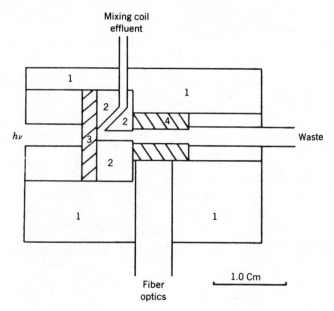

Figure 13. Flow cell for light-scattering detection in LC: 1, stainless steel detector block; 2, stainless steel entrance aperture; 3, Pyrex window; 4, Pyrex scattering cell. Reproduced with permission from Ref. 29.

particles should be larger than 110% of the wavelength. For the test case of several nonpolar lipids, the precipitating reagent is simply an ammonium sulfate solution, which changes the solubility of the former. The detector response thus depends on the ammonium sulfate concentration and its flow rate into the mixing chamber. The signal obtained is nonlinear relative to the injected quantity, approaching a square-root dependence for small samples. A chromatographic separation of nonpolar lipids in human serum is shown in Fig. 14. Cholesterol, its esters, and the triglycerides all show a response. The detectability is of the order of 0.5 μg for lipids. It can be expected that any of the traditional precipitation reactions can be adapted to this scheme.

Instead of relying on the formation of precipitates, the alternative is to vaporize the solvent completely. The residual particles represent the analytes as they elute, and can be monitored by light scattering (30). This is possible because most common LC solvents are highly volatile. The use of microcolumns (1-mm ID or smaller) also allows small amounts of solvent to be used. The main concern is solvent impurity, which, even at the 1-ppm level, can constitute a large background signal. A vaporization system for the LC effluent is shown in Fig.

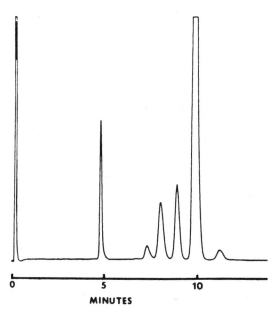

Figure 14. Chromatogram of nonpolar lipids in human serum with light-scattering detection. Conditions: column, C_{18} 10 μm; eluent, 2:1 acetone–methanol with postcolumn addition of ammonium sulfate in water; sample loop, 50 μL; flow, 1.0 mL/min. Reproduced with permission from Ref. 29.

15. Carbon dioxide is used as the nebulization gas because of its high heat capacity and low cost. After nebulization, the droplets pass through a drift tube at a temperature of 40–50°C. The solvent thus vaporizes and the only particles left are the nonvolatile impurities in the solvent and the analyte. A 1-mW HeNe laser at 633 nm irradiates the particles, and the scattered light is collected by a glass rod and transmitted to a photomultiplier tube. For water-containing solvents, the temperature of the drift tube can be raised to 80–85°C.

The intensity of the scattered light is a complex function of the particle size, RI, wavelength, polarization, and observation angle. The system in Fig. 15 has a wide distribution of particle sizes, and only a weighted average response is available. The signal is found to increase with the sample size to the 1.8 power. The detectability was found to be 0.55 μg injected, or a concentration of 40 ppm in the mobile phase. An advantage of this detector is that the contribution to band broadening is extremely small, of the order of 0.15 μL. Another advantage is that it is suitable for gradient elution, because all the solvent is vaporized. Figure 16 shows a separation of triglycerides with gradient elution.

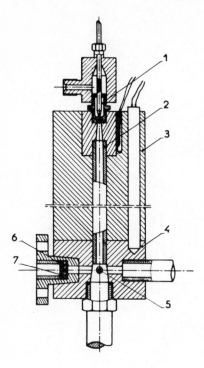

Figure 15. Detector block for light-scattering detector. (1) Nebulizer; (2) drift tube; (3) heated copper block; (4) light-scattering cell; (5) glass rod; (6) glass window; (7) diaphragms. Reproduced with permission from Ref. 30.

The base line is stable and the column efficiency is preserved. A commercial version of this detector has recently become available (31).

Quasi-electic light scattering can be used to probe particles in the 50 Å to 2 μm range in diameter. It is based on the slight Doppler shift in the scattered light due to Brownian motion. Highly monochromatic light is essential to these measurements. The scattering signal depends on the angle of observation, but in general either very small angles or very large angles are used. The optical arrangement for the former (32) is shown in Fig. 17. Changing from one mode to the other is quite easy. Only a 2-mW laser is needed to provide enough intensity. Because the important information is the frequency shift, photon-counting autocorrelation must be used after the photomultiplier tube. It was found that latex spheres in diameters of 0.1–0.3 μm can be measured in the concentration range of 5–100 μg/mL, still within the range of LC. The information that can be obtained includes particle size, molecular conformation, molecular weight, molecular weight distributions, and molecular rotations. Clearly the main applications are the studies of macromolecules and microemulsions in biochemistry and in polymer science, particularly in conjunction with gel-permeation chromatography (GPC). An example of the use of the low-angle light

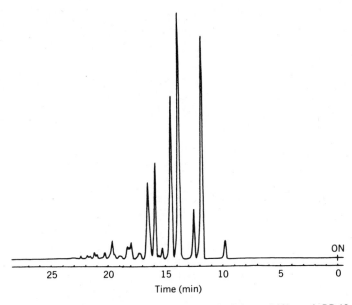

Figure 16. Separation of the triglycerides of soybean oil. Column, LiChrosorb RP-18 (5 μm), 200 × 2-mm ID; mobile phase, linear gradient from 33:67 acetone–acetonitrile to 99:1 acetone–acetonitrile in 25 min; flow rate 0.3 mL/min. Reproduced with permission from Ref. 30.

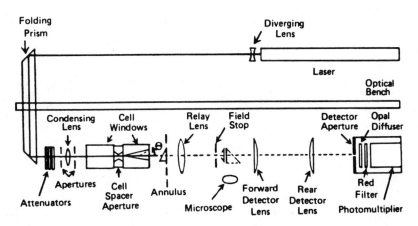

Figure 17. Optical arrangement for quasi-elastic light scattering at low scattering angles. Reproduced with permission from M. L. McConnell, *Anal. Chem.*, **53**, 1007A (1981). Copyright 1981 American Chemical Society.

scattering detector in GPC has been published (33). The experimental arrangement includes a UV detector and a RI detector in series with the light-scattering detector. When the molar absorptivity ε of the species is known or can be determined accurately, the responses of the three detectors can be used to determine the molecular weight. This is because a linear relationship exists between the molecular weight and the quantity $(UV)(LS)/(RI)^2\varepsilon$ for a given set of experimental conditions.

6. CIRCULAR DICHROISM

Circular dichroism (CD) is another property of chiral molecules. The same advantages in selectivity discussed in Chapter 6 for polarimetry also apply here. Again, the difficulty is that commercial CD spectrometers do not provide the sensitivity needed to study species at concentrations typical of an LC effluent. High-performance liquid chromatography coupled with circular dichroism (HPLC–CD) was demonstrated by interfacing an LC system to a CD spectrometer originally designed for static mode operation (34,35). A 3 μg limit of detection for L-tryptophan using HPLC–CD was reported (35). In contrast to polarimetry, having good polarizers is not the key to sensitivity in CD. CD is inherently an absorption measurement, and intensity fluctuations in the light source is the main contributor to noise. So if good intensity stability can be achieved in lasers, they can be used for CD. Normally, it is much simpler to use a fixed-wavelength laser. And, unless stop-flow techniques are used, single-wavelength detection is sufficient to provide a selective detection system for either a single species or a certain class of CD active compounds.

6.1. Theory

CD is defined as the difference in absorbance of left-circularly polarized light (LCPL) and right-circularly polarized light (RCPL). The CD quantity, $\Delta\varepsilon$, can be related to the molar absorptivities of LCPL and RCPL as

$$\Delta\varepsilon = \varepsilon_L - \varepsilon_R \tag{7}$$

Multiplication of both sides of Eq. 7 by path length (b, in cm) and concentration (C, in M) at the detector yields

$$\Delta A = \Delta\varepsilon bC = \varepsilon_L bC - \varepsilon_R bC = A_L - A_R \tag{8}$$

where absorbance, A_i, is defined in the conventional way, for the arbitrary subscript i, as

$$A_i = \log \left(\frac{I_{0,i}}{I_i} \right) = \varepsilon_i bC \tag{9}$$

with $I_{0,i}$ being the intensity of the incident beam and I_i being that of the transmitted beam, using the base 10 logarithm. Substituting Eq. 9 into Eq. 8 using $i = L$ for LCPL and $i = R$ for RCPL, we have

$$\Delta A = \log \left[\frac{I_{0,L}}{I_L} \right] - \log \left[\frac{I_{0,R}}{I_R} \right] \tag{10}$$

Because Eq. 7 dictates that $\varepsilon_L > \varepsilon_R$ for $\Delta \varepsilon$ to be positive, this requires $I_L < I_R$ (for $I_{0,L} \approx I_{0,R}$), or

$$\Delta I = I_R - I_L \tag{11}$$

Because $I_{0,L}$ and $I_{0,R}$ are not exactly the same in practice (owing to slight imbalance in the modulation), the difference can be arbitrarily expressed as

$$\Delta I_0 = I_{0,L} - I_{0,R} \tag{12}$$

This represents the offset from true null. Rearrangement of Eq. 10, substitution of Eqs. 11 and 12, and conversion to the natural logarithm gives,

$$\Delta A = \left[\frac{1}{2.303} \right] \ln \left[\left(1 - \frac{\Delta I_0}{I_{0,R}} \right) \left(\frac{1}{1 - (\Delta I / I_R)} \right) \right] \tag{13}$$

Taking the exponential of both sides, assuming $\Delta A \ll 1$, and multiplication of the right side of Eq. 13 yields

$$\Delta A = \left[\frac{1}{2.303} \right] \left[\frac{\Delta I}{I_R} - \frac{\Delta I_0}{I_{0,R}} - \frac{\Delta I_0 \Delta I}{I_R I_{0,R}} \right] \tag{14}$$

Using Eq. 9 and solving for I_R provides

$$I_R = \left[\frac{I_{0,R}}{1 + 2.303 A_R} \right] \tag{15}$$

Substituting Eq. 15 into Eq. 14, neglecting the insignificant term ($\Delta I_0 \Delta I / I_R I_{0,R}$), and solving for ΔI yields

$$\Delta I = \frac{2.303 \Delta A I_{0,R} - \Delta I_0}{1 + 2.303 A_R} \tag{16}$$

The definition of ΔI_0 in Eq. 12 was arbitrary; experimentally the sign can be either positive or negative. Also, for the purpose of Eq. 16 the subscript R can be neglected and absorbance "average" can be employed. These factors allow Eq. 16 to be expressed as

$$\Delta I = \frac{2.303 \Delta A I_0}{1 + 2.303 A} \pm \Delta I_0 (1 - 2.303 A) \tag{17}$$

This, then, is the complete description of the observed signal, ΔI, as it relates to the experimental parameters. Note that the signal, ΔI, is affected by the background absorbance (i.e., is attenuated) by the term $1 + 2.303 A$. Also note that the second term, containing ΔI_0, can have a marked effect if the background absorbance *is large and changing* with time. That is, it is the dynamic, concentration-dependent part of the second term, $2.303 A \Delta I_0$, that affects the signal, ΔI, in a chromatography context. By dividing the concentration-dependent ΔI_0 term, or "I_0 offset effect," by the numerator of the CD effect term, the ratio R can be calculated:

$$R = \frac{\Delta I_0 \varepsilon}{I_0 \Delta \varepsilon} \tag{18}$$

R is an indication of the error introduced into the CD measurement. An example of what Eq. 18 implies follows. In order for the CD measurement to be in error by only 1% because of the "I_0 offset effect," for $\varepsilon / \Delta \varepsilon = 10^3$, the modulation of the beams $I_{0,R}$ and $I_{0,L}$ must be able to maintain the power of these two beams to at least 1 part in 10^5 relative to true null. This includes the initial balancing of the beams or any subsequent "detector drift." Clearly, this is not an instrumental hardship for many inorganic complexes, but would be quite challenging for organic species that have a $\varepsilon / \Delta \varepsilon$ ratio typically near 1×10^5. It should be emphasized that the offset affects the *accuracy* but not necessarily the *detectability* of the measurement.

For conditions in which $2.303 A \leq 1 \times 10^{-2}$, that term can be neglected with a maximum of only 1% error in the accuracy of the measurement of ΔI in Eq. 17. If we assume the ΔI_0 term and the $2.303 A$ term can be neglected in Eq. 17, and substituting in $\Delta A = \Delta\varepsilon bC$, a simple expression for the detected CD signal is obtained:

$$\Delta I = 2.303\Delta\varepsilon bCI_0 \tag{19}$$

with the variables as defined earlier.

6.2. Detection System

The detection system is laser-based and modulated using polarization techniques; it is illustrated in Fig 18. The 488-nm light from the argon ion laser is sent through the center of a 33-cm focal length lens. The light from the lens is directed to an electrooptic modulator, through a Fresnel rhomb prism, comes to a focus in the detection cell (2-cm-long, 40-μL volume), and finally diverges to a larger diameter at the detector where the laser power measured was typically near 20 mW. The signal from the detector is sent to a high-frequency lock-in amplifier

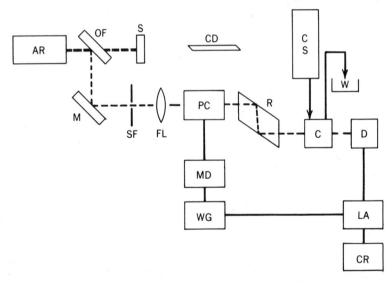

Figure 18. CD experimental configuration for conventional HPLC: AR, argon ion (488 nm) CW laser; OF, optical flat; S, beam stop; M, mirror; SF, pinhole spatial filter; FL, 33-cm focal-length lens; PC, Pockels cell; MD, modulation driver; WG, wave form generator; R, rhomb prism; C, detection cell; D, photodetector; LA, lock-in amplifier; CR, chart recorder; CS, chromatography system; W, waste.

and the output is sent to a chart recorder. The electrooptic modulator functions via a modulation driver operated at 500 kHz, which in turn is synchronized with the lock-in amplifier via a wave generator. By modulating the Pockels cell appropriately, on the first half cycle of the modulation frequency, RCPL is produced, and on the second half cycle, LCPL is produced. An optimum N/S ratio of 1.75×10^{-6} (500-kHz, 1-time constant) was obtained with the CD system. If the LOD is taken as twice the PPN, from Eq. 19, the differential absorbance LOD is $\Delta A_{LOD} = 1.5 \times 10^{-6}$ a.u. This constitutes an improvement by nearly a factor of 30 relative to previous work (34,35) for absolute CD absorbance detectability.

The demonstration of the HPLC–CD detection system for a mixture of three metal complexes is shown in Fig. 19. The CD chromatogram is shown in contrast to an absorbance chromatogram. Note that there are solvent disturbances in both chromatograms in Fig. 19 that occur at retention times earlier than the elution of $Co(NH_3)_5Cl^{2+}$. A LOD of 38 ng for $(+)$-$Co(en)_3^{3+}$ is calculated from Fig. 19 (240 g/mol, 10 μL injected, 1 s time constant). The "spikes" observed in the CD chromatogram at the retention times of the other two complexes cannot be attributed to CD activity. Because they are quite narrow, the "spikes" do not produce very much area as compared to the $(+)$-$Co(en)_3^{3+}$ peak; thus they can be tolerated. Most likely, these are artifacts due to thermal lensing at the ab-

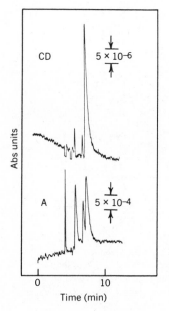

Figure 19. HPLC detection of $Co(NH_3)_5Cl^{2+}$, $Cr(NH_3)_6^{3+}$, and $(+)$-$Co(en)_3^{3+}$ in order of elution. CD: circular dichroism. A: absorbance. 10 μL injected, 0.67 mL/min, 5×10^{-4} M $(+)$-$Co(en)_3^{3+}$, 500-kHz modulation frequency, 1-s time constant, and a 2-cm path-length cell.

sorption maximum, in turn deflecting the laser beam relative to the photodiode. Utilization of a 10-s time constant compared to a 1-s time constant gives rise to some rounding of the analyte peak, yet a LOD of 20 ng was calculated, and the ΔA_{LOD} is better than 5.0×10^{-7} a.u.! A larger solvent peak is also observed. Note that when the sample injected was prepared in the same solution as used for the eluent no solvent peak was observed for our detection system. Although not shown, a $(-)$-Co(en)$_3^{3+}$ sample produced a peak that deflects in the opposite direction, as it should. For microbore chromatography (1-mm ID, 25-cm-long, 5-μm C$_{18}$ column) a cell 1 cm long and 2.6 μL in volume can be used. A LOD of 5.6 ng is achieved for the same compound, owing to the lower dilution factor in the microbore column, even though the absorbance LOD is degraded to 5.1×10^{-6} and the path length is shorter.

7. QUANTITATION WITHOUT STANDARDS

In the determination of impurities in truly unknown samples, if one knows in advance the number of components present and the amounts of each, then it is not necessary to test for each of the 6 million or so known compounds. The quantitative information can be used to limit the scope of the qualitative analysis, and total accountability is possible. In synthetic chemistry, even after the components have been identified, it may not be possible to obtain enough of the pure material to establish an analytical calibration curve for quantitation. Also, in complex samples, it may not be possible to achieve complete separation of the components. Even so, an unresolved chromatogram can be used for characterization if quantitation is achieved without identification.

The principles of quantitation without standards have been previously described (36,37). Briefly, if the detector responds to the eluent and to the analyte, the peak area, S_1, observed in a given chromatogram is related to this particular physical property of the eluent, to the physical property of the analyte, and to the concentration of the analyte. If, then, the same sample is eluted in a second eluent for which this physical property is different from that of the first eluent, a different peak area, S_2, will be observed. The two measured areas are then defined by two independent mathematical equations, which can be solved to determine the two unknowns, namely, the concentration of the analyte and its physical property. To provide useful sensitivities, the two eluents must show very different responses in the detector. This implies that not all detectors can be adapted for this quantitation scheme. Specifically, the detector must be able to maintain the same stable baseline when different eluents are used.

To apply this absolute quantitation scheme, the correct mathematical relationship must be derived to relate the peak areas to the corresponding concentrations. For the polarimeter (Chapter 6) the detector response (peak area) can be shown to be (38)

$$S_1K_1 = V_x\{[\alpha_x]\rho_x - [\alpha_1]\rho_1\} \tag{20}$$

where K_1 is a constant depending on the size of the injection loop and the eluent flow rate, V_x is the volume fractional concentration of the analyte, $[\alpha]$ is the specific rotation, ρ is the density, and subscripts x and 1 refer to the analyte and the eluent, respectively. Here, $[\alpha]_x\rho_x$ can be considered as a single unknown, being the special molecular physical property of x relevant to detector response in polarimetry.

For the RI detector, the correct expression is (36)

$$S_1K_1 = V_x \left(\frac{n_x^2 - 1}{n_x^2 + 2} - \frac{n_1^2 - 1}{n_1^2 + 2} \right) \tag{21}$$

where n is the refractive index.

For the absorption detector, it has been shown that (39)

$$S_1K_1 = V_x \left(\frac{\varepsilon_x}{v_x} - \frac{\varepsilon_1}{v_1} \right) \tag{22}$$

where ε is the molar absorptivity and v is the molar volume. Equation 22 is different from the standard form of Beer's law, which uses concentrations in moles per liter. The absorption detector is not very useful in general because it cannot be used for eluents with a high molar absorptivity due to loss of sensitivity. However, in conjunction with ion chromatography, the eluting ion is the relevant species and not the bulk eluent (e.g., water). The eluting ion is already at a low concentration typically, so that reasonable sensitivity based on the replacement of these eluting ions can be achieved. Because it is a charge replacement process,

$$S_1K_1 = N_x \left(\frac{\varepsilon_x}{m_x} - \frac{\varepsilon_1}{m_1} \right) \tag{23}$$

where N is the concentration in normality and m is the charge number of the species.

New instrumentation for conductivity detection (Section 1) provides a reasonably stable baseline even when an eluting ion with high equivalent conductance, λ, is used. Therefore the same quantitation scheme can be applied to ion chromatography using the following relationship (9):

$$S_1K_1 = N_x[\lambda_x - \lambda_1] \tag{24}$$

In the general case, one measures the peak areas, S, for the analyte in each of the two eluents, such that

$$S_1K_1 = C_x(F_x - F_1) \tag{25}$$

and

$$S_2K_2 = C_x(F_x - F_2) \tag{26}$$

with the subscript x designating the unknown and 1 and 2 the eluents. There are two methods of calibrating the response from the detector, so that one need not determine K_i or any physical properties of the eluent (F_i) in order to calculate the concentration of analyte injected onto the column. The only requirement is that the experimental conditions, for example, temperature, remain fixed for each of the eluents. The simplest method of calibration is to measure the peak area of each of the eluents eluted from the column using the other as the eluent at known concentrations C_1 and C_2. This will allow us to obtain areas S_a and S_b, such that

$$S_aK_1 = C_2(F_2 - F_1) \tag{27}$$

$$S_bK_2 = C_1(F_1 - F_2) \tag{28}$$

From Eqs. 25 through 28, using the derivation detailed in Ref. 36, we arrive at an expression for C_x:

$$C_x = \left(\frac{S_1C_2}{S_a} + \frac{S_2C_1}{S_b} \right) \tag{29}$$

One can also calculate the physical property of the analyte if one knows the physical properties of the eluent. The equation for F_x can be derived from Eqs. 15 and 17 of Ref. 36. With these substitutions, the equation becomes

$$F_x = \frac{F_2(S_1S_bC_2/S_2S_aC_1) + F_1}{(S_1S_bC_2/S_2S_aC_1) + 1} \tag{30}$$

The second method of calibrating the detector's response is to use two "calibrating" substances, 3 and 4. Thus one can calibrate even in the case where one eluent does not elute off the column using the other eluent within a reasonable amount of time. The two calibrating substances at known concentrations C_3 and C_4 allow us to obtain four additional relationships:

$$S_3 K_1 = C_3(F_3 - F_1) \tag{31}$$

$$S_4 K_2 = C_3(F_3 - F_2) \tag{32}$$

$$S_5 K_1 = C_4(F_4 - F_1) \tag{33}$$

$$S_6 K_2 = C_4(F_4 - F_2) \tag{34}$$

These equations plus Eqs. 25 and 26 can be manipulated, as shown in Ref. 39, to arrive at the following expression for C_x:

$$C_x = \left[\frac{S_1 - S_2 \dfrac{(S_3/C_3 - S_5/C_4)}{(S_4/C_3 - S_6/C_4)}}{S_3 - S_4 \dfrac{(S_3/C_3 - S_5/C_4)}{(S_4/C_3 - S_6/C_4)}} \right] C_3 \tag{35}$$

As before (39), we may solve for the physical property of the analyte if we know the physical properties of the calibrating substances 3 and 4, using the following equation:

$$F_x = \left[\frac{S_2/C_x - S_4/C_3}{S_4/C_3 - S_6/C_4} \right] (F_3 - F_4) + F_3 \tag{36}$$

To appreciate the potential of this quantitation scheme, it is necessary to define a figure of merit known as the dynamic reserve (DR). This is the ability of a detector to recognize a small change on top of a large background signal. This is because the two eluents must be chosen to differ greatly in the physical property being monitored. DR is quite different from the "dynamic range" concept often used in measurements, which is simply the ratio between the smallest and the largest signal that can be measured, independent of the background level. For the quantitation scheme described here, DR actually gives the detectability. For example, the absorbance detector for LC can measure a change of $\Delta A = 2 \times 10^{-4}$ a.u. when the background absorbance is unity (39). The dynamic reserve is simply the ratio of the two, so that $DR = 5 \times 10^3$. In other words, the analyte must be at a fractional concentration of at least 1 part in 5×10^3 (of that of the eluent) at the detector before a noticeable change in the signal is observed. We note that DR for the absorbance detector cannot be increased by using a larger background absorbance, because noise increases as well (in fact, more rapidly) to degrade the minimum detectable ΔA. So, quantitation without standards based on photometry (39) can only be used at analyte (fractional) concentrations of 1 part in 5×10^3 or higher. In normal LC situations, this is too high a concentration to be useful. The reason the scheme works in IC (39) is because the eluting ion is typically present at low concentration, for example, 10^{-3} M. Therefore the analyte need only be at 2×10^{-7} M to produce a fractional concentration of 1 part in 5×10^3. RI detector is a universal detector because the solvent provides the major contribution to the signal. One can measure $\Delta RI = 10^{-7}$ regardless of the background RI. Thus for a solvent–solute RI difference of 0.1, the DR is 10^6. This is why RI detectors can be used in this quantitation scheme. The best DR is obtained using polarimetry (38). When an optically active eluent is used, there can be a background rotation as large as 100°. By mechanical adjustment of the polarization analyzer, one can suppress this background signal to a level similar to that with an optically inactive eluent. A change of 4×10^{-6} degree (S/N = 3) can still be detected. The DR is then 2.5×10^7. In ion chromatography, the conductivity detector also functions as a universal detector (9). In the most favorable case, the background conductance can be 50 μmho without affecting the baseline noise and changes of 0.01 μmho can be measured. The DR is thus 5×10^3. Again, this DR becomes useful only because of the low initial concentration of the eluting ions.

ACKNOWLEDGMENTS

I thank the many co-workers in my laboratory who have contributed to various parts of this work, especially R. E. Synovec, D. R. Bobbitt, S. A. Wilson, and S. I. Mho, as well as the U.S. Department of Energy, Office of Basic Energy Sciences, for partial research support through contract No. W-7405-eng-82.

REFERENCES

1. H. Small, T. S. Stevens, and W. C. Bauman, *Anal. Chem.*, **47**, 1801 (1975).

2. D. T. Gjerde, J. S. Fritz, and G. Schmuckler, *J. Chromatogr.*, **186**, 509 (1979).

3. D. A. Skoog, *Principles of Instrumental Analysis*, 3rd ed., Saunders, New York, 1985, p. 704.

4. C. Duhne and O. de Ita. Sanchez, *Anal. Chem.*, **34**, 1074 (1962).

5. J. A. Dean, Ed., *Lange's Handbook of Chemistry*, 11th ed., McGraw-Hill, New York, 1973, p. 6.

6. J. S. Fritz, *LC Mag.*, **2**, 446 (1984).

7. D. T. Gjerde and J. S. Fritz, *Anal. Chem.*, **53**, 2324 (1981).

8. J. S. Fritz, D. T. Gjerde, and C. Pohlandt, *Ion Chromatography*, Huthig, New York, 1982, p. 99.

9. S. A. Wilson, E. S. Yeung, and D. R. Bobbitt, *Anal. Chem.*, **56**, 1457 (1984).

10. D. J. Popovich, J. B. Dixon, and B. J. Ehrlich, *J. Chromatogr. Sci.*, **17**, 643 (1979).

11. K. Siomos and L. G. Christophorou, *Chem. Phys. Lett.*, **72**, 43 (1980).

12. D. C. Locke, B. S. Dhingra, and A. D. Baker, *Anal. Chem.*, **54**, 447 (1982).

13. J. T. Schmermund and D. C. Locke, *Anal. Lett.*, **8**, 611 (1975).

14. E. Voigtman, A. Jurgensen, and J. D. Winefordner, *Anal. Chem.*, **53**, 1921 (1981).

15. E. Voigtman and J. D. Winefordner, *Anal. Chem.*, **54**, 1834 (1982).

16. S. Yamada, A. Hino, K. Kano, and T. Ogawa, *Anal. Chem.*, **55**, 1914 (1983).

17. S. Yamada, A. Hino, and T. Ogawa, *Anal. Chim. Acta.*, **156**, 273 (1984).

18. T. S. Krull and S. Jordan, *Amer. Lab.*, **12**, 21 (1980).

19. P. C. Uden and I. E. Bigley, *Anal. Chim. Acta*, **94**, 29 (1977).

20. S. Greenfield, H. McD. McGeachin, and P. B. Smith, *Talanta*, **22**, 1 (1975).

21. P. C. Uden, B. D. Quimby, R. M. Barnes, and W. G. Elliot, *Anal. Chim Acta*, **101**, 99 (1978).

22. M. Morita, T. Uehiro, and K. Fuwa, *Anal. Chem.*, **52**, 349 (1980).

23. K. Jinno, H. Tsuchida, S. Nakanishi, Y. Hirata, and C. Fujimoto, *Appl. Spectrosc.*, **37**, 258 (1983).

24. K. Jinno, S. Nakanishi, and T. Nagoshi, *Anal. Chem.*, **56**, 1977 (1984).

25. J. H. Knox and M. T. Gilbert, *J. Chromatogr.*, **186**, 405 (1979).

26. B. G. Julin, H. W. Vanderborn, and J. J. Kirkland, *J. Chromatogr.*, **112**, 443 (1975).

27. V. L. McGuffin and M. Novotny, *Anal. Chem.*, **53**, 946 (1981).

28. B. Chu, *Laser Light Scattering*, Academic, New York, 1974.

29. J. W. Jorgenson, S. L. Smith, and M. Novotny, *J. Chromatogr.*, **142**, 233 (1977).

30. A. Stolyhwo, H. Colin, and G. Guiochon, *J. Chromatogr.*, **265**, 1 (1983).

31. Anspec Model F2110 Mass Detector.

32. M. L. McConnell, *Anal. Chem.*, **53**, 1007A (1981).

33. S. Maezawa and T. Takagi, *J. Chromatogr.*, **280**, 124 (1983).

34. A. F. Drake, J. M. Gould, and S. F. Mason, *J. Chromatogr.*, **202**, 239 (1980).

35. S. A. Westwood, D. E. Games, and L. Sheen, *J. Chromatogr.*, **204**, 103 (1981).

36. R. E. Synovec and E. S. Yeung, *Anal. Chem.*, **55**, 1599 (1983).

37. R. E. Synovec and E. S. Yeung, *J. Chromatogr.*, **283**, 183 (1984).

38. D. R. Bobbitt and E. S. Yeung, *Anal. Chem.*, **56**, 1577 (1984).

39. S. A. Wilson and E. S. Yeung, *Anal. Chim. Acta*, **157**, 53 (1984).

INDEX

KRAUSKOPF LIBRARY
543.0894 D48   STACKS
/Detectors for liquid chromatography

3 1896 00006 2608